P9-DDG-425

Warped Passages

LISA RANDALL

Warped Passages

*Unraveling the Mysteries of the
Universe's Hidden Dimensions*

An Imprint of HarperCollinsPublishers

FIRST EDITION

Library of Congress Cataloging-in-Publication Data
has been applied for.

ISBN-10: 0-06-053108-8
ISBN-13: 978-0-06-053108-9

05 06 07 08 09 ❖/QWF 10 9 8 7 6 5 4

Contents

Preface and Acknowledgments

When I was a young girl, I loved the play and intellectual games in math problems or in books like *Alice in Wonderland*. But although reading was one of my favorite activities, books about science usually seemed more remote and less inviting to me—I never felt sufficiently engaged or challenged. The tone often seemed condescending to readers, overly worshipful of scientists, or boring. I felt the authors mystified results or glorified the men who found them, rather than describing science itself and the process by which scientists made their connections. That was the part I actually wanted to know.

As I learned more science, I grew to love it. I didn't always know that I would become a physicist and feel this way; no one I knew when I was young did science. But engaging with the unknown is irresistibly exciting. I found it thrilling to find connections between apparently disparate phenomena and to solve problems and predict surprising features of our world. As a physicist, I now understand that science is a living entity that continues to evolve. Not only the answers, but also the games and riddles and participation make it interesting.

When I decided to embark on this project, I envisioned a book that shares the excitement I feel about my work without compromising the presentation of the science. I hoped to convey the fascination of theoretical physics without simplifying the subject deceptively or presenting it as a collection of unchanging, finished monuments to be passively admired. Physics is far more creative and fun than people generally recognize. I wanted to share these aspects with people who hadn't necessarily arrived at this realization on their own.

There's a new world view pressing down upon us. Extra dimensions have changed the way physicists think about the universe. And because the connections of extra dimensions to the world could tie into many more well-established physics ideas, extra dimensions are a way to approach older, already-verified facts about the universe via new and intriguing pathways.

Some of the ideas I've included are abstract and speculative, but there's no reason why they shouldn't be understandable to anyone who is curious. I decided to let the fascination of theoretical physics speak for itself and chose not to over-emphasize history or personalities. I didn't want to give the misleading impression that all physicists are modeled on a single archetype or that any one particular type of person should be interested in physics. Based on my experiences and conversations, I'm pretty sure there are many readers who are smart, interested, and open enough to want more of the real thing.

This book doesn't skimp on the most advanced and intriguing theoretical ideas, but I've tried my best to make it self-contained. I've included both key conceptual advances and the physical phenomena to which they apply. The chapters are organized so that readers can tailor the book to their own backgrounds and interests. To help this process, I've bulleted the points that I'll refer to later on when I present more recent ideas about extra dimensions. I've also used bullets at the end of the extra-dimensional chapters to clarify what distinguishes each of the possible options for extra-dimensional universes.

Because the idea of extra dimensions is probably new to many readers, in the first few chapters I've explained what I mean when I use these words and why extra dimensions can exist but be invisible and intangible. After that, I've outlined the theoretical methods with which particle physicists approach their work to clarify the kind of thinking that enters into this admittedly very speculative research.

The recent work on extra dimensions relies on both more traditional and more modern theoretical physics concepts to motivate the questions it answers and its methods. In order to explain what is driving such research, I've included an extensive review of twentieth-century physics. Feel free to skim through this review if you like. But if you do, you'll miss a lot of good stuff!

The review begins with general relativity and quantum mechanics before turning to particle physics and the most important concepts that particle physicists employ today. I've presented some rather abstract ideas that are often neglected—in part because they are so abstract—but these concepts are now confirmed by experiment and enter into all research that we do today. Although not all of this material is essential for understanding the ideas you'll see later on about extra dimensions, I believe many readers will be glad to get a more complete picture.

After this, I've described some newer, more speculative notions that have been studied for the last thirty years—namely supersymmetry and string theory. Traditionally, physics has involved an interplay between theory and experiment. Supersymmetry is an extension of known particle physics concepts and has a good chance of being tested in forthcoming experiments. String theory is different. It is based solely on theoretical questions and ideas and isn't even completely mathematically formulated yet, so we can't yet be certain of its predictions. As for me, I'm an agnostic on this subject—I don't know what string theory will ultimately be or whether it will solve the questions of quantum mechanics and gravity it sets out to address. But string theory has been a rich resource for new ideas, some of which I've exploited in my own research on extra dimensions of space. These ideas exist independently of string theory, but string theory gives us a good reason to think some of their underlying assumptions could be right.

Having established the context, I'll finally return to the many exciting new developments about extra dimensions. They tell us remarkable things, such as that extra dimensions can be infinite in size yet remain unseen, or that we can be living in a three-spatial-dimensional sinkhole in a higher-dimensional universe. We now also know reasons why there can be unseen parallel worlds with very different properties from our own.

Throughout the text, I've explained physics concepts without equations. But for those who are interested in more mathematical detail, I've included a mathematical appendix. In the text itself, I've tried to expand the range of metaphors that are used to explain scientific concepts. A lot of the descriptive vocabulary everyone uses come from spatial analogies, but these often fail in the tiny realm of elementary particles and the hard-to-picture space with extra dimensions. It seemed to me that less conventional metaphors, even ones about art and food and personal relations, might work at least as well in explaining abstract ideas.

To make the transition for the new ideas in each chapter, I've begun the chapters with a brief story that isolates a key concept using more familiar metaphors and settings. I'm having fun with these stories, so go back to catch the references after you've read the chapter if you like. You might think of the stories as a two-dimensional narrative going "down" through the chapters and "horizontally" across the book. Or you might treat them as a sort of playful homework problem that lets you gauge when you've absorbed the ideas in a chapter.

Many friends and colleagues helped me accomplish my goals for this book. Although I often knew what I was after, I didn't always know when I had succeeded. A number of people deserve thanks for their generosity with their time, encouragement, and excitement and curiosity about the ideas I'm describing.

Several talented friends deserve particular thanks for their invaluable comments on the manuscript at various stages. Anna Christina Büchmann, a wonderful writer, gave beautifully detailed comments that helped me learn to complete the stories I was telling, both about physics and in general. She provided invaluable writing tips, always peppered with encouragement. Polly Shulman, another extremely talented friend, carefully read and commented on every chapter. I admire her logical and playful mind, and am very fortunate to have had her assistance. Lubos Motl, a brilliant physicist and dedicated science communicator (whose specious ideas about women in science we'll ignore), read everything, even before it was readable, and gave extraordinarily useful suggestions and encouragement at every stage. Tom Lewenson offered the important advice that only a skillful science writer could provide and contributed several critically important suggestions. Michael Gordin gave the perspective of a historian of science and a connoisseur of this type of literature. Jamie Robins gave insightful comments on more than one version of the manuscript. Esther Chiao gave useful comments on the manuscript and the extremely helpful perspective of a smart, interested reader with a background outside the sciences. And I'm delighted that Cormack

McCarthy volunteered valuable encouragement and suggestions in the final stages of this book.

Several people provided interesting stories and observations that helped me in the beginning stages of this project. Massimo Porrati is a storehouse of fascinating facts, some of which appear here. Gerald Holton's insights into early twentieth-century physics enriched my ideas about quantum mechanics and relativity. Jochen Brocks gave useful insights about what he liked in science writing and stimulated some writing ideas. Conversations with Chris Haskett and Andy Singleton helped me understand what non-physicists might hope to learn. Albion Lawrence made some valuable contributions that helped me sort out some difficult chapters. And John Swain passed along a couple of nice ways of presenting material.

Many colleagues gave valuable comments and suggestions. Among the many others to whom I am grateful, Bob Cahn, Csaba and Zsusanna Csaki, Paolo Creminelli, Joshua Erlich, Ami Katz, and Neil Weiner all read substantial portions of the book and provided insightful commentary. I also thank Allan Adams, Nima Arkani-Hamed, Martin Gremm, Jonathan Flynn, Melissa Franklin, David Kaplan, Andreas Karch, Joe Lykken, Peter Lu, Ann Nelson, Amanda Peet, Riccardo Rattazzi, Dan Shrag, Lee Smolin, and Darien Wood, who all gave useful comments and advice. Howard Georgi advised me and many of the physicists listed above about the effective theory way of thinking that is espoused in this book. I also thank Peter Bohacek, Wendy Chun, Enrique Rodriguez, Paul Graham, Victoria Gray, Paul Moorhouse, Curt McMullen, Liam Murphy Jeff Mrugan, Sesha Pretap, Dana Randall, Enrique Rodriguez, and Judith Surkis, who provided helpful criticism, suggestions, and encouragement. I am also grateful to Marjorie Caron, Tony Caron, Barry Ezarsky, Josh Feldman, Marsha Rosenberg, and other family members for helping me better understand my audience.

Greg Elliott and Jonathan Flynn executed the beautiful pictures contained in this book, and I'm extraordinarily grateful for their important contribution. I thank Rob Meyer and Laura Van Wyk for helping me obtain permissions for the many quotes throughout the book. I have made every effort to properly credit sources. If you think you have not been credited properly, please let me know.

I also want to thank my collaborators on my research that I describe in this book, particularly Raman Sundrum and Andreas Karch, who were both great to work with. And I'd like to acknowledge the contributions of the many physicists who have thought about these and related ideas, including those that I didn't have room to discuss.

I'd also like to express my appreciation to my Ecco Press editor, Dan Halpern, my Penguin editors, Stefan McGrath and Will Goodlad, and my copy editors in the U.S. and England, Lyman Lyons and John Woodruff, for their many helpful suggestions and for their support for this book. And I wish to thank my literary agent, John Brockman, as well as Katinka Matson, for their important commentary and advice, and for their invaluable help in getting this book launched. I'm also grateful to Harvard University and to the Radcliffe Institute for Advanced Study for providing some time to focus on this book, and to MIT, Princeton, Harvard, the National Science Foundation, the Department of Energy, and the Alfred P. Sloan Foundation for supporting my research.

Finally, I wish to thank my family: my parents, Richard Randall and Gladys Randall, and my sisters, Barbara Randall and Dana Randall, for backing my scientific career and for sharing their humor, thoughts, and encouragement over the years. Lynn Festa, Beth Lyman, Gene Lyman, and Jen Sacks were extremely supportive and I thank them all for their wonderful advice and suggestions along the way. And lastly, I'm so grateful to Stuart Hall for his insightful perspective, helpful comments, and unselfish support.

I thank you all and hope you find your contributions are repaid.

Lisa Randall

Cambridge, MA

April 2005

Introduction

Got to be good looking
'Cause he's so hard to see.   The Beatles

The universe has its secrets. Extra dimensions of space might be one of them. If so, the universe has been hiding those dimensions, protecting them, keeping them coyly under wraps. From a casual glance, you would never suspect a thing.

The disinformation campaign began back in the crib, which first

Figure 1. *A baby's three-dimensional world.*

introduced you to three spatial dimensions. Those were the two dimensions in which you crawled, plus the remaining one by which you climbed out. Since that time, physical laws—not to mention common sense—have bolstered the belief in three dimensions, quelling any suspicion that there might be more.

But spacetime could be dramatically different from anything you've ever imagined. No physical theory we know of dictates that there should be only three dimensions of space. Dismissing the possibility of extra dimensions before even considering their existence might be very premature. Just as "up-down" is a different direction from "left-right" or "forward-backward," other completely new dimensions could exist in our cosmos. Although we can't see them with our eyes or feel them with our fingertips, additional dimensions of space are a logical possibility.

Such hypothetical unseen dimensions don't yet have a name. But should they exist, they would be new directions along which something might travel. So when I need a name for an extra dimension, I'll sometimes call it a *passage*. (And when I explicitly discuss extra dimensions, I'll use chapter names with "passages" in the title.)

These passages could be flat, like the dimensions we are accustomed to. Or they could be warped, like reflections in a fun-house mirror. They might be tiny—far smaller than an atom—until recently, that's what anyone who believed in extra dimensions assumed. But new work has shown that extra dimensions might also be big, or even infinite in size, yet still be hard to see. Our senses register only three large dimensions, so an infinite extra dimension might sound incredible. But an infinite unseen dimension is one of many bizarre possibilities for what might exist in the cosmos, and in this book we'll see why.

Research into extra dimensions has also led to other remarkable concepts—ones that might fulfill a science fiction aficionado's fantasy—such as parallel universes, warped geometry, and three-dimensional sinkholes. I'm afraid such ideas might sound more like the province of novelists and lunatics than the focus of real scientific inquiry. But outlandish as they might seem at the moment, they are genuine scientific scenarios that could arise in an extra-dimensional world. (Don't worry if you are not yet familiar with these words or ideas; we'll introduce and investigate them later on.)

Why Consider Unseen Dimensions?

Even if physics with extra spatial dimensions permit these intriguing scenarios, you might still wonder why physicists concerned with making predictions about observable phenomena would bother to take them seriously. The answer is as dramatic as the idea of extra dimensions itself. Recent advances suggest that extra dimensions, not yet experienced and not yet entirely understood, might nonetheless resolve some of the most basic mysteries of our universe. Extra dimensions could have implications for the world we see, and ideas about them might ultimately reveal connections that we miss in three-dimensional space.

We wouldn't understand why Inuit and Chinese people share physical features, either, if we failed to include the dimension of time that lets us recognize their common ancestry. Similarly, the connections that can occur with additional dimensions of space might illuminate perplexing aspects of particle physics, shedding light on decades-old mysteries. Relationships between particle properties and forces that seemed inexplicable when space was shackled to three dimensions seem to fit together elegantly in a world with more dimensions of space.

Do I believe in extra dimensions? I confess I do. In the past, I've mostly viewed speculations about physics beyond what's been measured—including my own ideas—with fascination, but also with some degree of skepticism. I like to think this keeps me interested, but honest. Sometimes, however, an idea seems like it must contain a germ of truth. One day on my way to work about five years ago, as I was crossing the Charles River into Cambridge, I suddenly realized that I really believed that some form of extra dimensions must exist. I looked around and contemplated the many dimensions I couldn't see. I had the same shock of surprise at my altered worldview that I experienced when I realized that I, a native New Yorker, was rooting for the Red Sox during a playoff game against the Yankees— something else I never anticipated I'd do.

Greater familiarity with extra dimensions has only increased my confidence in their existence. Arguments against them have too many holes to be reliable, and physical theories without them leave too many questions unanswered. Furthermore, as we've explored extra dimensions in the last few years, we've expanded the range of possible extra-dimensional universes that can mimic our own, suggesting that we've identified only the tip of the iceberg. Even if extra dimensions don't conform precisely to the pictures I will present, I think they are very likely to be there, in one form or another, and their implications are bound to be surprising and remarkable.

You might be intrigued to know that there could be a vestige of extra dimensions hidden in your kitchen cabinet—on a nonstick frying pan coated with *quasicrystals*. Quasicrystals are fascinating structures whose underlying order is revealed only with extra dimensions. A crystal is a highly symmetric latticework of atoms and molecules with one basic element repeated many times. In three dimensions we know what structures crystals can form, and which patterns are possible. However, the arrangement of atoms and molecules in quasicrystals does not conform to any of these patterns.

An example of a quasicrystalline pattern is shown in Figure 2. It lacks the precise regularity you would see in a true crystal, which would look more like the kind of grid you would see on a piece of graph paper. The most elegant way of explaining the pattern of molecules in these strange

Figure 2. *This is a "Penrose tiling." It is a projection of a five-dimensional crystalline structure onto two dimensions.*

materials is with a projection—a sort of three-dimensional shadow—
of a higher-dimensional crystalline pattern, which reveals the symmetry
of the pattern in a higher-dimensional space. What looked like a completely inexplicable pattern in three dimensions reflects an ordered
structure in a higher-dimensional world. The nonstick frying pans that
are coated with quasicrystals exploit the structural differences between
the projections of higher-dimensional crystals in the pan's coating and
the more mundane structure of ordinary three-dimensional food. The
different arrangements of atoms, which keeps them from binding to
each other, is a tantalizing suggestion that extra dimensions exist and
explain some observable physical phenomena.

Overview

Just as extra dimensions help us understand the confusing arrangement of molecules in a quasicrystal, physicists today speculate that
theories of extra dimensions also will illuminate connections in particle physics and cosmology—connections that are difficult to understand with only three dimensions.

For thirty years, physicists have relied on a theory called the Standard Model of particle physics, which tells us about the fundamental
nature of matter and the forces through which elementary constituents
interact.* Physicists have tested the Standard Model by creating particles that have not been present in our world since the earliest seconds
of the universe, and they've found that the Standard Model describes
many of their properties extremely well. Yet the Standard Model
leaves some fundamental questions unanswered—questions so basic
that their resolution promises new insight into the building blocks of
our world and their interactions.

This book tells about how I and others searched for answers to
Standard Model puzzles and found ourselves in extra-dimensional
worlds. The new developments with extra dimensions will ultimately
take center stage, but I'll first introduce the supporting players—the
revolutionary physics advances of the twentieth century. The recent

---

*We'll discuss the Standard Model further in Chapter 7.

ideas that I discuss later are grounded in these stupendous break-throughs.

The review topics we'll encounter will, broadly, divide into three categories: early-twentieth-century physics, particle physics, and string theory. We'll investigate the key ideas of relativity and quantum mechanics, as well as the current state of particle physics and the problems that extra dimensions might address. We'll also consider the concepts that underlie string theory, which many physicists think is the leading contender for a theory that incorporates both quantum mechanics and gravity. String theory, which postulates that the most basic units in nature are not particles but fundamental, oscilllating strings, has provided much of the impetus for studying extra dimensions, because it requires more than three dimensions of space. And I will also describe the role of branes, membrane-like objects within string theory, which are as essential to the theory as strings themselves. We'll consider both the successes of these theories and the questions they leave open—the ones that motivate current research.

One of the chief mysteries is why gravity is so much weaker than the other known forces. Gravity might not feel weak when you're hiking up a mountain, but that's because the entire Earth is pulling on you. A tiny magnet can lift a paper clip, even though all the mass of the Earth is pulling it in the opposite direction. Why is gravity so defenseless against the small tug of a tiny magnet? In standard three-dimensional particle physics, the weakness of gravity is a huge puzzle. But extra dimensions might provide an answer. In 1998, my collaborator Raman Sundrum and I showed one reason this might be so.

Our proposal is based on warped geometry, a notion that arises in Einstein's theory of general relativity. According to this theory, space and time are integrated into a single spacetime fabric that gets distorted, or warped, by matter and energy. Raman and I applied this theory in a new, extra-dimensional context. We found a configuration in which spacetime warps so severely that even if gravity is strong in one region of space, it is feeble everywhere else.

And we found something even more remarkable. Although physicists have assumed for eighty years that extra dimensions must be tiny in order to explain why we haven't seen them, in 1999 Raman and I discovered that not only can warped space explain gravity's feebleness, but also that an invisible extra dimension can stretch out to infinity, provided it is suitably distorted in a curved spacetime. An extra dimension can be infinite in size– but nonetheless be hidden. (Not all physicists immediately accepted our proposal. But my non-physicist friends were more quickly convinced I was on to something—not because they fully grasped the physics, but because when I attended a conference banquet after speaking about my work, Stephen Hawking saved me a seat.)

I will explain the physical principles underlying these and other theoretical developments and the new notions about space that make them conceivable. And later on, we'll also encounter an even weirder possibility, which the physicist Andreas Karch and I discovered a year later: we could be living in a three-dimensional pocket of space, even though the rest of the universe behaves as if it is higher-dimensional. This result opens a host of new possibilities for the fabric of spacetime, which could consist of distinct regions, each appearing to contain a different number of dimensions. Not only are we not in the center of the universe, as Copernicus shocked the world by suggesting five hundred years ago, but we just might be living in an isolated neighborhood with three spatial dimensions that's part of a higher-dimensional cosmos.

The newly-studied membrane-like objects called branes are important components of the rich higher-dimensional landscapes. If extra dimensions are a physicist's playground, then *braneworlds*— hypothesized universes in which we live on a brane—are the tantalizing, multi-layered, multi-faceted jungle gyms.* This book will take you to braneworlds and universes with curled-up, warped, large, and infinite dimensions, some of which contain a single brane and others of which have multiple branes housing unseen worlds. All of these are within the realm of possibility.

The Excitement of the Unknown

The postulated braneworlds are a theoretical leap of faith, and the ideas they contain are speculative. However, as with the stock market, riskier ventures might fail but they could also reward you with greater returns.

*For British readers, a child's climbing frame.

Imagine the sight of the snow under a ski chairlift on the first sunny day after a storm, when untracked powder tempts you from below. You know that no matter what, once you hit the snow, it's going to be a great day. Some runs will be steep and full of bumps, some will be easy cruisers, and some will be tricky routes through trees. But even if you take the occasional wrong turn, most of the day will be wonderfully rewarding.

For me, model building—which is what physicists call the search for theories that might underlie current observations—has this same irresistible appeal. Model building is adventure travel through concepts and ideas. Sometimes new ideas are obvious, and sometimes they are tricky to find and negotiate. However, even when we don't know where they're heading, interesting new models often explore untouched, delightful terrain.

We will not know right away which of the theories gets it right about our place in the universe. For some of them, we might never know. But, incredibly, that is not true for all extra-dimensional theories. The most exciting feature of any extra-dimensional theory that explains the weakness of gravity is that if it is correct, we will soon find out. Experiments that study very energetic particles could discover evidence supporting these proposals and the extra dimensions they contain within the next five years—as soon as the Large Hadron Collider (LHC), a very high energy *particle collider* near Geneva, is up and running.

This collider, which turns on in 2007, will bang together tremendously energetic particles that could turn into new types of matter we have never seen before. If any of these extra-dimensional theories is right, it could leave visible signs at the LHC. The evidence would include particles called *Kaluza-Klein modes*, which travel in the extra dimensions yet leave traces of their existence here in the familiar three dimensions. Kaluza-Klein modes would be fingerprints of extra dimensions in our three-dimensional world. And if we're very lucky, experiments will register other clues as well, perhaps even higher-dimensional black holes.

The detectors that will record these objects will be large and impressive—so much so that working on them will require climbing gear like harnesses and helmets. In fact, I once took advantage of this gear when I went glacier hiking in Switzerland close to the European

8

Organization for Particle Research (CERN), the physics center that will house the LHC. These enormous detectors will record particle properties that physicists will use to reconstruct what passed through.

Admittedly, the evidence for extra dimensions will be somewhat indirect, and we will have to piece together various clues. But that is true of almost all recent physics discoveries. As physics evolved in the twentieth century, it moved away from things that can be directly observed with the naked eye to things that can be "seen" only through measurements coupled with a theoretical train of logic. For example, quarks, components of the proton and neutron familiar from high-school physics, never appear in isolation; we find them by following the trail of evidence they leave behind them as they influence other particles. It's the same with the intriguing kinds of stuff known as dark energy and dark matter. We don't know where most of the energy in the universe comes from or the nature of most of the matter that the universe contains. Yet we know that dark matter and dark energy exist in the universe, not because we've detected them directly, but because they have noticeable effects on matter that surrounds them. Like quarks or dark matter and dark energy, whose existence we only indirectly ascertain, extra dimensions will not appear to us directly. Nonetheless, signatures of extra dimensions, even when indirect, could ultimately reveal their existence.

Let me say at the outset that obviously not all new ideas prove correct, and that many physicists are skeptical about any new theories. The theories I present here are no exception. But speculation is the only way to make progress in our understanding. Even if it turns out that the details don't all align with reality, a new theoretical idea can still illuminate physical principles at work in the true theory of the cosmos. I'm fairly certain that the ideas about extra dimensions we'll encounter in this book contain more than a germ of truth.

When engaging with the unknown and working with speculative ideas, I find it comforting to recall that the discovery of fundamental structure has always come as a surprise and been met with skepticism and resistance. Oddly enough, not just the general populace, but sometimes even the very people who suggest underlying structures have been reluctant to believe in them at first.

James Clerk Maxwell, for example, who developed the classical theory of electricity and magnetism, didn't believe in the existence of fundamental units of charge such as electrons. George Stoney, who at the end of the nineteenth century proposed the electron as a fundamental unit of charge, didn't believe that scientists would ever isolate electrons from the atoms of which they are components. (In fact, all it takes is heat or an electric field.) Dmitri Mendeleev, creator of the periodic table, resisted the notion of valence, which his table encoded. Max Planck, who proposed that the energy carried by light was discontinuous, didn't believe in the reality of the light quanta that were implicit in his own idea. Albert Einstein, who suggested these quanta of light, didn't know that their mechanical properties would permit them to be identified as particles—the photons we now know them to be. Not everyone with correct new ideas has denied their connection to reality, however. Many ideas, whether believed-in or mistrusted, have turned out to be true.

Is there more waiting to be discovered? For the answer to that question, I turn to the all-too-mortal words of George Gamow, the prominent nuclear physicist and science popularizer. In 1945 he wrote, "Instead of a rather large number of 'indivisible atoms' of classical physics, we are now left with only three essentially different entities; nucleons, electrons, and neutrinos . . . Thus it seems that we have actually hit the bottom in our search for the basic elements from which matter is formed." When Gamow wrote this, he had no idea that the nucleons are composites of quarks, which would be discovered within thirty years!

Wouldn't it be strange if we turn out to be the first people for whom the search for further underlying structure ceased to be fruitful? So strange, in fact, that it seems hardly credible? Inconsistencies in existing theories tell us they can't be the final word. Earlier generations had neither the tools nor the motivations of today's physicists for exploring the extra-dimensional arenas that this book will describe. Extra dimensions, or whatever underlies the Standard Model of particle physics, would be a discovery of major importance.

When it comes to the world around us, is there any choice but to explore?

I

Entryway Passages:
Demystifying Dimensions

You can go your own way.
Go your own way.

Fleetwood Mac

"Ike, I'm not so sure about this story I'm writing. I'm considering adding more dimensions. What do you think of that idea?"

"Athena, your big brother knows very little about fixing stories. But odds are it won't hurt to add new dimensions. Do you plan to add new characters, or flesh out your current ones some more?"

"Neither; that's not what I meant. I plan to introduce new dimensions—as in new dimensions of space."

"You're kidding, right? You're going to write about alternative realities—like places where people have alternative spiritual experiences or where they go when they die, or when they have near-death experiences?* I didn't think you went in for that sort of thing."

"Come on, Ike. You know I don't. I'm talking about different spatial dimensions—not different spiritual planes!"

"But how can different spatial dimensions change anything? Why would using paper with different dimensions—11" × 8" instead of 12" × 9", for example—make any difference at all?"

"Stop teasing. That's not what I'm talking about either. I'm really planning to introduce new dimensions of space, just like the dimensions we see, but along entirely new directions."

*Questions I've actually been asked.

"Dimensions we don't see? I thought three dimensions is all there are."

"Hang on, Ike. We'll soon see about that."

The word "dimension," like so many words that describe space or motion through it, has many interpretations—and by now I think I've heard them all. Because we see things in spatial pictures we tend to describe many concepts, including time and thought, in spatial terms. This means that many words that apply to space have multiple meanings. And when we employ such words for technical purposes, the alternative uses of the words can make their definitions sound confusing.

The phrase "extra dimensions" is especially baffling because even when we apply those words to space, that space is beyond our sensory experience. Things that are difficult to visualize are generally harder to describe. We're just not physiologically designed to process more than three dimensions of space. Light, gravity, and all our tools for making observations present a world that appears to contain only three dimensions of space.

Because we don't directly perceive extra dimensions—even if they exist—some people fear that trying to grasp them will make their head hurt. At least, that's what a BBC newscaster once said to me during an interview. However, it's not thinking about extra dimensions but trying to picture them that threatens to be unsettling. Trying to draw a higher-dimensional world inevitably leads to complications.

Thinking about extra dimensions is another thing altogether. We are perfectly capable of considering their existence. And when my colleagues and I use the words "dimensions," and "extra dimensions," we have precise ideas in mind. So before taking another step forward or exploring how new ideas fit into our picture of the universe—note the spatial phrases—I will explain the words "dimensions" and "extra dimensions" and what I will mean when I use them later on.

We'll soon see that when there are more than three dimensions, words (and equations) can be worth a thousand pictures.

What Are Dimensions?

Working with spaces that have many dimensions is actually something everyone does every day, although admittedly most of us don't think of it that way. But consider all the dimensions that enter into your calculations when you make an important decision, like buying a house. You might consider the size, the schools nearby, the proximity to places of interest, the architecture, the noise level—and the list goes on. You need to optimize in a multidimensional context, enumerating all your desires and needs.

The number of dimensions is the number of quantities you need to know to completely pin down a point in a space. The multidimensional space might be an abstract one, such as the space of features you are looking for in a house, or it might be concrete, like the real physical space we will soon consider. But when buying a house, you can think of the number of dimensions as the number of quantities you would record in each entry in a database—the number of quantities you find worth investigating.

A more frivolous example applies dimensions to people. When you peg someone as one-dimensional, you actually have something rather specific in mind: you mean that the person has only a single interest. For example, Sam, who does nothing but sit at home watching sports, can be described with just one piece of information. If you felt so inclined, you could picture this information as a dot on a one-dimensional graph: Sam's proclivity to watch sports, for example. In drawing this graph you need to specify your units so that someone else can understand what the distance along this single axis means. Figure 3 shows a plot with Sam as a point along a horizontal axis. This plot represents the number of hours Sam spends per week watching sports on TV. (Fortunately, Sam won't be insulted by this

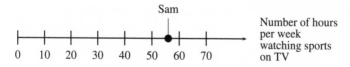

Figure 3. *The one-dimensional Sam plot.*

example; he is not among the multidimensional readers of this book.)

Let's explore this notion a little further. Icarus Rushmore III (Ike in the above story), a Boston resident, is a more complex character. In fact, he is three dimensional. Ike is twenty-one, drives fast cars, and loses money at Wonderland, a town near Boston with a dog-racing track. In Figure 4 I've plotted Ike. Although I've drawn it on the two-dimensional surface of a piece of paper, the three axes tell us that Ike is definitely three-dimensional.*

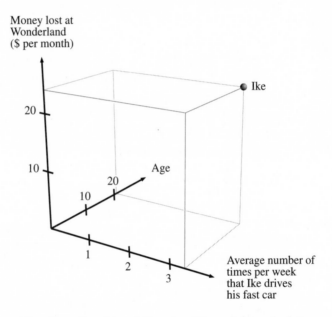

Figure 4. *The three-dimensional Ike plot. The solid notched lines are the coordinate axes of the three-dimensional plot. The point that is labeled Ike corresponds to a 21-year-old boy who loses 24 dollars at Wonderland every month and drives his fast car (on average) 3.3 times a week.*

When we describe most people, however, we usually assign them more than one, or even three, characteristics. Athena, Ike's sister, is

---

*If you're picky, you'll object that Sam too has an age and therefore another dimension. However, I've assumed that Sam has been the same way for years so his age isn't relevant.

an eleven-year-old who reads avidly, excels at math, keeps abreast of current events, and raises pet owls. You might want to plot this too (though why, exactly, I'm not really sure). In that case, Athena would have to be plotted as a point in a five-dimensional space with axes corresponding to age, number of books read per week, average math test score, number of minutes spent reading the newspaper per day, and number of owls she owns. However, I'm having trouble drawing such a graph. It would require a five-dimensional space, which is very hard to draw. Even computer programs only have 3D graphics.

Nonetheless, in an abstract sense, there exists a five-dimensional space with a collection of five numbers, such as (11, 3, 100, 45, 4), which tells us that Athena is eleven, that she reads three books on the average each week, that she never gets a math question wrong, that she reads the newspaper for forty-five minutes each day, and that she has four owls at the moment. With these five numbers, I've described Athena. If you knew her, you could recognize her from this point in five dimensions.

The number of dimensions for each of the three people above was the number of attributes I used to identify them: one for Sam, three for Ike, and five for Athena. Real people, of course, are generally more difficult to capture with so few items of information.

In the following chapters, we'll use dimensionality to explore not people, but space itself. By "space" I mean the region in which matter exists and physical processes take place. A *space of a particular dimension* is a space requiring a particular number of quantities to specify a point. In one dimension, that would be a point on a plot with a single x axis; in two dimensions, a point on a plot with an x and a y axis; in three dimensions, it would be a point on a plot with an x, a y, and a z axis.[1,*] Those axes are shown in Figure 5.

In three-dimensional space, three numbers are all you ever need to know your precise location. The numbers you specify might be latitude, longitude, and altitude; or length, width, and height; or you might have a different way to choose your three numbers. The critical thing is that three dimensions means you need precisely three numbers.

*This and other superscript numbers (1, 2, . . .) refer to the Math Notes at the end of the book.

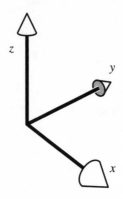

Figure 5. *The three coordinate axes that we use for three-dimensional space.*

In two-dimensional space you need two numbers, and in higher-dimensional space you need more.

More dimensions means freedom to move in a greater number of completely different directions. A point in a four-dimensional space simply requires one additional axis—again, difficult to draw. But it should not be hard to imagine its existence. We'll think about it using words and mathematical terms.

String theory suggests even more dimensions: it postulates six or seven extra spatial dimensions, meaning that six or seven additional coordinates are needed to plot a point. And very recent work in string theory has shown that there could be even more dimensions than that. In this book, I'll keep an open mind and entertain the possibility of any number of extra dimensions. It is too soon to say how many dimensions the universe actually contains. Many of the concepts about extra dimensions that I will describe apply to any number of extra dimensions. In the rare cases when that isn't true, I will make sure that it is clear.

Describing a physical space involves more than just identifying points, however. You need also to specify a *metric*, which establishes the measurement scale, or the physical distance between two points. These are the markings along the axis of a graph. It's not enough to know that the distance between a pair of points is 17 unless you know whether 17 means 17 centimeters, 17 miles, or 17 light-years. A metric is required to tell us how to measure distance: what the distance between two points on a graph corresponds to in the world that the graph represents. A metric gives a measuring rod that reveals your choice of units in order to set the scale, just like on a map, where a half-inch might represent one mile, or as in the metric system, which gives us a meter stick we all agree on.

But that is not all a metric specifies. It also tells us whether space bends or curls around, like the surface of a balloon when it is blown up into a sphere. The metric contains all the information about the shape of space. A metric for curved space tells us about both distances and angles. Just as an inch can represent different distances, an angle can correspond to different shapes. I'll go into this later on when we explore the connection between curved space and gravity. For now, let's just say that the surface of a sphere is not the same as the surface of a flat piece of paper. Triangles on one don't look like triangles on the other, and the difference between these two-dimensional spaces can be seen in their metrics.[2]

As physics has evolved, so has the amount of information stored in the metric. When Einstein developed relativity, he recognized that a fourth dimension—time—is inseparable from the three dimensions of space. Time, too, needs a scale, so Einstein formulated gravity by using a metric for four-dimensional *spacetime*, adding the dimension of time to the three dimensions of space.

And more recent developments have shown that additional spatial dimensions might also exist. In that case, the true spacetime metric will involve more than three dimensions of space. The number of dimensions and the metric for those dimensions is how one describes such a multidimensional space. But before we explore metrics and metrics for multidimensional spaces any further, let's think more about the meaning of the term "multidimensional space."

Playful Passages Through Extra Dimensions

In Roald Dahl's *Charlie and the Chocolate Factory*, Willy Wonka introduced visitors to his "Wonkavator." In his words, "An elevator can only go up and down, but a Wonkavator goes sideways and slantways and longways and backways and frontways and squareways and any other ways that you can think of . . ."* Really, what he had was a device that moved in any direction, so long as it was a direction in the three dimensions we know. It was a nice, imaginative idea.

However, the Wonkavator didn't really go any way "you can think of." Willy Wonka was remiss in that he neglected extra-dimensional passages. Extra dimensions are other directions entirely. They are hard to describe, but they may be easier to understand by analogy.

In 1884, to explain the notion of extra dimensions, the English mathematician Edwin A. Abbott wrote a novel called *Flatland*.† It takes place in a fictitious two-dimensional universe—the Flatland of the title—where two-dimensional beings (of various geometric shapes) reside. Abbott shows us why Flatlanders, who live their whole lives in two dimensions—on a table top, for example—are as mystified by three dimensions as people in our world are by the idea of four.

For us, more than three dimensions requires a stretch of the imagination, but in Flatland three dimensions are beyond its inhabitants' comprehension. Everyone thinks it is obvious that the universe holds no more than their two perceived dimensions. Flatlanders are as insistent about this as most people here are about three.

The book's narrator, A. Square (the namesake of the author, Edwin A[2]), is introduced to the reality of a third dimension. In the first stage of his education, while he is still confined to Flatland, he watches a three-dimensional sphere travel vertically through his two-dimensional world. Because A. Square is confined to Flatland, he sees a series of disks that increase and then decrease in size, which are slices of the sphere as it passes through A. Square's plane (see Figure 6).

This is initially perplexing to the two-dimensional narrator, who has never imagined more than two dimensions and has never contemplated a three-dimensional object like a sphere. It is not until A. Square has been lifted out of Flatland into the surrounding three-dimensional world that he can truly imagine a sphere. From his new perspective, he recognizes the sphere as the shape made by gluing together the two-dimensional slices he witnessed. Even in his two-dimensional

*Roald Dahl, *Charlie and the Chocolate Factory* (London: Puffin Books, 1998).
†The full title is *Flatland: A Romance of Many Dimensions*.

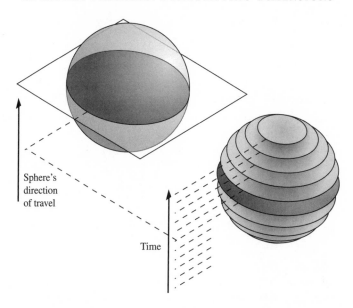

Sphere's
direction
of travel

Time

Figure 6. *If a sphere passes through a plane, a two-dimensional observer would see a disk. The sequence of disks that the observer sees over time comprises the sphere.*

world, A. Square could have plotted the disks he sees as a function of time (as in Figure 6) to construct the sphere. But it wasn't until his trip through a third dimension opened his eyes that he fully comprehended the sphere and its third spatial dimension.

By analogy, we know that if a *hypersphere* (a sphere with four spatial dimensions) were to pass through our universe, it would appear to us as a time sequence of three-dimensional spheres that increase, then decrease, in size.[3] Unfortunately, we don't have the opportunity to journey through an extra dimension. We will never see a static hypersphere in its entirety. Nonetheless, we can make deductions about how objects look in spaces of different dimensions—even dimensions that we don't see. We can confidently deduce that our perception of a hypersphere passing through three dimensions would look like a series of three-dimensional spheres.

As another example, let's imagine the construction of a *hypercube*—a generalization of a cube to more than three dimensions. A line segment of one dimension consists of two points connected by a straight, one-dimensional line. We can generalize this in two dimensions to a square by putting one of these one-dimensional line segments above another and connecting them with two additional segments. We can generalize further in three dimensions to a cube, which we can construct by placing one two-dimensional square above the other and connecting them with four additional squares, one on each edge of the original squares (see Figure 7).

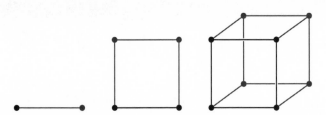

Figure 7. *How we put together lower-dimensional objects to make higher-dimensional ones. We connect two points to make a line segment, two line segments to make a square, two squares to make a cube, and (not pictured since it's too difficult to draw) two cubes to make a hypercube.*

We can generalize in four dimensions to a hypercube, and in five dimensions to something for which we don't yet have a name. Even though we three-dimensional mortals have never seen these two objects, we can generalize the procedure that worked in lower dimensions. To construct a hypercube (also known as a tesseract), put one cube above the other, and connect them by adding six additional cubes, connecting the faces of the two original cubes. This construction is an abstraction and difficult to draw, but that doesn't make the hypercube any less real.

In high school, I spent a summer at math camp (which was far more entertaining than you might think), where we were shown a film version of *Flatland.** At the end, the narrator, in a delightful British accent, tried futilely to point to the third dimension that was inaccessible to Flatlanders, saying, "Upward, not Northward." Unfortunately, we have the same frustration if we try to point to a fourth spatial

---

*This animated film, directed by Eric Martin, featured the voices of Dudley Moore and other members of the British theatrical comedy group *Beyond the Fringe.* It was very entertaining.

dimension, a passage. But just as Flatlanders didn't see or travel through the third dimension, even though it existed in Abbott's story, our not having yet seen another dimension doesn't mean there is none. So although we haven't yet observed or traveled through such a dimension, the subtext throughout *Warped Passages* will be, "Not Northward, but Forward along a passage." Who knows what exists that we haven't yet seen?

Three from Two

For the rest of this chapter, rather than thinking about spaces that have more than three dimensions, I will talk about how, with our limited visual capacity, we go about thinking and drawing three dimensions using two-dimensional images. Understanding how we perform this translation from two-dimensional images to three-dimensional reality will be useful later on when interpreting lower-dimensional "pictures" of higher-dimensional worlds. Think of this section as a warm-up exercise for wrapping your mind around extra dimensions. It might be good to remember that you cope with dimensionality all the time in ordinary life. It really isn't that unfamiliar.

Often all we can see are parts of the surface of things, the surface being only the exterior. This exterior has two dimensions, even though it curves through three-dimensional space, because you only need two numbers to identify any point. We deduce that the surface isn't three-dimensional because it has no thickness.

When we look at pictures, movies, computer screens, or the figures in this book, we are generally looking at two-dimensional, not three-dimensional representations. But we can nonetheless deduce the three-dimensional reality that is being portrayed.

We can use two-dimensional information to construct three dimensions. This involves suppressing information in making two-dimensional representations while trying to keep enough information to reproduce essential elements of the original object. Let's now reflect on the methods we often use to reduce higher-dimensional objects to lower dimensions—slicing, projection, holography, and sometimes just ignoring the dimension—and how we work backwards to deduce the three-dimensional objects they represent.

The least complicated way of seeing beyond the surface is to make slices. Each slice is two-dimensional, but the combination of the slices forms a real three-dimensional object. For example, when you order ham at the deli, the three-dimensional lump of ham is readily exchanged for many two-dimensional slices.* By stacking all the slices you could reconstruct the full three-dimensional shape.

This book is three-dimensional. However, its pages have only two dimensions. The union of the two-dimensional pages comprises the book.† We could illustrate this union of pages in many ways. One is shown in Figure 8, in which we view the book edge on. In this picture we've again played with dimensionality, since each line represents a page. So long as we all know that the lines represent two-dimensional pages, this illustration should be clear. Later on, we'll use a similar shorthand when we depict objects in multidimensional worlds.

Slicing is only one way to replace higher dimensions with lower

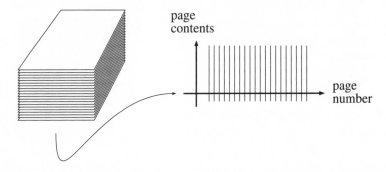

Figure 8. *A three-dimensional book is made up of two-dimensional pages.*

---

*Slices of ham do have some thickness, so they are in reality thin, but three-dimensional. Their size in this extra dimension is so small that it is a good approximation to think of them as two-dimensional. However, even with arbitrarily thin two-dimensional slices, we can imagine putting them together to make a three-dimensional object in this way.

†Again, for the pages to be truly two-dimensional they would have to be infinitely thin slices with no thickness at all in the third dimension. For now, though, two dimensions is a fine approximation for pages as thin as these.

ones. *Projection*, a technical term borrowed from geometry, is another. A projection gives a definite prescription for creating a lower-dimensional representation of an object. A shadow on a wall is an example of a two-dimensional projection of a three-dimensional object. Figure 9 illustrates how information is lost when we (or rabbits) make a projection. Points on the shadow are identified by only two coordinates, left-right or up-down along the wall. But the object that is projected also has a third spatial dimension that the projection doesn't retain.

Figure 9. *A projection carries less information than the higher-dimensional object.*

The simplest way to make a projection is to just ignore one dimension. For example, Figure 10 shows a cube in three dimensions being projected onto two dimensions. The projections can take many forms, the simplest of which is a square.

To return to our earlier examples of the graphs of Ike and Athena, we might make a two-dimensional plot of Ike by neglecting his driving fast cars. And we might not really want to know the number of owls Athena raises, and might therefore make a four-dimensional rather than a five-dimensional plot. Disregarding Athena's owls is a projection.

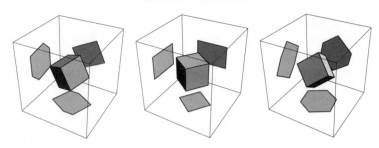

Figure 10. *Projections of a cube. Notice that the projection can be a square, as we see in the middle diagram, but that projections can also take other shapes.*

A projection discards information from the original, higher-dimensional object (see Figure 9). However, when we make a lower-dimensional picture using a projection, we sometime include information to help retain some of what was lost. The additional information might be shading or color, as in a painting or photograph. It might be a number, as in a topographic map to illustrate height. Or there might be no label at all, in which case the two-dimensional characterization simply offers less information.

Without both our eyes, which work together to let us reconstruct three dimensions, everything we see would be projections. Depth perception is tougher when you close one of your eyes. A single eye constructs a two-dimensional projection of three-dimensional reality. You need two eyes to reproduce three dimensions.

I am nearsighted in one eye and farsighted in the other, so I don't properly combine the images from both eyes unless I'm wearing glasses—which is rarely the case. Although I was told I should have trouble reconstructing three dimensions, I don't usually notice any problem: things still look three-dimensional to me. That is because I rely on shading and perspective (and my familiarity with the world) to reconstruct three-dimensional images.

But one day in the desert, a friend and I were trying to reach a distant cliff. My friend kept telling me that we could walk directly there, and I couldn't understand why he was insisting that we should walk straight through a piece of rock. It turned out the rock that I thought projected directly from the cliff, so that it would completely block our way, was in fact located much closer to us, in front of the cliff. The rock I had thought would bar our path wasn't actually attached to the cliff at all. This misunderstanding occurred because we were near the cliff around noon, when there were no shadows, and I had no way to construct the third dimension that would have told me how the distant cliffs and rocks were lined up. I wasn't really conscious of my compensating strategy of using shading and perspective until then, when it failed.

Painting and drawing have always required artists to reduce what they see to projected images. Medieval art did this in the simplest manner. Figure 11 shows a mosaic image of a city as a two-dimensional projection. This mosaic doesn't tell us anything about a third dimension; there are no labels or indications of its existence.

Since medieval times, painters have developed ways to make projections that partially redress painting's loss of a dimension. One approach that opposes the medieval flattening of space is the method used by the cubists in the twentieth century. A cubist painting (for example Picasso's *Portrait of Dora Maar*, Figure 12) presents several projections simultaneously, each from a different angle, and thereby conveys the subject's three-dimensionality.

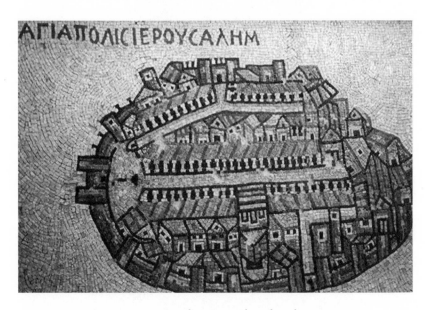

Figure 11. *A two-dimensional medieval mosaic.*

Figure 12. Portrait of Dora Maar, *a cubist painting by Picasso.*

Figure 13. *Dali's* Crucifixion (Corpus Hypercubus).

Most Western painters since the Renaissance, however, have used perspective and shading to create the illusion of a third dimension. One of the essential skills in painting is the ability to reduce a three-dimensional world to a two-dimensional representation that allows the observer to reverse the process and reconstitute the initial three-dimensional scene or object. We are acculturated to know how to decode the images, even though not all of the three-dimensional information is there.

Artists have even tried representing higher-dimensional objects on two-dimensional surfaces. For example, Salvador Dali's *Crucifixion (Corpus Hypercubus)* (see Figure 13) shows the cross as an opened-up hypercube. A hypercube consists of eight cubes attached in four-dimensional space. These are the cubes he has drawn. I've shown a few projections of a hypercube in Figure 14.

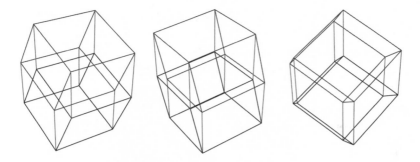

Figure 14. *Projections of a hypercube.*

I have already mentioned a physics example: quasicrystals, which look like the projection of a higher-dimensional crystal into our three-dimensional world. Projections can also be used for practical, not just artistic purposes. Medicine contains many examples where three-dimensional objects are projected onto two dimensions. An X-ray always records a two-dimensional projection. CAT (computer-assisted tomography) scans combine multiple X-ray images to reconstruct a more informative three-dimensional representation. With X-rays taken from sufficiently many angles, one can use interpolation to piece together full three-dimensional images. An MRI (magnetic resonance imaging) scan, on the other hand, reconstructs a three-dimensional object from slices.

A holographic image is another way to record three dimensions on a two-dimensional surface. Although a holographic image is recorded on a lower-dimensional surface, it actually carries all the information of the original higher-dimensional space. You probably have an example of this technique in your wallet: the three-dimensional-looking image on your credit card is a hologram.

A holographic image records relationships between light in different places, so that the full higher-dimensional image can be recovered. This principle is much the same as that used in a good stereo, which lets you hear where instruments were being played in relation to each other when they were recorded. With the information stored in a hologram, the eye can truly reconstruct the three-dimensional object it represents.

These methods tell us how we might get more information from a lower-dimensional image. But maybe all you really need is less information. Sometimes you just don't care about all three dimensions. For example, something might be so thin in the third dimension that nothing interesting happens in this direction: even though the ink on this paper is really three-dimensional, we lose nothing by thinking of it as two-dimensional. Unless we look at the page under a microscope, we simply don't have the necessary resolution to see the ink's thickness. A wire looks one-dimensional even though, on closer inspection, you can see it has a two-dimensional cross-section and therefore three dimensions in all.

Effective Theories

There is nothing wrong with ignoring an extra dimension that's too small to be seen. Not only the visual effects, but also the physical effects of tiny, undetectable processes can usually be ignored. Scientists often average over or ignore (often unwittingly) physical processes that occur on immeasurably small scales when formulating their theories or setting up their calculations. Newton's laws of motion work at the distances and speeds he could observe. He didn't need the details of general relativity to make successful predictions. When biologists study a cell, they don't need to know about quarks inside the proton.

Selecting relevant information and suppressing details is the sort of pragmatic fudging everyone does every day. It's a way of coping with too much information. For almost anything you see, hear, taste, smell, or touch, you have the choice between examining details by scrutinizing very closely, and looking at the "big picture" with its other priorities. Whether you are staring at a painting, tasting wine, reading philosophy, or planning your next trip, you automatically parcel your thoughts into the categories of interest—be they sizes or flavors or ideas—and the categories that you don't find relevant at the time. When appropriate, you ignore some details so that you can focus on the issue of interest, and not obscure it with inessential details.

This procedure of disregarding small-scale information should be familiar because it's actually a conceptual leap people make all the time. Take New Yorkers, for example. New Yorkers living in the thick of the city see the details and variation within Manhattan. To them, downtown is funkier, older, with narrower, more crooked streets. Uptown has more real estate that was designed for human beings to actually live in, as well as Central Park and most of the museums. Although such distinctions are blurred from far away, within the city they are very real.

But now think about how people far away see New York. To them, it's a dot on a map. An important dot, perhaps, a dot with a distinctive character; but from outside New York, a dot nonetheless. Even with all their variety, New Yorkers are in a single category when viewed from the Midwest or Kazakhstan, for example. When I mentioned this analogy to my cousin who lives downtown (in the West Village, to be precise), he confirmed my point by balking at the suggestion of grouping together New Yorkers living uptown and downtown. Nonetheless, as any non-New Yorker could tell him, the distinctions are too small to matter to people not living in their midst.

It is common practice in physics to formalize this intuition, and organize categories in terms of the distance or energy that is relevant. Physicists accept this practice and have given it a name—*effective theory*. The effective theory concentrates on the particles and forces that have "effects" at the distance in question. Rather than describing particles and interactions in terms of unmeasurable parameters that describe ultra-high-energy behavior, we formulate observations in terms of the things that are actually relevant to the scales we might detect. The effective theory at any single distance scale doesn't go into the details of an underlying short-distance physical theory; it only asks about things you could hope to measure or see. If something is beyond the resolution of the scales at which you are working, you don't need its detailed structure. This practice is not scientific fraud, but a way of disregarding the clutter of superfluous information. It is an "effective" way to obtain accurate answers efficiently.

Everyone, including physicists, is happy to return to a three-dimensional universe when higher-dimensional details are beyond our resolution. Just as physicists will often treat a wire as if it is one-dimensional, we will also describe a higher-dimensional universe in lower-dimensional terms when the extra dimensions are minuscule and higher-dimensional details are too tiny to matter. Such a lower-dimensional description would summarize the observable effects of all possible higher-dimensional theories in which the extra dimensions are too tiny to see. For many purposes, such a lower-dimensional description is adequate, independent of the number, size, and shape of the additional dimensions.

The lower-dimensional quantities are not providing the fundamental description, but they are a convenient way of organizing observations and predictions. If you do know the short-distance details, or the microstructure, of a theory, you can use them to derive the quantities that appear in the low-energy description. Otherwise, those quantities are just unknowns to be experimentally determined.

The following chapter elaborates these ideas and considers the consequences of tiny rolled-up extra dimensions. The dimensions we'll consider first are minuscule, too tiny to make any difference at all. Later on, when we return to extra dimensions, we'll explore both the large and the infinite dimensions that recently radically revised this picture.

2

Restricted Passages:
Rolled-up Extra Dimensions

No way out
None whatsoever.

Jefferson Starship

Athena awoke with a start. The previous day she had read Alice in Wonderland *and* Flatland *in order to seek some inspiration about dimensions. But that night she had the strangest dream, which, when fully conscious, she recognized as the result of having read the two books on the same day.* *

Athena dreamed she had turned into Alice, slipped into a rabbit hole, and met the resident Rabbit, who had pushed her out into an unfamiliar world. Athena had thought it a rather rude way to convey a guest. Even so, she had eagerly looked forward to her upcoming adventure in Wonderland.

Athena was in for a disappointment, however. The resident Rabbit, who was fond of puns, had sent her instead to OneDLand, a strange, not so wonderful, one-dimensional world. Athena looked around— or, I should say, to her left and right—and discovered that all she could see were two points—one to her left and another to her right (but in a prettier color, she thought).

In OneDLand, all the one-dimensional people with their one-dimensional possessions were lined up along this single dimension like long, thin beads strung out along a thread. But even with her limited

---

*Or perhaps this story is a result of my having begun my education at the perhaps questionably named Lewis Carroll School, P.S. 179, in Queens.

purview, Athena knew there must be more to OneDLand than met her eyes because of the outrageous din that met her ears. A Red Queen was well hidden behind a dot, but Athena couldn't miss her strident yells: "This is the most ridiculous chess game I have ever seen! I can't move any pieces, not even to castle!" Athena was relieved when she realized her one-dimensional existence shielded her from the wrath of the Red Queen.

But Athena's cozy universe did not last long. Slipping through a gap in OneDLand, she returned to the dreamworld's rabbit hole, which had an elevator that could take her to hypothetical, other-dimensional universes. Almost immediately, the Rabbit announced, "Next stop: TwoDLand—a two-dimensional world." Athena didn't think "TwoDLand" a very nice name, but she cautiously entered all the same.

Athena needn't have been so hesitant. Almost everything in TwoD-Land looked the same as in OneDLand. She did notice one differ-ence—a vial labeled "Drink me." Bored with one dimension, Athena promptly obeyed. She quickly shrank to a tiny size, and as she became smaller, a second dimension came into view. This second dimension was not very big—it was wrapped around in a fairly small circle. Her surroundings now resembled the surface of an extremely long tube. A Dodo was racing around this circular dimension, but he wanted to stop. So he kindly offered Athena, who looked rather hungry, some cake.

When Athena ate a morsel of the Dodo's dreamcake, she started to grow. After only a few bites (she was quite sure of this, as she was still rather hungry), the cake very nearly disappeared; all that remained was a very tiny crumb. At least Athena thought there was a crumb, but she could see it only when she squinted very hard. And the cake wasn't the only thing that had vanished from view: when Athena returned to her usual size, the entire second dimension had dis-appeared.

She thought to herself, "TwoDLand is very odd indeed. I'd best be getting home." Her return journey was not without further adven-tures, but those will be kept for another time.

Even if we don't know *why* three spatial dimensions are special, we can ask *how*. How is it possible that the universe could appear to have only three dimensions of space if the fundamental underlying spacetime contains more? If Athena is in a two-dimensional world, why does she sometimes see only one? If string theory is the correct description of nature, and there are nine dimensions of space (plus one of time), what has become of the missing six spatial dimensions? Why aren't they visible? Do they have any discernible impact on the world we see?

The last three questions are central to this book. However, the first order of business is to determine whether there is any way in which the evidence of extra dimensions can be hidden so that Athena's two-dimensional world would appear as one-dimensional, or a universe with extra dimensions would appear to have the three-spatial-dimensional structure we observe around us. If we're to accept the idea of a world with extra dimensions, whatever theory they come from, there must be a good explanation for why we have not yet detected even the slightest trace of their existence.

This chapter is about extremely small *compactified*, or rolled-up, dimensions. They don't extend for ever, like the three familiar dimensions; instead, they quickly loop back on themselves, like a tightly wound spool of thread. No two objects could be separated very far along a compactified dimension; any attempt at a long-distance excursion would instead turn into a journey that went round and round, like the Dodo's laps. Such compactified dimensions could be so small that we wouldn't ever notice their existence. Indeed, we'll see that if tiny rolled-up dimensions exist, they will be quite a challenge to detect.

Rolled-up Dimensions in Physics

String theory, the most promising candidate for a theory combining quantum mechanics and gravity, gives a concrete reason to think about extra dimensions: the only coherent versions of string theory that we know of are laden with these surprising appendages. However, although the arrival of string theory in the physics world improved the respectability of extra dimensions, the idea of extra dimensions originated much earlier.

Back in the early twentieth century, Einstein's theory of relativity opened the door to the possibility of extra dimensions of space. His theory of relativity describes gravity, but it doesn't tell us why we experience the particular gravity we do. Einstein's theory does not favor any particular number of spatial dimensions. It works equally well for three or four or ten. Why, then, do there seem to be only three?

In 1919, close on the heels of Einstein's theory of general relativity (completed in 1915), the Polish mathematician Theodor Kaluza recognized this possibility in Einstein's theory and boldly proposed a fourth spatial dimension, a new unseen dimension of space.* He suggested that the extra dimension somehow might be distinguished from the three familiar infinite ones, though he didn't specify how. Kaluza's goal with this extra dimension was to unify the forces of gravity and electromagnetism. Although the details of that failed unification attempt are irrelevant here, the extra dimension that he had so brazenly introduced is very relevant indeed.

Kaluza wrote his paper in 1919. Einstein, who was the referee evaluating it for publication in a scientific journal, wavered about the merits of the idea. Einstein delayed the publication of Kaluza's paper for two years, but eventually acknowledged its originality. Yet Einstein still wanted to know what this dimension was. Where was it and why was it different? How far did it extend?

These are the obvious questions to ask. They might be some of the very same questions that are bothering you. No one responded to Einstein until 1926, when the Swedish mathematician Oskar Klein addressed his questions. Klein proposed that the extra dimension would be curled up in the form of a circle, and that it would be extremely small, just 10^{-33} cm,† one tenth of a millionth of a trillionth

---

*We will specify spatial dimensions in this and the following chapter. After introducing relativity, we will switch to spacetime, and consider time as an additional dimension.

†I will sometimes use scientific notation for very large or very small numbers. When a power of ten has a negative exponent, as in 10^{-33}, it indicates a decimal number; for example, 10^{-33} is the number 0.000,000,000,000,000,000,000,000,000,000,001.

of a trillionth of a centimeter. This tiny rolled-up dimension would be everywhere: each point in space would have its own minuscule circle, 10^{-33} cm in size.

This small quantity represents the Planck length, a quantity that will be relevant later when we discuss gravity in more detail. Klein picked the Planck length because it is the only length that could naturally appear in a quantum theory of gravity, and gravity is connected to the shape of space. For now, all you need to know about the Planck length is that it is extraordinarily, unfathomably small—far smaller than anything we would ever have a chance of detecting. It is about twenty-four orders of magnitude† smaller than an atom and nineteen orders of magnitude smaller than a proton. It's easy to overlook anything as tiny as that.

There are many examples in daily life of objects whose extent in one of the three familiar dimensions is too small to be noticed. The paint on a wall, or a clothesline viewed from far away, are examples of things that seem to extend in fewer than three dimensions. We overlook the paint's depth and the clothesline's thickness. To a casual observer, the paint looks as if it has only two dimensions, and the clothesline appears to have only one, even though we know that actually both have three. The only way to see the three-dimensional structure of such things is to look up close, or with sufficiently fine resolution. If we stretched a hose across a football field and viewed it from a helicopter above, as is illustrated in Figure 15, the hose would look one-dimensional. But up close, you can resolve the two dimensions of the hose's surface and the three-dimensional volume it encloses.

For Klein, though, the thing that was undiscernibly small was not the thickness of an object, but a dimension itself. So what does it mean for a dimension to be small? What would a universe with a

---

This is an extremely tiny number and would be too cumbersome to write in full each time it occurs. A number with a positive exponent, such as 10^{33}, has 33 zeroes after a 1, 1,000,000,000,000,000,000,000,000,000,000,000, which is an enormous number that would also be difficult to write in full each time. I will often give a number in both scientific notation and in words the first time I use it.

†An order of magnitude is a factor of ten. Twenty-four orders of magnitude is 1,000,000,000,000,000,000,000,000, or one trillion trillion.

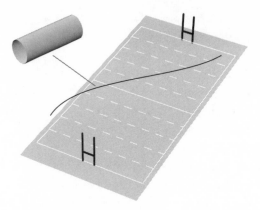

Figure 15. *When you view a hose spread over a football field from above, it looks like it has one dimension. But when you view it up close, you see that the surface has two dimensions and the volume it encloses has three.*

curled-up dimension look like to someone living inside it? Once again, the answer to these questions depends entirely on the size of the curled-up dimension. Let's consider an example to see what the world would look like to conscious beings that are small or big compared with the size of a rolled-up extra dimension. Because drawing four or more spatial dimensions is impossible, the first picture I'll present of a universe with a small, compactified dimension will have only two dimensions, with one of them rolled up tightly to a very small size (see Figure 16).

Figure 16. *When one dimension is curled up, a two-dimensional universe looks one-dimensional.*

Imagine again a garden hose, which can be thought of as a long sheet of rubber rolled up into a tube with a small circular cross-section. This time, we'll think of the hose as the entire universe (not an object inside the universe).* If the universe were shaped like this garden hose, we would have one very long dimension and one very small, rolled-up dimension—exactly what we want.

For a little creature—a flat bug, say—that lived in the garden-hose universe, the universe would look two-dimensional. (In this scenario, our bug has to stick to the surface of the hose—the two-dimensional universe doesn't include the interior, which is three-dimensional.) The bug could crawl in two directions: along the length of the hose or around it. Like the Dodo, who could run laps in its two-dimensional universe, a bug that started somewhere along the hose could crawl around and eventually return to where it started. Because the second dimension is small, the bug wouldn't travel very far before it returned.

If a population of bugs living on the hose experienced forces, such as the electric force or gravity, those forces would be able to attract or repel bugs in any direction on the surface of the hose. Bugs could be separated from one another either along the length of the hose or around the hose's circumference, and would experience any force that was present on the hose. Once there is sufficient resolution to distinguish distances as small as the diameter of the hose, forces and objects exhibit both of the dimensions they actually have.

However, if our bug could observe its surroundings, it would notice that the two dimensions were very different. The one along the length of the hose would be very big. It could even be infinitely long. The other dimension, on the other hand, would be very small. Two bugs could never get very far from each other in the direction around the hose. And a bug that tried to take a long trip in that direction would quickly end up back where it started. A thoughtful bug that liked to stretch its legs would know that its universe was two-dimensional, and that one dimension extended a long way while the other was very small and rolled up into a circle.

But the bug's perspective is nothing like the one that creatures like

---

*The garden hose has always been a popular analogy to illustrate rolled-up dimensions. I learned it at math camp and it has most recently been described in Brian Greene's *Elegant Universe* (Norton, 1999; Vintage, 2000). I'll use this same analogy since it's so good and because I want to expand on it in the following section (and in later chapters), in which I'll also include sprinklers to explain extra-dimensional gravity.

us would have in Klein's universe, in which the extra dimension is rolled up to an extremely small size, 10^{-33} cm. Unlike the bug, we are not small enough to detect—never mind travel in—a dimension of such a tiny size.

So to complete our analogy, suppose that something much bigger than a bug, capable only of much coarser resolution and therefore unable to detect small objects or structure, lived in the garden-hose universe. Since the lens through which this bigger being views the world blurs any details that are as small as the hose's diameter, from the vantage point of this bigger being the extra dimensions would be invisible. It would see only a single dimension. Someone would see that the garden-hose universe had more than a single dimension only if he had sufficiently sharp vision to register something as small as the width of the hose. If his vision is too fuzzy to register that width, all he'll ever notice is a line.

Moreover, physical effects wouldn't betray the extra dimension's existence. Big beings in the garden-hose universe would fill out the entire second, small dimension and would never know that this dimension was there. Without the ability to detect structure or variations along the extra dimension, such as wiggles or undulations of matter or energy, they could never register its existence. Any variations along the second dimension would be completely washed out, much as any variation in the thickness of a piece of paper on the scale of its atomic structure is something you don't ever notice.

The two-dimensional world in which the dreaming Athena found herself was very much like the garden-hose universe. Because Athena had the opportunities to be both big and small relative to TwoDLand's width, she could observe this universe from both the perspective of someone bigger and that of someone smaller than its second dimension. To the big Athena, TwoDLand and OneDLand appeared the same in every respect. Only the small Athena could tell the difference. Similarly, in the garden-hose universe a being would be ignorant of an additional spatial dimension if it were too tiny for it to see.

Let's now return to the Kaluza-Klein universe, which has the three spatial dimensions we know about, supplemented by an extra one that's unseen. We can again use Figure 16 to think about this situation. Ideally, I would draw four spatial dimensions, but unfortunately that's not possible (even a pop-up book wouldn't suffice). However, since the three infinite dimensions that constitute our space are all qualitatively the same, I really need only draw just one representative dimension. That leaves me free to use the other dimension to represent the unseen extra dimension. The other dimension shown here is the one that's curled up—the one that's fundamentally different from the other three.

Just as with our two-dimensional garden-hose universe, a four-dimensional Kaluza-Klein universe with a single tiny, rolled-up dimension would appear to us to have one dimension fewer than the four it actually has. Because we wouldn't know about the additional spatial dimension unless we could detect evidence of structure on its minute scale, the Kaluza-Klein universe would look three-dimensional. Rolled-up, or compactified, extra dimensions will never be detected if they are sufficiently tiny. Later on, we'll investigate just *how* tiny, but for now, rest assured that the Planck length is well below the threshold of detectability.

In life, and in physics, we only register those details that actually matter to us. If you cannot observe detailed structure, you might as well pretend it isn't there. In physics, this disregard of local detail is embodied in the effective theory idea of the previous chapter. In an effective theory, all that matters are the things that you can actually perceive. In the example above, we would use a three-dimensional effective theory where the information about extra dimensions is suppressed.

Although the curled-up dimension of the Kaluza-Klein universe is not far away, it's so small that any variation within it is imperceptible. Just as differences among New Yorkers don't really matter to people outside, the structure in the extra dimensions of the universe is irrelevant when its details vary on such minuscule a scale. Even if fundamentally there turn out to be many more dimensions than we acknowledge in our daily lives, everything we see can still be described in terms of only the dimensions we observe. Extremely small extra dimensions change nothing about the way we view the world, or even about how we do most physics calculations. Even if additional dimensions exist, if we are incapable of seeing or experiencing them, we can ignore them and still correctly describe what we see. Later on we'll see modifications to this simple picture for which this won't always be true, but those will involve additional assumptions.

We can understand one further important point about a rolled-up dimension from Figure 17, which illustrates the hose, or universe with one dimension, rolled up into a circle. Focus on any point along the infinite dimension. Notice that at each and every point there sits the entire compact space, namely the circle. The hose consists of all these circles glued together, like the slices I talked about in Chapter 1.

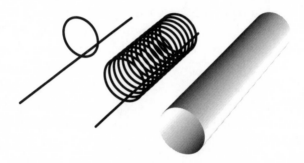

Figure 17. *In a two-dimensional universe, when a dimension is curled up there is a circle at every point along the infinite dimension of space.*

Figure 18 presents a different example: here there are two infinite dimensions rather than one, plus a single additional dimension curled up into a circle. In this case, there is a circle at each and every

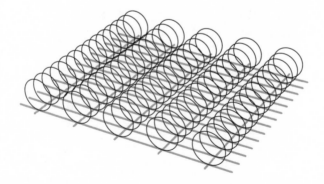

Figure 18. *In a three-dimensional universe, if one of the three dimensions is curled up you have a circle at every point in the plane.*

point in the two-dimensional space. And if there were three infinite dimensions, the rolled-up dimensions would exist at every point in three-dimensional space. You might liken the points in extra-dimensional space to the cells in your body, each of which carries your entire DNA sequence. Similarly, each point in our three-dimensional space could host an entire compactified circle.

So far, we've only considered a single additional dimension, which is rolled up into a circle. But everything we've said would hold true even if that curled-up dimension took some other shape—any shape at all. And it would also be true if there were two or more tiny, rolled-up dimensions of any shape at all. Any and all dimensions that are sufficiently small would be completely invisible to us.

Let us consider an example with two rolled-up dimensions. There are many possible shapes that these rolled-up dimensions could take. We'll choose a *torus*, a donut-like shape in which the two additional dimensions are both simultaneously rolled into a circle. This is illustrated in Figure 19. If both circles—the one that winds through the donut hole and the one that winds around the donut itself—are sufficiently small, the additional two rolled-up dimensions would never be seen.

Figure 19. *When two out of four dimensions are curled into a donut, you have a donut at every point in space.*

But that's just one example. With more dimensions there are a huge number of conceivable *compact spaces*—spaces with rolled-up dimensions, distinguished by the precise manner in which the dimensions are rolled up. One category of compact spaces important to string theory are the *Calabi-Yau manifolds*, named after the Italian mathematician Eugenio Calabi, who first proposed these particular shapes, and the Chinese-born Harvard mathematician Shing-Tung Yau, who showed that they are mathematically possible. These geometric shapes roll up and wind together extra dimensions in a very special way. The dimensions are curled up into a small size, as with all compactifications, but they are tangled in a way that is more complicated and difficult to draw.[4]

Whatever shape the rolled-up extra dimensions take, and however many there are, at each point along the infinite dimensions there would be a small compact space containing all the curled-up dimensions. So, for example, if string theorists are right, everywhere in visible space— at the tip of your nose, at the North Pole of Venus, at the spot above the tennis court where your racket hit the ball the last time you served—there would be a six-dimensional Calabi-Yau manifold of invisibly tiny size. The higher-dimensional geometry would be present at every point in space.

String theorists often suggest—as Klein did—that curled-up dimensions are as small as the Planck length, 10^{-33} cm. Planck-length-size compact dimensions would be extraordinarily well hidden; there is almost certainly no way for us to detect something so small. Therefore, Planck-length extra dimensions would very likely leave no visible trace of their existence. So even if we live in a universe with Planck-length extra dimensions, we would still register only the three familiar dimensions. The universe could have many such tiny dimensions, but we might never have the resolving power to find out.

Newton's Gravitational Force Law with Extra Dimensions

It is nice to have a pictorial, descriptive explanation for why the extra dimensions are hidden when compactified, or rolled up, to a very minute size. But it's a good idea to check that the laws of physics accord with this intuition.

Let's take a look at Newton's gravitational force law, the well-established form of the gravitational force law that Newton proposed in the seventeenth century. This law tells us how the gravitational force depends on the distance between two massive objects.* It's known as an *inverse square law*, which means that the strength of gravity decreases with distance proportionally to the distance squared. For example, if you double the distance between two objects, the strength of their gravitational attraction goes down by a factor of four. If the separation is increased to three times its original value, gravitational attraction decreases by a factor of nine. The inverse square law of gravity is one of the oldest and most important laws of physics. Among other things, it is the reason that planets have the type of orbits they do. Any viable physical theory of gravity must reproduce the inverse square law or it would be bound to fail.

The way in which the gravitational force law depends on distance, which is encoded in Newton's inverse square law, is intimately connected to the number of spatial dimensions. This is because the number of dimensions determines how quickly gravity diffuses as it spreads out in space.

Let's reflect on the connection, which will be very relevant to us later on when we consider extra dimensions. We'll do this by imagining a water supply whose water can be directed through either a hose or a sprinkler. We'll assume that both the hose and the sprinkler have the same amount of water running through them, and that they can each water a certain flower in a garden (see Figure 20). When the water goes through the hose, which is directed at the flower, the flower will get all the water. The distance from the base of the hose to the nozzle directed at the flower is irrelevant, because all the water must end up on the flower, no matter how far away the hose happened to be.

However, suppose instead that the same water is directed through a sprinkler that simultaneously waters many flowers. That is, the sprinkler sends out water in a circle, reaching all the flowers a certain distance away. Because the water will now be distributed among everything at that distance, the original flower will no longer get all

---

*In this book a "massive" object means an object with mass. A massive object is to be distinguished from a "massless" object, which has zero mass (and travels at the speed of light).

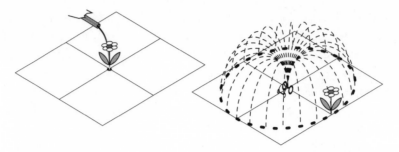

Figure 20. *The amount of water delivered to a flower by a sprinkler that sprinkles water around a circle is less than the amount delivered directly by a hose.*

the water. Moreover, the farther away the flower is from the source, the more greenery the sprinkler will water, and the more widely distributed the water will be (see Figure 21). That's because you can fit more plants on a circle three meters in circumference, say, than a circle just one meter in circumference. Because the water is more widely spread out, a farther away flower receives less water.

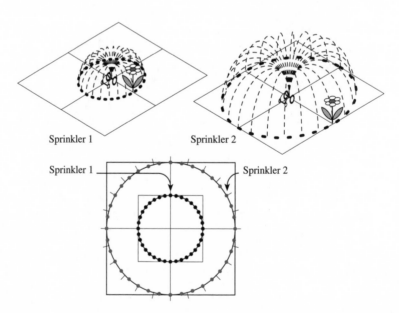

Figure 21. *When a sprinkler delivers water around a circle of larger radius, the water is spaced out more and the flower receives less water.*

Similarly, anything that is shared uniformly in more than one direction will have a smaller impact on any particular thing that is farther away—whether that thing is a flower or, as we will soon see, an object experiencing the force of gravity. Gravity, like water, is more widely distributed when it is farther away.

With this example, we can also see why the distribution depends so strongly on the number of dimensions in which water (or gravity) is spread. The water from the two-dimensional sprinkler is spread out with distance, unlike the water from the one-dimensional hose, which is not spread out at all. Now imagine a sprinkler that spreads its water over the surface of a sphere, and not just around a circle. (Such a sprinkler would look something like a dandelion gone to seed.) Here, the water will spread out with distance much more quickly.

Let's now apply this reasoning to gravity, and derive the precise distance dependence of the gravitational force in three dimensions. Newton's gravitational force law follows from two facts: that gravity acts equally in all directions, and that there are three dimensions of space. Let's now imagine a planet, which attracts any mass in its vicinity. Because the gravitational force is the same in all directions, the strength of the gravitational attraction that the planet exerts on another massive object—a moon, for example—will depend not on direction, but on the distance between them.

To pictorially represent the strength of the gravitational force, the left of Figure 22 shows radial lines extending outwards from the planet's center, resembling water spreading out from a sprinkler. The density of these lines determines the strength of gravitational attraction that the planet exerts on anything in its vicinity. More force

Figure 22. *Gravitational force lines emitted from a massive object, such as a planet. The same number of lines intersect a sphere of any radius; therefore, the force lines are more diffuse and gravity is weaker the farther you are from the massive object at the center.*

lines passing through an object would mean a greater gravitational attraction, and fewer force lines would mean a smaller gravitational attraction.

Notice that the same number of force lines intersect a spherical shell drawn any distance away, no matter how far or near (center and right of Figure 22). The number of force lines never changes. But because the force lines are spread out among all the points on the sphere's surface, the force at a greater distance is necessarily weaker. The precise dilution factor is determined by the quantitative measure of how widely distributed the force lines are at any given distance.

A fixed number of force lines passes through a sphere's surface, whatever its distance from the mass. The area of that sphere's surface is proportional to its radius squared: the surface area is equal to a number multiplied by the square of the radius. Because the fixed number of gravitational force lines is spread out over the sphere's surface, the gravitational force has to decrease as the square of the radius. This spreading out of the gravitational field is the origin of the inverse square law for gravity.

Newton's Law with Compact Dimensions

So we now know that in three dimensions, gravity should obey an inverse square law. Notice that the argument seems to depend critically on the fact that there are three spatial dimensions. Had there been only two dimensions, gravity would have been spread out only over a circle, and the force of gravity would have decreased with distance at a slower rate. Had there been more than three dimensions, the surface area of a hypersphere would have grown far more rapidly with the separation between the planet and its moon, and the force would have fallen off that much more quickly. It seems that only three spatial dimensions yields the inverse square distance dependence. But if that is the case, how can theories with extra dimensions yield Newton's inverse square law for gravity?

It is very interesting to see how compactified dimensions resolve this potential conflict. The essence of the logic is that force lines cannot spread arbitrarily far into the compact dimensions because those compact dimensions have finite size. Although force lines initially spread out in all dimensions, when they have spread beyond the extra dimensions' sizes they have no choice but to spread out solely in the directions of the infinite dimensions.

This can be illustrated once again with our hose example. Imagine that water enters the hose through a small pinhole in a cap covering the end of the hose (see Figure 23). Water directed through the puncture will not immediately travel directly down the hose, but will first spread throughout the tube's cross-section. Nonetheless, it should be clear that if you were at the other end of the hose, watering your flower, the way the water entered would make no difference at all. Although the water would first spread in more than one direction, it would quickly reach the inside surface of the hose and flow once again as if there were only one direction. This is essentially what happens to gravitational field lines in small, compactified dimensions.

Figure 23. *Water entering a garden hose through a pinhole at the end first spreads in three dimensions before traveling only along the single long dimension of the hose.*

As before, we can imagine a fixed number of force lines emanating from a massive sphere. At a distance smaller than the extra dimensions' size, these force lines will spread out equally in all directions. If you could measure gravity on that small scale, you would measure the consequences of higher-dimensional gravity. The force lines would spread the way water does as it enters the hose through the pinhole and spreads throughout the hose's interior.

However, at distances greater than the extra dimensions' sizes, the force lines can spread only in the infinite directions (see Figure 24). In the small, compact dimensions, the force lines will hit the edge of space, so they can't spread out any farther that way. They have to bend, and the only way for them to go is in the direction of the large

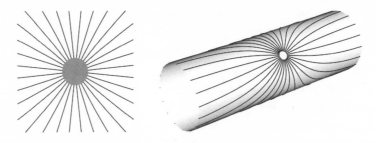

Figure 24. *Gravitational force lines emitted from a massive object when a dimension is curled up. The force lines spread radially over short distances, but over long distances they extend only along the infinite dimension.*

dimensions. Therefore, at distances greater than the sizes of the extra dimensions, it's just as if the extra dimensions didn't exist, and the force law reverts to Newton's inverse square law—the one we observe. This means that even from a quantitative point of view, you won't know there are extra dimensions if you measure the gravitational force only between objects with separations greater than the curled-up dimensions' size. The distance dependence reflects extra dimensions only in the tiny region inside the compact space.

Other Ways to Bound Dimensions?

We've now established that when extra dimensions are sufficiently small, they are invisible and have no detectable consequences on the length scales we observe. For a long time, string theorists assumed that extra dimensions were Planck-length dimensions, but recently some of us have questioned this assumption.

No one understands string theory well enough to say definitively what the sizes of extra dimensions will turn out to be. Sizes comparable to the Planck length are possible, but any dimension too small to observe is also in the running. The Planck length is so tiny that even considerably larger curled-up dimensions might well escape notice. An important question for the study of extra dimensions is just how big these dimensions can be, given that we haven't seen them yet.

The questions we'll address in this book include how big extra dimensions can be, whether these dimensions have any discernible effect on elementary particles, and how experiments might probe them. We will see that the existence of extra dimensions can significantly change the rules by which we do particle physics and, furthermore, that some of these changes will have experimentally observable consequences.

An even more radical question we'll investigate is whether additional dimensions have to be small. We don't see tiny dimensions, but do dimensions have to be small to be invisible? Could an extra dimension possibly extend for ever without our seeing it? If so, extra dimensions would have to be very different from the dimensions we've looked at. So far I've presented only the simplest possibility. We'll see later why even the radical possibility of an infinite extra dimension cannot be excluded if it is sufficiently different from the three familiar infinite dimensions.

The next chapter will address yet another question that might have occurred to you: why can't small extra dimensions just be intervals, not curled up into a ball but instead bounded between two "walls"? This possibility didn't occur to anyone right away—but why not? The reason is that imagining an end to space entails knowing what is happening there. Would things fall off the end of the universe, as old pictures of the flat Earth seemed to imply? Or would they be reflected back? Or would they never get there? The need to specify what would happen at the end means that you have to know what scientists call *boundary conditions*. If space ends, where and on what does it end?

Branes—membrane-like objects in higher-dimensional space—provide the necessary boundary conditions for worlds that "end." As we will see in the following chapter, branes can make a world (or many worlds) of difference.

3

Exclusive Passages:
Branes, Braneworlds, and the Bulk

I'm gonna stick like glue,
Stick, because I'm stuck on you.

Elvis Presley

Unlike the studious Athena, Ike rarely read any books. He generally preferred playing with games, gadgets, and cars. But Ike hated driving in Boston, where the drivers were reckless, the roads were badly signposted, and the highways were invariably under construction. Ike always ended up stuck in traffic, which he found especially frustrating when he could see a nearly empty freeway overhead. Though the empty road would be tempting, Ike would have no way to quickly reach it since, unlike Athena's owls, he couldn't fly. For Ike trapped on slow roads in Boston, the third dimension was no use at all.

Until very recently, few self-respecting physicists considered extra dimensions worth thinking about. They were too speculative and too foreign: no one could say anything definitive about them. But in the last few years, extra dimensions have found their fortunes rising. No longer shunned as undesirable gatecrashers, they've evolved into highly sought-after, stimulating company. They owe their newfound respectability to branes and to the many genuinely new theoretical possibilities that these fascinating constructs have introduced.

Branes took the physics community by storm in 1995, when the physicist Joe Polchinski of the Kavli Institute for Theoretical Physics

(KITP) in Santa Barbara established that they were essential to string theory. But even before then, physicists had proposed branelike objects. One such example was a *p-brane* (so called by p-layful p-hysicists), an object that extends infinitely far in only some dimensions, which physicists derived mathematically using Einstein's theory of general relativity. Particle physics had also suggested mechanisms for confining particles on branelike surfaces. But string theory branes were the first known type of brane that could trap forces as well as particles, and we'll soon see that is part of what makes them so interesting. Like Ike stuck on a two-dimensional road in three-dimensional space, particles and forces can be trapped on lower-dimensional surfaces called branes, even if the universe has many other dimensions to explore. If string theory accurately describes the world in which we live, physicists have no choice but to acknowledge the potential existence of such branes.

The world of branes is an exciting new landscape that has revolutionized our understanding of gravity, particle physics, and cosmology. Branes might really exist in the cosmos, and there is no good reason that we couldn't be living on one. Branes might even play an important role in determining the physical properties of our universe and ultimately explain observable phenomena. If they do, branes and extra dimensions will be here to stay.

Branes as Slices

In Chapter 1 we looked at one way of thinking about the two-dimensional world of Flatland: as a two-dimensional slice of a three-dimensional space. In Abbott's novel, the character A. Square took a journey beyond two-dimensional Flatland, into the third dimension, and recognized that Flatland was a mere slice of the bigger three-dimensional world.

Upon his return, A. Square suggested—logically enough—that the three-dimensional world he had seen might also be a mere slice: a three-dimensional slice of an even higher-dimensional space. By "slice," of course, I don't mean merely a paper-thin, two-dimensional membrane, but the logical extension of such a thing—a generalized membrane, if you like. You might think of the three-dimensional slices that A. Square suggested as three-dimensional chunks in four-dimensional space.

But his three-dimensional guide promptly dismissed A. Square's speculation about three-dimensional slices. Like almost everyone we know, this unimaginative inhabitant of three dimensions believed in only the three dimensions of space he could see; he couldn't even contemplate a fourth.

Branes have introduced mathematical notions into physics that are similar to those described in *Flatland* over a century ago. Physicists have now returned to the idea that the three-dimensional world that surrounds us could be a three-dimensional slice of a higher-dimensional world. A brane is a distinct region of spacetime that extends through only a (possibly multidimensional) slice of space. The word "membrane" motivated the choice of the word "brane" because membranes, like branes, are layers that either surround or run through a substance. Some branes are "slices" inside the space, but others are "slices" that bound space, like slices of bread in a sandwich.

Either way, a brane is a domain that has fewer dimensions than the full higher-dimensional space that surrounds or borders it.[5] Note that membranes have two dimensions, but branes can have any number of dimensions. Although the branes that will most interest us have three spatial dimensions, the word "brane" refers to all "slices" of this sort; some branes have three spatial dimensions, but other branes have more (or fewer).[6] We'll use *3-branes* to refer to branes with three dimensions, *4-branes* to refer to those with four, and so on.

Boundary Branes and Embedded Branes

In the previous chapter I explained why we might not see extra dimensions. They could be curled up into sizes so small that evidence of their existence never would appear. The key point was that the extra dimensions would be small. None of the reasons for the invisibility of dimensions relied on the fact that extra dimensions were curled up.

This suggests an alternative possibility: perhaps dimensions are not rolled up, but simply terminate within a finite distance. Because

dimensions that disappear into nothing are potentially dangerous—
you wouldn't want pieces of the universe to fall off the ends—there
must be boundaries for the finite dimensions that tell them where and
how to end. The question is, what happens to particles and energy
when they reach these boundaries?

The answer is that they encounter a brane. In a higher-dimensional
world, branes would be the boundaries of the full higher-dimensional
space, known as the *bulk*. Unlike a brane, the bulk extends in all
directions. The bulk spans every dimension, both on and off the
brane (see Figure 25). The bulk is therefore "bulky," whereas, in
comparison, the brane is flat (in some dimensions), like a pancake. If
branes bordered the bulk in certain directions, some of the bulk's
dimensions would be parallel to the brane, while other dimensions
would lead off it. If the brane is the boundary, the dimensions off the
brane would extend only to one side.

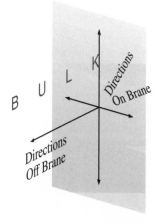

Figure 25. *A brane is a lower-dimensional surface with directions along it
and directions that lead away from it, into the higher-dimensional bulk.*

To understand the nature of finite dimensions that end on branes,
let us consider a very long thin pipe. Within the pipe there are three
dimensions: one long and two short. To make the analogy to flat
branes most straightforward, let's imagine that our pipe has a square
cross-section. An infinitely long pipe of this type would have four
infinitely long straight walls. If the pipe were a universe in its own

right, it would be one with three dimensions, two of which are bounded on either side by walls and one that extends infinitely far.

We know that a long thin pipe when viewed from afar (or with insufficient resolution) looks one-dimensional, much like the garden hose of the previous chapter. But we can also ask, as we did before with the garden-hose universe, how the pipe universe—consisting of the pipe and its interior—would appear to a conscious being living inside.

As you might suspect, this would depend on the being's resolution. A small fly that could move around within the square pipe would experience it as three-dimensional. Unlike the two-dimensional garden-hose example, we are assuming that the fly can move inside the pipe, and not just on its exterior. Nonetheless, as with the garden hose, the fly would experience the one long dimension differently than the other two. In one direction the fly could go arbitrarily far (assuming that our pipe is very long or infinite), whereas in the other two directions the fly could only go a short distance—the width of the pipe.

But there is a difference between the garden-hose universe and the pipe universe, aside from the number of dimensions each has. Unlike the bug of the previous chapter, the fly in the pipe travels inside it. Thus the fly sometimes encounters walls. It can go back and forth, or up and down, and reach a boundary. The bug on the hose, on the other hand, would never reach such a boundary: instead, it would only go round and round.

When the fly reaches the boundary of its pipe universe, there have to be rules that govern how it behaves. The walls of the pipe determine that behavior. The fly might hit the wall and splat into it; or the pipe might be reflective, so that the fly bounces off. If the pipe were a true universe bounded by branes, then the branes, which would be two-dimensional, would determine what happens when a particle, or anything else that could carry energy, reaches them.

When things get to a boundary brane, they bounce back, just as billiard balls bounce from the edges of the table or light bounces back from a mirror. This is an example of what physicists call a *reflective boundary condition*. If things bounce back from a brane, energy is not lost; it doesn't get absorbed in the branes or leak away. Nothing goes beyond the branes. The boundary branes are the "ends of the world."

In a multidimensional universe, branes serve the role of the boundary walls in the pipe-universe example above. Like walls, such branes would have lower dimension than the full space; a boundary always has a lower dimension than the object it bounds. That is as true for the boundary of space as it is for the crust that is the boundary of a loaf of bread. It is also true for the walls in your house, which have one lower dimension than the room they enclose: the room is three-dimensional, whereas any individual wall (when we ignore its thickness) spans only two dimensions.

Although so far in this section I have concentrated on branes that sit at boundaries, branes don't always sit at the edge of the bulk. They could conceivably exist anywhere in space. In particular, branes might sit somewhere away from the boundary, inside of space. If a boundary brane is like a thin heel at the end of a loaf of bread, such a non-boundary brane would be like a thin slice of bread within the loaf. A non-boundary brane would still be a lower-dimensional object, like the ones we have already considered. But non-boundary branes would have higher-dimensional bulk space on either side.

In the next section, we'll see that whatever the number of dimensions of the bulk or of the brane, and no matter whether branes are inside a space or at a boundary, branes can trap particles and forces along them. This makes the region of space they occupy very special.

Trapped on Branes

It is very unlikely that you will explore all the space available to you. There are probably places that you wish you had visited and voyages you'll never take—into outer space or the depths of the sea, for example. You haven't been to these places, but, in principle, you could go. There is no physical law that makes it impossible.

If, however, you lived inside a black hole, your travel opportunities would be far more severely constrained, more restricted even than those of women in Saudi Arabia. The black hole (until it decayed away) would keep you (or rather, the mutilated, black hole version of you) trapped in the interior, and you would never be able to escape.

There are many more familiar examples of things with restricted freedom of movement for which there are regions of space that are truly inaccessible. A charge on a wire and a bead on an abacus are both objects that live in a three-dimensional world, but travel in only one of its dimensions. There are also commonplace things that are confined to two-dimensional surfaces. Water droplets on a shower curtain travel only along the curtain's two-dimensional surface (see Figure 26). Bacteria trapped between microscope slides also experience only two-dimensional motion. Another example is Sam Loyd's "fifteen" game, the annoying game consisting of a little plastic tray with letters on tiles that you push around until they are correctly arranged in a square and say something like LOOK/YOUF/INIS/HED (see Figure 27). Unless you cheat, the letters stay within their plastic enclosure; they can never move in a third dimension.

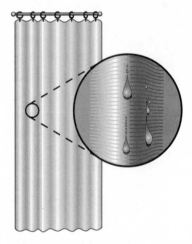

Figure 26. *Drops of water stuck on a two-dimensional shower curtain in a three-dimensional room.*

Branes, like shower curtains and Loyd's fifteen game, trap things on lower-dimensional surfaces. They introduce the possibility that in a world with additional dimensions, not all matter is free to travel everywhere. Just as the water droplets on the curtain are bound to a two-dimensional surface, particles or strings can be confined to a three-dimensional brane sitting inside a higher-dimensional world. But unlike the droplets on the curtain, they are truly trapped. And

Figure 27. *Sam Loyd's "fifteen" game.*

unlike the fifteen game, branes are not arbitrary. They are natural players in a higher-dimensional world.

Particles confined to branes are truly trapped on those branes by physical laws. Brane-bound objects never venture into the extra dimensions that extend off the brane. Not all particles will be trapped on branes; some particles might be free to travel throughout the bulk. But what distinguishes theories with branes from multidimensional theories without them are the particles on the branes—the ones that don't travel through all the dimensions.

In principle, branes and the bulk could have any number of dimensions, so long as a brane never has more dimensions than the bulk. The *dimensionality of a brane* is the number of dimensions in which brane-confined particles are permitted to travel. Although there are many possibilities, the branes that will be most interesting to us later on will be the three-dimensional ones. We don't know why three dimensions should appear to be so special. But branes with three spatial dimensions could be relevant to our world because they could extend along the three spatial dimensions we know. Such branes could appear in a bulk space with any number of dimensions that is more than three—four, five, or more dimensions.

Even if the universe does have many dimensions, if the particles and forces with which we are familiar are trapped on a brane that extends in three dimensions, they would still behave as if they lived in only three. Particles confined to branes would travel only along the brane.

And if light were also stuck to the brane, light rays would spread out only along the brane. In a three-dimensional brane, light would behave exactly as it would in a truly three-dimensional universe.

Furthermore, forces trapped on a brane influence only particles confined to this same brane. The material of which we are composed, such as nuclei and electrons, and the forces through which these building blocks interact, such as the electric force, might be confined on a three-dimensional brane. Brane-bound forces would spread out only along their brane, and brane-bound particles would be exchanged and would travel solely along the dimensions of the brane.

So if you lived in such a three-dimensional brane, you would be able to travel freely along its dimensions, much as you do in three dimensions now. Anything confined within a three-dimensional brane would look just the same as it would if the world were truly three-dimensional. The other dimensions would exist adjacent to the brane, but things stuck to a three-dimensional brane would never penetrate the higher-dimensional bulk.

But although forces and matter can be stuck on a brane, brane-worlds are interesting precisely because we know that not everything is confined to a single brane. Gravity, for example, is never confined to a brane. According to general relativity, gravity is woven into the framework of space and time. That means that gravity must be exerted throughout space and in every dimension. If it could be confined to a single brane, we would have to abandon general relativity.

Fortunately this is not the case. Even if branes exist, gravity will be felt everywhere, on and off branes. This is important because it means that braneworlds have to interact with the bulk, even if only via gravity. Because gravity extends into the bulk, and everything interacts via gravity, braneworlds will always be connected to the extra dimensions. Braneworlds do not exist in isolation: they are part of a larger whole with which they interact. In addition to gravity, there could conceivably exist other particles and forces in the bulk. If there are, such particles could also interact with particles confined to a brane and connect brane-bound particles to the higher-dimensional bulk.

The string theory branes that we will briefly consider later on have specific properties aside from the ones I have mentioned: they can carry particular charges, and they will respond in particular ways when something pushes on them. However, I will rarely bring in such detailed properties later on when I talk about branes. It will be enough to know the properties we have considered in this chapter: branes are lower-dimensional surfaces that can house forces and particles, and they can be the boundaries of higher-dimensional space.

Braneworlds: Blueprints for a Jungle Gym of Branes

Because branes could trap most particles and forces, the universe we live in could conceivably be housed on a three-dimensional brane, floating in an extra-dimensional sea. Gravity would extend into the extra dimensions, but stars, planets, people, and everything else that we sense could be confined to a three-dimensional brane. We would then be living on a brane. A brane might be our habitat. The concept of braneworlds is based on this assumption (see Figure 28).

If there can be one brane suspended in a higher-dimensional spacetime, there is no denying the possibility of many more. Braneworld scenarios often involve more than a single brane. We don't yet

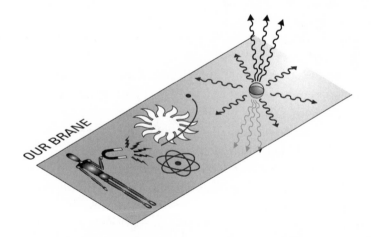

Figure 28. *We could be living on a brane. That is, the matter we are made of, photons, and other Standard Model particles can all be on the brane. But gravity is always everywhere—on the brane and in the bulk, as is illustrated by the squiggly lines.*

know the number or types of branes that could be present in the cosmos. *Multiverse* is a name that is sometimes attached to theories with more than one brane (see Figure 29). People often use the word to describe a cosmos with non-interacting or only weakly interacting pieces.

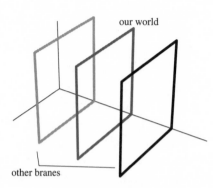

Figure 29. *The universe can contain multiple branes that interact only via gravity or don't interact at all. Such set-ups are sometimes called multiverses.*

I find the term "multiverse" a bit strange, since a universe is defined as the whole that is the unity of its parts. It is possible, however, to have different branes that are too far apart ever to communicate with one another, or that can communicate with one another only weakly, through mediating particles that travel between them. Particles on distant branes, then, would experience entirely different forces, and brane-bound particles would never have direct contact with particles bound on another brane. So when there is more than one brane with no force in common aside from gravity, I will sometimes refer to the universe housing them both as a multiverse.

Thinking about branes makes you aware of just how little we know about the space in which we live. The universe might be a magnificent composition linking intermittent branes. Even if we know the basic ingredients, in a multiverse populated by more than one brane, exotic new scenarios for the geometry of space are conceivable as well as myriad possibilities for how the particles we know and don't know are distributed among them. A single deck of cards can yield many different hands. There are scores of possibilities.

Other branes might be parallel to ours and might house parallel worlds. But many other types of braneworld might exist too. Branes could intersect and particles could be trapped at the intersections. Branes could have different dimensionality. They could curve. They could move. They could wrap around unseen invisible dimensions. Let your imagination run wild and draw any picture you like. It is not impossible that such a geometry exists in the cosmos.

In a world in which branes are embedded in a higher-dimensional bulk, there could be some particles that explore the higher dimensions and others that stay trapped on branes. If the bulk separates one brane from another, some particles can be on the first brane, some on the other, and some in the middle. Theories tell us about many ways in which particles and forces might be distributed among different branes and the bulk. Even for branes derived from string theory, we don't yet know why string theory should single out any particular allocation of particles and forces. Braneworlds introduce new physical scenarios that might describe both the world we think we know and other worlds we don't know on other branes we don't know, separated from our world in unseen dimensions.

New forces confined to distant branes might exist. New particles with which we will never directly interact might propagate on such other branes. Additional stuff accounting for dark matter and dark energy—the matter and energy that we surmise from their gravitational effects but whose identity is a mystery—might be distributed among different branes, or even in both the bulk and on other branes. And gravity might even influence particles differently as you go from one brane to the next.

If there is life on another brane, those beings, imprisoned in an entirely different environment, most likely experience entirely different forces that are detected by different senses. Our senses are attuned to the chemistry, light, and sound surrounding us. Because fundamental forces and particles are likely to be different, the creatures of other branes, should they exist, are unlikely to bear much resemblance to the life of our brane. The other branes will probably be nothing like our own. The only necessarily shared force is gravity, and even gravity's influence can vary.

The consequences of a braneworld will depend on the number and types of branes, and where they are located. Unfortunately for the curious, particles and forces confined to distant branes are not required to influence us very strongly. They might merely determine what travels in the bulk, and emit weak signals which might never even reach us. Therefore many conceivable braneworlds will be very difficult to detect, even if they do exist. After all, gravity is the only interaction that we know for sure is shared between the stuff on our brane and the stuff on any other brane, and gravity is an extremely weak force. Without direct evidence, other branes will remain cloistered in the realm of theory and conjecture.

But some of the braneworlds I will present could lead to detectable signals. The detectable braneworlds are the ones that have implications for the physical features of our world. Even though the proliferation of possible braneworlds is in some respects frustrating, it is really quite exciting. Not only might branes help resolve long-standing problems in particle physics, but if we're lucky, and one of the scenarios that I will describe is correct, evidence for braneworlds should appear in experiments with elementary particle physics very soon. We might really be living on a brane—and we might actually know it within a decade.

As of now, we do not know which, if any, of the many possibilities is the true description of the universe. I will therefore keep all options open, so as not to omit anything interesting. Whatever scenario turns out to describe our world, the ones I will present introduce new and fascinating ideas that no one would previously have thought possible.

4

Approaches to Theoretical Physics

She's a model and she's looking good.

Kraftwerk

"Hey, Athena, is that Casablanca *you're watching?"*
"Sure is. Want to join me? This is such a great scene."

> *You must remember this,*
> *A kiss is just a kiss.*
> *A sigh is just a sigh.*
> *The fundamental things apply as time goes by.*

"Hang on, Ike. Don't you think that last line's a little weird? It's supposed to be so romantic, but it almost sounds as if it's about physics."
"Athena, if you think that's strange, you've got to hear the opening verse of the original:"

> *This day and age we're living in*
> *Give cause for apprehension,*
> *With speed and new invention.*
> *And things like fourth dimension,*
> *Yet we get a trifle weary*
> *With Mr. Einstein's theory . . .*

"Ike, you don't really expect me to believe that, do you? Next thing I know you'll tell me Rick and Ilsa escape into the seventh dimension! Why don't we forget I ever said anything and just sit back and watch the movie?"

Einstein introduced general relativity in the early twentieth century, and by 1931 Rudy Vallee had recorded Ike's (true) version of Herman Hupfeld's song. However, by the time Sam played the tune in *Casablanca*, the omitted lyrics—as well as the science of spacetime—were all but forgotten in popular culture. And although Theodor Kaluza introduced the idea of an extra dimension back in 1919,* physicists didn't take the idea all that seriously until very recently.

Now that we've seen *what* dimensions are and *how* dimensions could escape our notice, we are almost set to ask what triggered this renewed interest in extra dimensions. *Why* should physicists believe that they might actually exist in the real physical world? That will require a far longer explanation—one that involves some of the most significant physics developments of the past century. In the next few chapters, before launching into a description of possible extra-dimensional universes, I will review these developments and why they serve as precursors for more recent theories. We will look at the major paradigm shifts that happened in the early twentieth century (quantum mechanics, general relativity), the essence of particle physics today (the Standard Model, symmetry, symmetry breaking, the hierarchy problem), and new ideas for approaching currently unresolved problems (supersymmetry, string theory, extra dimensions, and branes).

However, before we plunge into these subjects, this chapter will take a brief journey inside matter in order to set the physical stage. And because understanding where we're heading also requires some familiarity with the types of reasoning that today's theorists employ, we'll consider the theoretical approaches that are critical to more recent developments.

At first I thought "the fundamental things apply" was a clever choice of song quote. But on further reflection the words sounded so much like physics that I decided to check that my memory wasn't playing tricks on me, as sometimes happens with song lyrics—even those you think are burnt into your head. I was rather surprised (and amused) when I discovered that the song was more rooted in physics

---

*Only a year after the last time before 2004 that the Red Sox won the World Series—quite a while ago.

than I had ever imagined. I certainly hadn't realized that the "time going by" was supposed to be the fourth dimension!

Physical insights can work like this discovery; small clues sometimes reveal unanticipated connections. When you're lucky, what you find is better than what you were looking for—but you have to be looking in the right place. In physics, once you discover relationships, even by following tenuous leads, you look for meaning in the way you think best. That might involve educated guesses or it might involve trying to deduce the mathematical consequences of a theory you think you trust.

In the next section we'll consider the modern methods used to pursue such clues: model building—my forte—and the alternative approach to fundamental high-energy physics, namely, string theory. String theorists try to derive universal predictions from a definite theory, whereas model builders try to find ways to solve particular physical problems and then to build up theories from these starting points. Model builders and string theorists both seek more comprehensive theories with more explanatory power. They aim to answer similar questions, but they approach them in different ways. Research sometimes involves educated guesses, as with model building, and sometimes it involves deducing logical consequences of the ultimate theory you already believe to be correct, as with the string theory approach. We'll soon see that the recent research on extra dimensions successfully combines elements of both methods.

Model Building

Although I was first drawn to math and science by the certainty they promised, today I find the unanswered questions and the unexpected connections at least as attractive. The principles contained in quantum mechanics, relativity, and the Standard Model stretch the imagination, but they barely scratch the surface of the remarkable ideas engrossing physicists today. We know that something new is required because of the deficiencies in existing ideas. Those shortfalls are harbingers of novel physical phenomena that should emerge when we do more precise experiments.

Particle physicists try to find the laws of nature that explain how elementary particles behave. These particles, and the physical laws they obey, are components of what physicists call a *theory*—a definite set of elements and principles with rules and equations for predicting how those elements interact. When I speak about theories in this book, I'll be using the word in this sense; I won't mean "rough speculations", as in more colloquial usage.

Ideally, physicists would love to find a theory capable of explaining all observations, one that uses the sparest possible set of rules and the fewest possible fundamental ingredients. The ultimate goal for some physicists is a simple, elegant, unifying theory—one that can be used to predict the result of any particle physics experiment.

The quest for such a unifying theory is an ambitious—some might say audacious—task. Yet in some respects it mirrors the search for simplicity that began long ago. In ancient Greece, Plato imagined perfect forms, such as geometric shapes and ideal beings, that earthly objects only approximate. Aristotle also believed in ideal forms, but he thought that only observations can reveal the ideals that physical objects resemble. Religions also often postulate a more perfect or more unified state that is removed from, but somehow connected to, reality. The story of the fall from the Garden of Eden supposes an idealized prior world. Although the questions and methods of modern physics are very different from those of our ancestors, physicists, too, are seeking a simpler universe, not in philosophy or religion but in the fundamental ingredients that constitute our world.

However, there is an obvious impediment to finding an elegant theory that we can connect to our world: when we look around us, we see very little of the simplicity that such a theory should embody. The problem is that the world is complex. It takes a lot of work to connect a simple, spare formulation to the more complicated real world. A unified theory, while being simple and elegant, must some-how accommodate enough structure for it to match observations. We would like to believe that there is a perspective from which everything is elegant and predictable. Yet the universe is not as pure, simple, and ordered as the theories with which we hope to describe it.

Particle physicists negotiate the terrain connecting theory to obser-vations with two distinct methodologies. Some theorists follow a "top-down" approach: they start with the theory they believe to be correct—for example, string theorists start with string theory—and try to derive its consequences so that they can connect it to the much more disordered world we observe. Model builders, on the other hand, follow a "bottom-up" approach: they try to deduce an underlying theory by making connections among observed elementary particles and their interactions. They search for clues in physical phenomena. They make models, which are sample theories that may or may not prove correct. Both approaches have their merits and their deficiencies, and the best route to progress is not always apparent.

The conflict between the two scientific approaches is interesting because it reflects two very different ways of doing science. This division is the latest incarnation of a long debate in science. Do you follow the Platonic approach, which tries to gain insights from more fundamental truth, or the Aristotelian approach, rooted in empirical observations? Do you take the top-down or the bottom-up route?

The choice could also be phrased as "Old Einstein vs. Young Einstein." As a young man, Einstein rooted his work in experiments and physical reality. Even his so-called thought experiments were grounded in physical situations. Einstein changed his approach after learning the value of mathematics when he developed general relativity. He found that mathematical advances were crucial to completing his theory, which led him to use more theoretical methods later in his career. Looking to Einstein won't resolve the issue, however. Despite his successful application of mathematics to general relativity, his later mathematical search for a unified theory never reached fruition.

As Einstein's research demonstrated, there are different types of scientific truth and different ways of finding them. One is based in observations; this is how we learned about quasars and pulsars, for example. The other is based on abstract principles and logic: for example, Karl Schwarzschild first derived black holes as a mathematical consequence of general relativity. Ultimately, we would like these to converge—black holes have now been deduced from both the mathematical description of observations and from pure theory—but in the first phases of investigation, the advances we make based on the two types of truth are rarely the same. And in the case of string theory, the principles and equations are not nearly so well laid out as are those of general relativity, making deriving its consequences that much harder.

When string theory first rose to prominence, it sharply divided the particle physics world. I was a graduate student in the mid-1980s when the "string revolution" first split the world of particle physics asunder. At that time, one community of physicists decided to devote themselves wholeheartedly to the ethereal, mathematical realm of string theory.

String theory's basic premise is that strings—not particles—are the most fundamental objects of nature. The particles we observe in the world around us are mere consequences of strings: they arise from the different vibrational modes of an oscillating string, much as different musical notes arise from a vibrating violin string. String theory gained favor because physicists were looking for a theory that consistently includes quantum mechanics and general relativity and that can make predictions down to the tiniest conceivable distance scales. To many people, string theory looked like the most promising candidate.

However, another group of physicists decided to stay in touch with the relatively low-energy world that experiments could explore. I was at Harvard, and the particle physicists there—which included the excellent model builders Howard Georgi and Sheldon Glashow, along with many talented postdoctoral fellows and students, were among the stalwarts who continued with the model building approach.

Early on, the battles between the merits of the two opposing viewpoints—string theory and model building—were fierce, with each side claiming better footing on the road to truth. Model builders thought that string theorists were in mathematical dreamland, whereas string theorists thought that model builders were wasting their time and ignoring the truth.

Because of the many brilliant model builders at Harvard, and because I relished the challenges of model building, when I first entered the world of particle physics I stayed within that camp. String theory is a magnificent theory which has already led to profound mathematical and physical insights, and it might well contain the correct ingredients to ultimately describe nature. But finding the connection between string theory and the real world is a daunting task. The problem is that string theory is defined at an energy scale that is about ten million billion times larger than those we can experimentally explore with our current instruments. We still don't even know what will happen when the energy of particle colliders increases by a factor of ten!

An enormous theoretical gulf separates string theory, as it is currently understood, from predictions that describe our world. String theory's equations describe objects that are so incredibly tiny and possess such extraordinarily high energy that any detectors we could imagine making with conceivable technologies would be unlikely ever to see them. Not only is it mathematically tremendously challenging to derive string theory's consequences and predictions, it is not even always clear how to organize string theory's ingredients and determine which mathematical problem to solve. It is too easy to get lost in a thicket of detail.

String theory can lead to a plethora of possible predictions at distances we actually see—the particles that are predicted depend on the as yet undetermined configuration of fundamental ingredients in the theory. Without some speculative assumptions, string theory looks like it contains more particles, more forces, and more dimensions than we see in our world. We need to know what separates the extra particles, forces, and dimensions from the visible ones. We don't yet know what physical features, if any, favor one configuration over another, or even how to find a single manifestation of string theory that conforms to our world. We would have to be very lucky to extract all the correct physical principles that will make the predictions of string theory match what we see.

For example, string theory's invisible extra dimensions have to be different from the three that we see. The gravity of string theory is more complex than the gravity we see around us—the force that caused Newton's apple to fall on his head. Instead, string theory's gravity operates in six or seven additional dimensions of space. Fascinating and remarkable as string theory is, puzzling features such as its extra dimensions obscure its connection to the visible universe. What distinguishes those extra dimensions from the visible ones? Why aren't they all the same? Discovering how and why nature hides string theory's extra dimensions would be a stunning achievement, making it worthwhile to investigate all possible ways in which this might happen.

So far, however, all attempts to make string theory realistic have had something of the flavor of cosmetic surgery. In order to make its predictions conform to our world, theorists have to find ways to cut away the pieces that shouldn't be there, removing particles and tucking dimensions demurely away. Although the resulting sets of particles come tantalizingly close to the correct set, you can nonetheless tell that they aren't quite right. Elegance might well be the hallmark of a correct theory, but we can only really judge a theory's beauty once we've fully understood all its implications. String theory is captivating at first, but ultimately string theorists have to address these fundamental problems.

When exploring mountainous territory without a map, you can rarely tell what the most direct route to your destination will turn out to be. In the world of ideas, as in complex terrain, the best path to follow is not always clear at the outset. Even if string theory does ultimately unify all the known forces and particles, we don't yet know whether it contains a single peak representing a particular set of particles, forces, and interactions, or a more complicated landscape with many possible implications. If the paths were smooth, well-signposted grids, route-finding would be simple. But that is rarely the case.

So, the approach to advancing beyond the Standard Model that I will emphasize is model building. The term "model" might evoke a small-scale battleship or castle you built in your childhood. Or you might think of numerical simulations on a computer that are meant to reproduce known dynamics—how a population grows, for example, or how water moves in the ocean. Modeling in particle physics is not the same as either of these definitions. However, it's not entirely different from the use of the word in magazines or fashion shows: models, both on runways* and in physics, demonstrate imaginative creations and come in a variety of shapes and forms. And the beautiful ones get all the attention.

Needless to say, the similarities end there. Particle physics models

---

*Catwalks in the UK.

are guesses at alternative physical theories that might underlie the Standard Model. If you think of a unified theory as the summit of a mountain, model builders are trailblazers who are trying to find the path that connects the solid ground below, consisting of well-established physical theories, to the peak—the path that will ultimately tie new ideas together. Although model builders acknowledge the fascination of string theory and the possibility that it could turn out to be true, they are not as certain as string theorists that they know what theory they will find if they ever get to the top.

As we will see in Chapter 7, the Standard Model is a definite physical theory with a fixed set of particles and forces that reside in a four-dimensional world. Models that go beyond the Standard Model incorporate its ingredients and mimic its consequences at energies that have already been explored, but they also contain new forces, new particles, and new interactions that can be seen only at shorter distances. Physicists propose these models to address current puzzles. Models might suggest different behaviors for known or conjectured particles, behaviors determined by a new set of equations that follow from a model's assumptions. Or they might suggest a new spatial setting, such as the ones we'll explore with extra dimensions or branes.

Even when we fully understand a theory and its implications, that theory can be implemented in different ways, which might have different physical consequences for the real world in which we live. For example, even when we know how particles and forces interact in principle, we still need to know which particular particles and forces exist in the real world. Models allow us to sample the possibilities.

Different assumptions and physical concepts distinguish theories, as do the distance or energy scales at which a theory's principles apply. Models are a way of getting at the heart of such distinguishing features. They let you explore a theory's potential implications. If you think of a theory as general instructions for making a cake, a model would be a precise recipe. The theory would say to add sugar, but the model would specify whether to add half a cup or two cups. The theory would say that raisins are optional, but the model would tell you to be sensible and leave them out.

Model builders look at the unresolved aspects of the Standard Model and try to use known theoretical ingredients to address its inadequacies. The model building approach is fueled by the instinct that the energies for which string theory makes definite predictions are too far away from those we can observe. Model builders try to see the big picture so they can find the pieces that could be relevant to our world.

We model builders pragmatically admit that we can't derive everything at once. Instead of trying to derive string theory's consequences, we try to figure out which ingredients of the underlying physical theory will explain known observations and reveal relationships among experimental discoveries. A model's assumptions could be part of the ultimate underlying theory, or they might illuminate new relationships even before we understand their deeper theoretical underpinnings.

Physics always strives to predict the largest number of physical quantities from the smallest number of assumptions, but that doesn't mean that we always manage to identify the most fundamental theories right away. Advances have often been made before everything was understood at the most fundamental level. For example, physicists understood the notions of temperature and pressure and employed them in thermodynamics and engine design long before anyone had explained these ideas at a more fundamental microscopic level as the result of the random motion of large numbers of atoms and molecules.

Because models relate to physical "phenomena," (meaning experimental observations) model builders with stronger ties to experiment are sometimes called phenomenologists. "Phenomenology" is a poor choice of word, however. It does not do justice to data analysis, which in today's complex scientific world is deeply embedded in theory. Model building is far more tied to interpretation and mathematical analysis than phenomenology, in the philosophical sense of the word, would suggest.

The best models do, however, have an invaluable feature. They yield definite predictions for physical phenomena, giving experimenters a way to verify or contradict a model's claims. High-energy experiments are not merely searching for new particles—they are testing models and looking for clues to better ones. Any proposed particle physics model will involve new physical principles and new physical laws that apply at measurable energies. It will therefore predict new particles and testable relationships among them. Finding these particles and measuring their properties can confirm or rule out proposed ideas. The goal of high-energy experiments is to shed light on underlying physical laws and the conceptual framework that gives them their explanatory power.

Only some models will prove correct, but models are the best way to investigate possibilities and build up a reservoir of compelling ingredients. And if string theory is right, we might eventually learn how some models follow as consequences of it, much as thermodynamics was rooted in atomic theory. However, for about a decade the two communities were sharply divided. As Albion Lawrence, a young string theorist from Brandeis University, commented recently when he and I were discussing this schism, "One of the tragedies is that string theory and model building were distinct intellectual subjects. Model builders and string theorists weren't talking to each other for years. I always thought of string theory as the granddaddy of all models."

Both string theorists and model builders are searching for a tractable, elegant route that connects theory to the observed world. Any theory will be truly compelling and likely to be correct only if this path, and not just the view from the top, manifests this elegance. Model builders, who start from the bottom, run the risk of many false starts, but string theorists, who start at the top, run the risk of finding themselves at the edge of a precipitous, isolated cliff, too remote for them to find their way back to base camp.

You might say that we are all searching for the language of the universe. But whereas string theorists focus on the inner logic of the grammar, model builders focus on the nouns and phrases that they think are most useful. If particle physicists were in Florence learning Italian, the model builders would know how to ask for lodging and acquire the vocabulary that would be essential to finding their way around, but they might talk funny and never fully comprehend the *Inferno*. String theorists, on the other hand, might aspire to grasp the subtleties of Italian literature—but run the risk of starving to death before learning how to ask for dinner!

Fortunately, things have now changed. These days, theory and low-energy phenomena bolster each other's progress, and many of us now think about string theory and experimentally oriented physics simultaneously. I have continued to follow the model building approach in my research, but I now also incorporate ideas from string theory. I think we're ultimately most likely to make advances by combining the best of both methods.

Albion points out that "the distinction is becoming fuzzy again, catalyzed in large part by the study of extra dimensions. People are talking to each other." The communities are no longer so rigidly defined, and there is more common ground. There has been a renewed convergence of purpose and ideas. Both scientifically and socially, there are now strong overlaps between model builders and string theorists.

One of the beautiful aspects of the extra-dimensional theories I will describe is that ideas from both camps converged to produce them. String theory's extra dimensions might be a nuisance, but they might also prove to be an opportunity for finding new ways of addressing old problems. We can certainly ask where the extra dimensions are, and why we haven't seen them. But we might also ask whether these unseen dimensions could have any import in our world. These dimensions might help explain underlying relationships that are relevant to observed phenomena. Model builders relish the challenge of connecting notions such as extra dimensions to observable quantities such as relations among particle masses. And, if we're lucky, the insights based on extra-dimensional models might successfully address one of the biggest problems facing string theory: its experimental inaccessibility. Model builders have used theoretical elements derived from string theory to attack questions in particle physics. And those models, including the ones that have extra dimensions, will have testable consequences.

When we investigate extra-dimensional models later on, we will see that the model building approach in conjunction with string theory has generated major new insights into particle physics, the evolution of the universe, gravity, and string theory. With the string theorist's knowledge of grammar and the model builder's vocabulary, the two together have begun to write quite a reasonable phrase book.

The Heart of Matter

Ultimately, the ideas we will consider encompass the entire universe. However, these ideas are rooted in particle physics and in string theory—theories that aspire to describe the smallest components of matter. So before setting out on our journey to the extreme theoretical territory these theories address, we'll now take a brief trip into matter down to its smallest parts. On this guided tour of the atom, take note of matter's basic building blocks and the sizes of the objects that different physical theories deal with. They should provide a few landmarks that you can use to orient yourself later on and help you to recognize the components with which each area of physics concerns itself.

The basic premise in most of physics is that elementary particles constitute the building blocks of matter. Peel away the layers, and inside you will always ultimately find elementary particles. Particle physicists study a universe in which these objects are the smallest elements. String theory takes this assumption one step further and postulates that those particles are the oscillations of elementary strings. But even string theorists believe that matter is composed of particles—the unbreakable entities at its core.

It might be difficult to believe that everything is composed of particles; they certainly are not evident to the naked eye. But that is because of the coarse resolving power of our senses, which cannot directly detect anything anywhere nearly as tiny as an atom. Nonetheless, even though we can't directly view them, elementary particles are the elementary building blocks of matter. Just as the images on your computer or TV are composed of tiny dots, even though they present images that appear to be continuous, matter is composed of atoms, which are in turn composed of elementary particles. Physical objects around us appear to be continuous and uniform, but in reality they are not.

Before physicists could look inside matter and deduce its composition, they needed technological advances to create sensitive measuring instruments. But every time they developed more accurate technological tools, *structure*—more elementary constituents—

emerged. And every time physicists had access to tools that could probe still smaller sizes, they discovered yet more fundamental ingredients: *substructure*, constituents of the previously known structural elements.

The goal of particle physics is to discover matter's most basic constituents and the most fundamental physical laws obeyed by those constituents. We study small distance scales because elementary particles interact at these scales, and it's easier to disentangle fundamental forces. At large scales, the basic ingredients are bound into composite objects, which makes fundamental physical laws difficult to disentangle and therefore more obscure. Small distance scales are interesting because new principles and connections apply there.

Matter is not simply a Russian doll with smaller copies of similar entities inside. Smaller distances reveal truly novel phenomena. Even the workings of the human body—the heart and the circulation of the blood, for example—were badly misconstrued until scientists such as William Harvey cut people open in the 1600s and looked inside. Recent experiments have done the same thing with matter, exploring smaller distances where new worlds operate via more fundamental physical laws. And just as the blood's circulation has important consequences for all human activity, the fundamental physical laws have important consequences for us on larger scales.

We now know that all matter is made up of *atoms*, which combine through chemical processes into *molecules*. Atoms are very small, about an angstrom, or one-hundredth of a millionth of a centimeter in size. But atoms are not fundamental: they consist of a central, positively charged *nucleus* which is surrounded by negatively charged *electrons* (see Figure 30). The nucleus is far smaller than the atom, occupying only about one hundred thousandth of the atom's size. And the positively charged nucleus is itself composite: it is made from positively charged *protons* and neutral (uncharged) *neutrons*, collectively known as *nucleons*, which are not very much smaller than the nucleus in size. This was the picture of matter that scientists held before the 1960s, and is very likely the blueprint you learned about in school.

This template for the atom is correct, although, as we will see later, quantum mechanics gives a more interesting picture of an electron's

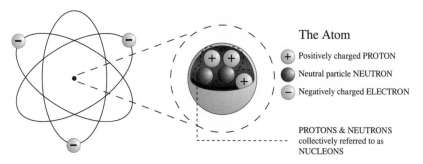

The Atom

+ Positively charged PROTON

● Neutral particle NEUTRON

− Negatively charged ELECTRON

PROTONS & NEUTRONS
collectively referred to as
NUCLEONS

Figure 30. *The atom consists of electrons circulating around a tiny nucleus. The nucleus is composed of positively charged protons and charge-neutral neutrons.*

orbits than any picture you can draw. But we now know that even the proton and neutron are not fundamental. Contrary to Gamow's quote in the introduction, the proton and neutron contain substructure, more fundamental ingredients known as *quarks*. The proton contains two *up quarks* and one *down quark*, while the neutron contains two down quarks and one up quark (see Figure 31). These

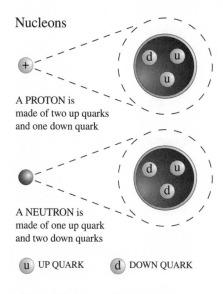

Nucleons

+

A PROTON is
made of two up quarks
and one down quark

●

A NEUTRON is
made of one up quark
and two down quarks

u UP QUARK   d DOWN QUARK

Figure 31. *The proton and neutron are composed of more elementary quarks bound together through the strong force.*

quarks are bound together through a nuclear force known as the *strong force*. The electron, the other component of the atom, is different. So far as we can tell, it is fundamental: the electron cannot be divided into smaller particles and contains no substructure within.

The Nobel Prize-winning physicist Stephen Weinberg coined the term "Standard Model" to label the well-established particle physics theory that describes the interactions of these fundamental building blocks of matter—the electron, the up quark, and the down quark—as well as other fundamental particles that we will get to momentarily. The Standard Model also describes three of the four forces through which the elementary particles interact: electromagnetism, the weak force, and the strong force. (It usually omits gravity.)

Although gravity and electromagnetism were known for hundreds of years, no one understood the last two less familiar forces until the second half of the twentieth century. Those weak and strong forces act on fundamental particles and are important for nuclear processes. They permit quarks to bind together and nuclei to decay, for example.

If we wanted, we could also include gravity in the Standard Model. We usually don't though, because gravity is far too weak a force to be of any consequence at the distance scales that are relevant to particle physics at experimentally accessible energies. At very high energies and very small distances, our usual notions about gravity break down; this is relevant to string theory, but it does not happen on measurable distance scales. When studying elementary particles, gravity is important only in certain extensions of the Standard Model, such as the extra-dimensional models we will consider later on. For all other predictions about elementary particles, we can forget about gravity.

Now that we've entered the world of fundamental particles, let's look around a little and take stock of our neighbors. The up quark, the down quark, and the electron lie at the core of matter. However, we now know that there also exist additional, heavier quarks and other heavier electron-like particles that are nowhere to be found in ordinary material.

For example, whereas the electron has a mass of about one-half of one-thousandth that of a proton, a particle called the *muon*, with precisely the same charge as the electron, has a mass that is two hundred times greater than the electron's. A particle called the *tau*, which also has the same charge, has a mass that is ten times greater still. And experiments at high-energy colliders have discovered even heavier particles in the past thirty years. To produce them, physicists needed the large amount of highly concentrated energy that today's high-energy particle colliders can create.

I realize that this section was billed as a tour inside matter, but the particles I am talking about now are not inside the stable objects of the material world. Although all known matter consists of elementary particles, heavier elementary particles are not constituents of matter. You won't find them in your shoelaces, on your table top, or on Mars, or in any other physical object that we know about. But these particles are currently created today at high-energy collider experiments, and they were a part of the early universe immediately after the Big Bang.

Nonetheless, these heavy particles are essential components of the Standard Model. They interact through the same forces as the more familiar particles do, and will very likely play a role in a deeper understanding of matter's most basic physical laws. I've listed the Standard Model particles in Figures 32 and 33. I've included neutrinos and

First generation	up 3 MeV	down 7 MeV	electron neutrino ~ 0	electron 0.5 MeV
Second generation	charm 1.2 GeV	strange 120 MeV	muon neutrino ~ 0	muon 106 MeV
Third generation	top 174 GeV	bottom 4.3 GeV	tau neutrino ~ 0	tau 1.8 GeV

Figure 32. *The matter particles of the Standard Model and their masses. Particles in the same column have identical charges but different masses.*

	electromagnetism	weak force	strong force
Force-carrying gauge bosons	photon massless	weak gauge bosons $W^{\pm}$ Z 80 Gev 91 GeV	gluons massless

Figure 33. *The force-carrying gauge bosons of the Standard Model, their masses, and the forces they communicate.*

force-carrying gauge bosons, which I'll tell you more about in Chapter 7 when I discuss all the elements of the Standard Model in detail.

No one knows why the heavy Standard Model particles exist. The questions of their purpose, what role they play in the ultimate underlying theory, and why their masses are so different from those of the constituents of more familiar matter are some of the major mysteries facing the Standard Model. And these are only a few of the puzzles that the Standard Model leaves unresolved. Why, for example, are there four forces and no others? Could there be others we haven't yet detected? And why is gravity so much weaker than the other known forces?

The Standard Model also leaves open a more theoretical question, the one that string theory hopes to address: how do we reconcile quantum mechanics and gravity consistently at all distance scales? This question differs from the others in that it doesn't concern currently visible phenomena, but is instead a question about the intrinsic limitations of particle physics.

Both types of unanswered question—those that concern visible and purely theoretical phenomena—give us reasons to look beyond the Standard Model. Despite the Standard Model's power and success, we're confident that more fundamental structure awaits discovery and that the search for more fundamental principles will be rewarded. As the composer Steve Reich elegantly put it in the *New York Times* (when making an analogy to a piece he wrote), "First there were just atoms, then there were protons and neutrons, then there were quarks, and now we're talking about string theory. It seems like every 20, 30, 40, 50 years a trapdoor opens and another level of reality opens up."*

Experiments at current and future particle colliders are no longer looking for the ingredients of the Standard Model—those have all been found. The Standard Model nicely organizes these particles according to their interactions, and the full complement of Standard Model particles is now known. Instead, experimenters are looking for particles that should be even more interesting. Current theoretical models include the Standard Model ingredients, but add new elements

---

*Quoted in Anne Midgette, "At 3 score and 10, the music deepens," *New York Times*, 28 January 2005.

to address some of the questions that the Standard Model leaves unresolved. We hope that current and future experiments will provide clues that will allow us to distinguish among them and find the true underlying nature of matter.

Although we have experimental and theoretical hints about the nature of a more fundamental theory, we are unlikely to know what is the correct description of nature until higher-energy experiments (that probe shorter distances) provide the answer. As we will see later on, theoretical clues tell us that experiments in the next decade will almost certainly discover something new. It probably won't be definitive evidence of string theory, which will be very difficult to discover, but we'll see that it could be something as exotic as new relations in spacetime, or new and as yet unseen extra dimensions—new phenomena that feature in string theory as well as other particle physics theories. And despite the broad scope of our collective imagination, these experiments also have the potential to reveal something that no one has yet thought of. My colleagues and I are very curious about what that will be.

Preview

We know about the structure of matter we just visited as a result of the critical physics developments of the last century. These stupendous advances are essential to any more comprehensive theory of the world we might come up with and were also major achievements in themselves.

Starting in the next chapter, we'll review those developments. Theories grow out of the observations and deficiencies of progenitor theories, and you can better appreciate the role of more recent advances by becoming acquainted with these remarkable earlier developments. Figure 34 indicates some of the ways in which the theories we will discuss interconnect. We'll see how each of these theories was built using the lessons from older ones and how newer theories filled in gaps that were detected only after the older theories were complete.

We'll begin with the two revolutionary ideas of the early twentieth century: relativity and quantum mechanics, through which we learned

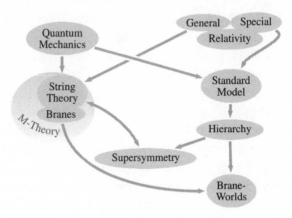

Figure 34. *The fields of physics we will encounter and how they are connected.*

about the shape of the universe and the objects it contains, and the composition and structure of the atom. We'll then introduce the Standard Model of particle physics, which was developed in the 1960s and 1970s to predict the interactions of the elementary particles we just encountered. We'll also consider the most important principles and concepts in particle physics: symmetry, symmetry breaking, and scale dependence of physical quantities, through which we've learned a great deal about how matter's most elementary components create the structures we see.

However, despite its many successes, the Standard Model of particle physics leaves some fundamental questions unanswered— questions so basic that their resolution promises new insight into the building blocks of our world. Chapter 10 presents one of the most interesting and mysterious aspects of the Standard Model: the origin of the elementary particles' masses. We'll see that we almost certainly need a more profound physical theory than the Standard Model if we are to explain the masses of known particles and the weakness of gravity.

Extra-dimensional models address such particle physics problems, but they also use ideas from string theory. After discussing the basics of particle physics, we'll introduce the fundamental motivation and concepts in string theory. We won't derive models directly from string theory, but string theory contains some of the elements that are used when developing extra-dimensional models.

This review covers a lot of ground because research on extra dimensions ties together many theoretical advances in the two major strands of particle physics—model building and string theory. Some familiarity with many of the most interesting recent developments in these fields will help you to better understand the motivations and the methods underlying the development of extra-dimensional models.

However, in case you want to jump around, I will end each of the review chapters with a bulleted list of vital concepts that we will refer to later on when we return to extra-dimensional model building. The bullets will serve as a short cut, a summary, in case you want to skip a chapter or if you want to focus on the material we'll turn to later on. I might occasionally refer to points that aren't in the bullets, but the bullets will review the key ideas that are essential to the major results in the later part of the book.

In Chapter 17 we'll start to explore extra-dimensional brane-worlds—theories that propose that the matter of which our universe is composed is confined to a brane. Braneworld ideas have provided new insights into general relativity, particle physics, and string theory. The different braneworlds I'll present make different assumptions and explain different phenomena. I'll summarize the distinctive features of each model with bullets at the end of these chapters as well. We don't yet know which, if any, of these ideas correctly describe nature. But it's entirely conceivable that we'll ultimately discover that branes are a part of the cosmos, and that we—along with other, parallel universes—are confined to them.

One thing I have learned from this research is that the universe often has more imagination than we do. Sometimes its properties are so unexpected that we stumble across them only by accident. Discovering such surprises can be amazing. Our known physical laws turn out to have startling consequences.

Let's now begin our exploration of what those laws are.

5

Relativity: The Evolution of Einstein's Gravity

The laws of gravity are very, very strict.
And you're just bending them for your own benefit.

Billy Bragg

Icarus (Ike) Rushmore III couldn't wait to show Dieter his new Porsche. But as proud as he was of his car, he was even more excited about his Global Positioning System (GPS) that he had recently designed and installed himself.

Ike wanted to impress Dieter, so he convinced his friend to drive with him to the local track. They got in the car, Ike programmed in their destination, and the two of them set off. But to Ike's chagrin, they ended up in the wrong place—the GPS system didn't work nearly as well as he had thought it would. Dieter's first thought was that Ike must have made some ridiculous error, like confusing meters and feet. But Ike didn't believe he could have made such a stupid mistake, and he bet Dieter that wasn't the problem.

The next day, Ike and Dieter did some troubleshooting. But to their dismay, when they went for a drive the GPS was even worse than before. Ike and Dieter searched again for the problem and finally, after a frustrating week, Dieter had an epiphany. He did a quick calculation and made the startling discovery that without accounting for general relativity, Ike's GPS system would build up errors at the rate of more than 10 km each day. Ike didn't think his Porsche was fast enough to warrant relativistic calculations, but Dieter explained that the GPS signals—not the car—travel at the speed of light. Dieter modified the software to account for the changing gravitational field the GPS signals had to pass through. Ike's system then worked as well as the readily available commercial variety. Relieved, Ike and Dieter began to plan a road trip.

At the beginning of the last century the British physicist Lord Kelvin said, "There is nothing new to be discovered in physics now. All that remains is more and more precise measurement."* Lord Kelvin was famously incorrect: very soon after he uttered those words, relativity and quantum mechanics revolutionized physics and blossomed into the different areas of physics that people work on today. Lord Kelvin's more profound statement, that "scientific wealth tends to accumulate according to the law of compound interest,"† is certainly true, however, and is especially appropriate to these revolutionary developments.

This chapter explores the science of gravity, and how it evolved from the impressive achievement of Newton's laws to the revolutionary advances of Einstein's theory of relativity. Newton's laws of motion are the classical physics laws that scientists used for centuries to compute mechanical motion, including motion caused by gravity. Newton's laws are magnificent, and they allow us to make predictions of motion that work spectacularly well—well enough to send men to the Moon and satellites into orbit, well enough to keep the superfast trains in Europe on the tracks when rounding corners, well enough to prompt the search for the eighth planet, Neptune, based on peculiarities in Uranus's orbit. But alas, not well enough for an accurate GPS system.

Incredibly, the GPS system now in use requires Einstein's theory of general relativity to achieve its one-meter accuracy. Determinations of the variation in snow depth on Mars using laser ranging data from orbiting spacecraft also incorporate general relativity, and yield values with an unbelievable precision of 10 cm. Certainly, at the time it was developed, no one—not even Einstein—anticipated such practical applications of a theory as abstract as general relativity.

---

*An address to a group of physicists at the British Association for the Advancement of Science in 1900.
†Presidential Address to British Association, 1871.

This chapter will explore Einstein's theory of gravity, a spectacularly accurate theory that applies to a wide range of systems. We'll begin by briefly reviewing Newton's gravitational theory, which works fine for the energies and speeds we encounter in daily life. We'll then move on to the extreme limits in which it fails: namely, very high speed (close to the speed of light) and very large mass or energy. In these limits, Newtonian gravity is superseded by Einstein's theory of relativity. With Einstein's general relativity, space (and spacetime) evolved from a static stage to a dynamical entity that can move and curve and have a rich life of its own. We'll consider this theory, the clues that led to its development, and some of the experimental tests that convince physicists that it's right.

Newtonian Gravity

Gravity is the force that keeps your feet on the ground and is the source of the acceleration that returns a tossed ball to Earth. In the late sixteenth century, Galileo showed that this acceleration is the same for all objects on the surface of the Earth, no matter what their mass.

However, this acceleration does depend on how far the object is from the Earth's center. More generally, the strength of gravity depends on the distance between the two masses—gravity's pull is weaker when objects are farther apart. And when what creates the gravitational attraction is not the Earth, but some other object, gravity's strength will depend on the mass of that object.

Isaac Newton developed the gravitational force law that summarizes how gravity depends on mass and distance. Newton's law says that the force of gravity between two masses is proportional to the mass of each of them. They could be anything: the Earth and a ball, the Sun and Jupiter, a basketball and a soccer ball, or any two objects you please. The more massive the objects, the greater the gravitational attraction.

Newton's gravitational force law also says how the gravitational force depends on the distance between the two objects. As discussed in Chapter 2, the law says that the force between two objects is proportional to the inverse square of their separation. This inverse square law was where the famous apple entered in.* Newton could deduce the acceleration due to the Earth's gravitational pull on an apple located near the Earth's surface and compare it with the acceleration induced on the Moon, which is located sixty times further away than the Earth's surface is from its center. The acceleration of the Moon due to the earth's gravity is 3,600 times smaller (3,600 is the square of 60) than the acceleration of the apple. This is in accordance with the gravitational force decreasing as the square of the distance from the Earth's center.[7]

However, even when we know the dependence of the gravitational attraction on mass and distance, we still need another piece of information before we can determine the overall strength of gravitational attraction. The missing piece is a number, called *Newton's gravitational constant*, that factors into the calculation of any classical gravitational force. Gravity is very weak, and this is reflected in the tiny size of Newton's constant, to which all gravitational effects are proportional.

The Earth's gravitational pull or the gravitational attraction between the Sun and the planets might seem pretty big. But that's only because the Earth, the Sun, and the planets are so massive. Newton's constant is very small, and the gravitational attraction between elementary particles is an extremely weak force. This feebleness of gravity is itself a big puzzle that we will return to later on.

Although his theory was correct, Newton delayed its publication for twenty years, until 1687, while he tried to justify a critical assumption of his theory: that the Earth's gravitational pull was exerted as if its mass were all concentrated at the center. While Newton was hard at work developing calculus to solve this problem, Edmund Halley, Christopher Wren, Robert Hooke, and Newton himself made tremendous progress in determining the gravitational force law by analyzing the motion of the planets, whose orbits Johannes Kepler had measured and found to be elliptical.

These men all made major contributions to the problem of planetary motion, but it is Newton who gets credited with the inverse square

*The story might be apocryphal, but the reasoning is not.

law. That is because Newton ultimately showed that elliptical orbits would arise as a result of a central force (that of the Sun) only if the inverse square law was true, and he showed with calculus that the mass of a spherical body did in fact act as if it were concentrated at the center. Newton did, however, acknowledge the significance of others' contributions in his words, "If I have seen further, it is because I have stood upon the shoulders of giants."* (However, rumor has it that he said this only because of his intense dislike for Hooke, who was very short.)

In high school physics, we learned Newton's laws and calculated the behavior of interesting (if somewhat contrived) systems. I remember my outrage when our teacher, Mr Baumel, informed us that the gravitational theory we had just learned was wrong. Why teach us a theory that we know to be incorrect? In my high school view of the world, the whole merit of science was that it could be true and reliable, and could make accurate and factual predictions.

But Mr Baumel was simplifying, perhaps for dramatic effect. Newton's theory was not wrong: it was merely an approximation, one that works incredibly well in most circumstances. For a large range of parameters (speed, distance, mass, and so on), it predicts gravitational forces quite accurately. The more precise underlying theory is relativity, and you only make measurably different predictions with relativity when you are dealing with extremely high speeds or large amounts of mass or energy. Newton's law predicts the motion of a ball admirably well, since neither of the above criteria apply. To use relativity to predict the motion of a ball would be pure silliness.

In fact, Einstein himself initially thought of special relativity merely as an improvement on Newtonian physics—not as a radical paradigm shift. This, of course, grossly underplays the ultimate significance of his work.

*Letter from Isaac Newton to Robert Hooke, 5 February 1675.

Special Relativity

A very reasonable thing to expect from physical laws is that they should be the same for everyone. No one could blame us for questioning their validity and utility if people in different countries or sitting on moving trains or flying on an airplane experienced different physical laws. Physical laws should be fundamental and hold true for any observer. Any differences in calculations should be due to differences in environment, not the physical laws. It would be very strange indeed to have universal physical laws that required a particular vantage point. The particular quantities you might measure could depend on your reference frame, but the laws that govern these quantities should not. Einstein's formulation of special relativity ensures that this is the case.

In fact, it's somewhat ironic that Einstein's work on gravity is referred to as "the theory of relativity." The essential point that drove both special and general relativity was that physical laws should apply for everyone, independent of their reference frame. In fact, Einstein would have preferred the term *Invariantentheorie*.* In a letter Einstein wrote in 1921 in reply to a correspondent who had suggested he reconsider the name, he admitted that the term "relativity" was unfortunate.† But by that time, the term was too well entrenched for him to attempt to change it.

Einstein's first insight about reference frames and relativity came from thinking about electromagnetism. The well-known theory of electromagnetism from the nineteenth century was based on Maxwell's laws, which describe the behavior of electromagnetism and electromagnetic waves. The laws gave correct results, but everyone initially falsely interpreted the predictions in terms of the motion of an *aether*, a hypothesized invisible substance whose vibrations were supposed to be electromagnetic waves. Einstein realized that if there were an aether, there would also be a preferred observational vantage

---

*Gerald Holton, *Einstein, History, and Other Passions* (Cambridge, MA: Harvard University Press, 2000).
†Letter to E. Zschimmer, 30 September 1921.

point, or frame of reference: the one in which the aether is at rest. He reasoned that the same physical laws should apply to people who are moving at constant velocity* with respect to each other and with respect to someone at rest—that is, in frames of reference that physicists refer to as inertial frames. By requiring that *all* physical laws, including those of electromagnetism, should hold for observers in all inertial reference frames, Einstein was led to abandon the idea of the aether and, ultimately, to formulate special relativity.

Einstein's theory of special relativity, with its radical revision of the concepts of space and time, was a major leap. Peter Galison,† a physicist and historian of science, suggests that it was not only the aether theory that put Einstein on the right track, but Einstein's job at the time. Galison reasoned that Einstein, who grew up in Germany and worked at the patent office in Bern, Switzerland, must have had time and time coordination on his mind. Anyone who has traveled in Europe knows that precision is valued highly in countries such as Switzerland and Germany, which has the happy consequence that passengers can count on the trains to run on time. Einstein worked in the patent office between 1902 and 1905, during an era when train travel was becoming increasingly important, and coordinating time was at the forefront of new technology. In the early 1900s, Einstein was very likely thinking about real-world problems, such as how to coordinate the time at one train station with that at another.

Of course, Einstein did not need to develop relativity to solve the problem of coordinating real trains. (For those of us accustomed to the frequently delayed American trains, coordinated time might sound exotic in any case.‡) But coordinating time raised some interesting questions. Time coordination of relativistically moving trains is not a straightforward problem. If I were to coordinate my watch with someone on a moving train, I would need to account for the time delay of a signal traveling between us because light has a finite speed. Coordinating my watch with that of the person sitting next to me

---

*Velocity gives both speed and direction.
†Peter Galison, *Einstein's Clocks, Poincaré's Maps: Empires of Time* (New York: W.W. Norton, 2003).
‡Don't get me wrong—I like trains. But I wish they were better supported in the U.S.

would not be the same as coordinating watches with someone far away.*

Einstein's critical insight, the one that led him to special relativity, was that ideas about time had to be reformulated. According to Einstein, space and time could no longer be considered independently. Although they are not the same thing—time and space are clearly different—the quantities you measure depend on the speed at which you are traveling. Special relativity was the result of this insight.

Bizarre as they are, one can derive all of Einstein's novel consequences of special relativity from two postulates. To state them, we need to understand the meaning of *inertial frames*—a particular category of reference frames. Let's first choose any frame of reference that moves at constant velocity (speed and direction); the one that's at rest is often a good one. The inertial frames would then be those that are moving at fixed velocity with respect to that first one—someone running or driving by at constant speed, for example.

Einstein's postulates then state that:

The laws of physics are the same in all inertial frames.
The speed of light, c, is the same in any inertial frame.

The two postulates tell us that Newton's laws are incomplete. Once we accept Einstein's postulates, we have no choice but to replace Newton's laws with new physical laws that are consistent with these rules.[8] The laws of special relativity that follow lead to all the surprising consequences you might have heard of, such as time dilation, the observer dependence of simultaneity, and Lorentz contraction of a moving object. The new laws should look very much like the old classical physics laws when applied to objects moving at speeds that are small compared with the speed of light. But when applied to something moving very fast, at or near the speed of light, the difference

---

*Although American trains don't always coordinate time very well, Amtrak does appear to acknowledge special relativity when they say, "time and the space to use it" in their advertising slogan for the Acela, the high-speed train that travels the Northeast corridor. However, "time" and "space" are not precisely interchangeable. Although the slogan "space and the time to use it" does describe my more heavily delayed train rides, the phrase wouldn't be a very compelling advertisement for a high-speed train.

between the Newtonian and special relativity formulations should become apparent.

For example, in Newtonian mechanics speeds are simply added together. A car driving towards yours on the freeway approaches you at a speed that's the sum of its speed and yours. Similarly, if someone throws a ball at you from the platform while you are on a moving train, the ball's speed appears to be the sum of the speed of the ball itself plus the speed of the moving train. (A former student of mine, Witek Skiba, can attest to this fact. Witek was nearly knocked out when he was hit by a ball that someone threw at the approaching train he was riding.)

According to Newtonian physics, the speed of a beam of light directed at a moving train should be the sum of the speed of light and the speed of the moving train. But this can't be true if the speed of light is constant, as Einstein's second postulate asserts. If the speed of light is always the same, then the speed of the beam aimed at the moving train will be identical to the speed of a light beam that approaches you when you're standing still on the ground. Even though it runs counter to the intuition gained from your experience of the slow speeds you encounter in daily life, light speed is constant, and in special relativity speeds don't simply add up as they do in Newtonian physics. Instead, you add speeds according to a relativistic formula that follows from Einstein's postulates.

Many of special relativity's implications don't jibe with our familiar notions of time and space. Special relativity treats time and space differently than they had been treated before in Newtonian mechanics, and this is what gives rise to many of its counterintuitive results. Time and space measurements depend on speed and get mixed up in systems that move relative to each other. Nonetheless, surprising as they are, once you accept the two postulates then a different notion of space and time is an inevitable consequence.

Here's one argument why. Imagine two identical ships with identical masts. One ship is docked by the shore, while the other is moving away. Also imagine that the captains of the two ships synchronized their watches when the first ship sailed off.

Now suppose that the two captains do a rather odd thing: each decides to measure time on her ship by placing a mirror at the top of the mast and a second mirror at the bottom, shining a light from the bottom mirror to the top one, and measuring the number of times light hits the mirror and returns. As a practical matter, of course, this would be absurd, since light would cycle up and down far too frequently to count. But bear with me, and imagine that the captains can count extraordinarily fast; I'll be using this somewhat contrived example to argue that time stretches out on the moving ship.

If each captain knows how long it takes for light to cycle once, she can calculate the passage of time by multiplying the light-cycle time by the number of times light cycles up and down between the mirrors. Now suppose, though, that instead of using her own stationary mirror clock, the captain on the docked ship measures time by the number of times the light on the moving ship hits the mast's mirror and returns.

Now from the perspective of the captain on the moving ship, the light simply goes straight up and down. However, from the perspective of the captain on the docked ship, the light has to travel farther (in order to cover the distance traveled by the moving ship—see Figure 35). But—and this is the counterintuitive part—the speed of light is constant. It is the same for the light sent to the top of the mast on the docked ship as it is for the light sent to the top of the mast on the moving ship. Since speed measures distance traveled over time, and the speed of light for the moving ship is the same as the speed of light for the stationary one, the moving mirror clock has to "tick" at a slower rate to compensate for the longer distance the moving light

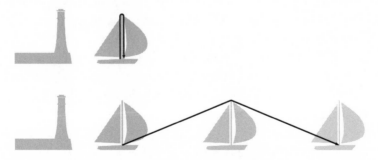

Figure 35. *The path of a light beam that bounces off the top of a mast of a stationary ship and of a moving one. The stationary observer (in a boat by the shore or in a lighthouse) would see a longer path in the second case.*

has to travel. This very counterintuitive conclusion—that moving and stationary clocks must tick at different rates—follows from the fact that the speed of light in a moving reference frame is the same as the speed of light in a stationary one. And although this is a funny way to measure time, the same conclusion—that moving clocks run slower—would hold true independently of how time is measured. If the captains had watches on, they would observe the same thing (again, with the caveat that for normal speeds, the effect would be tiny).

While the above example is artificial, the phenomenon described produces genuinely measurable effects. For example, special relativity gives rise to the different time experienced by fast-moving objects— the phenomenon known as time dilation.

Physicists measure time dilation when they study elementary particles produced at colliders or in the atmosphere, which travel at relativistic speeds—speeds approaching that of light. For example, the elementary particle called a muon has the same charge as an electron, but is heavier and can decay (that is, it can turn into other, lighter particles). The muon's lifetime, the time before it decays, is only 2 microseconds. If a moving muon had the same lifetime as a stationary one, it would be able to travel only about 600 meters before it disappeared. But muons manage to make it all the way through our atmosphere, and in colliders, to the edges of large detectors, because their near-light-speed velocity makes them appear to us much longer-lived. In the atmosphere, muons travel at least ten times further than they would in a universe based on Newtonian principles. The very fact that we see these muons at all shows us that time dilation (and special relativity) gives rise to true physical effects.

Special relativity is important both because it was a dramatic deviation from classical physics and because it was essential to the development of general relativity and quantum field theory, both of which play a significant role in more recent developments. Because I won't use specific special relativity predictions when I discuss particle physics and extra-dimensional models later on, I'll resist the urge to go into all the fascinating consequences of special relativity, such as why simultaneity depends on whether an observer is moving and how the sizes of moving objects are different from when they are at rest.

Instead, we'll delve into another dramatic development, namely general relativity, which will be critical when we consider string theory and extra dimensions later on.

The Principle of Equivalence:
General Relativity Begins

Einstein wrote down his theory of special relativity in 1905. In 1907, while working on a paper that summarized his recent work on the subject, he found himself already questioning whether the theory could apply to all situations. He noticed two major omissions. For one thing, physical laws looked the same only in certain special inertial reference frames—those that moved at fixed velocity with respect to each other.

In special relativity, these inertial reference frames occupied a privileged position. The theory left out any reference frame that was accelerating. If you pressed the accelerator while driving your car, you would no longer be in one of the special reference frames where the laws of special relativity apply. That's what's "special" in special relativity: the "special" inertial frames are only a small subset of all possible reference frames. For someone convinced that no one's reference frame is special, it was a big problem that the theory singled out inertial reference frames.

Einstein's second misgiving concerned gravity. Although he had figured out how objects respond to gravity in some situations, he still hadn't come up with formulas for determining the gravitational field in the first place. The form of the gravitational force law was known in some simpler settings, but Einstein wasn't yet able to deduce the field for every possible distribution of matter.

Between 1905 and 1915, in a sometimes grueling exploration, Einstein addressed these problems. The result was general relativity. He centered his new theory around the *equivalence principle*, which states that the effects of acceleration cannot be distinguished from those of gravity. All the laws of physics would look the same to an accelerating observer as they would to a stationary observer placed in a gravitational field that accelerates everything in the stationary frame with an acceleration of the same magnitude—but in the opposite direction—as the original observer's acceleration. In other words, you wouldn't have any way of distinguishing uniform acceleration from standing still in a gravitational field. According to the principle of equivalence, there is no measurement that would distinguish between these two situations. An observer could never know which situation he was in.

The equivalence principle follows from the equivalence of *inertial* and *gravitational mass*, two quantities that in principle could have been different from each other. Inertial mass determines how an object will respond to any force—how much the object would accelerate if you applied that force. The role of inertial mass is summarized in Newton's second law of motion, $F = ma$, which says that if you apply a force of magnitude F to an object with mass m, you will produce an acceleration a. Newton's famous second law tells us that a given force produces smaller acceleration on an object that has bigger inertial mass, which is probably very familiar to you from experience. (If you shove a footstool, it will go further and faster than if you shove a grand piano just as hard.) Notice that this law applies for any sort of force—the force of electromagnetism, for example. It can apply in situations that have nothing whatsoever to do with gravity.

Gravitational mass, on the other hand, is the mass that enters the gravitational force law and determines the strength of gravitational attraction. As we saw, the strength of the Newtonian gravitational force is proportional to the two masses that get attracted to each other. These masses are gravitational mass. Gravitational mass and the inertial mass that enters Newton's second force law turn out to be the same, and that's why we can safely give them the same name: mass. But in principle they could have been different, and we would have had to call one "mass" and the other "ssam." Fortunately, we don't need to do that.

The mysterious fact that the two masses are the same has deep implications, which it took an Einstein to recognize and develop. The gravitational force law states that the strength of gravity is proportional to mass, and Newton's law tells us how much acceleration would be generated by that (or any other) force. Because the strength of gravity is proportional to the same mass that determines the amount of acceleration, the two laws together tell us that even though the *force* depends on mass through $F = ma$, the acceleration induced by gravity is entirely independent of the mass that gets accelerated.

The acceleration of gravity that any object experiences must be the same for anyone or anything separated by the same distance from another object. This is the claim that Galileo allegedly verified by dropping objects off the Tower of Pisa,* demonstrating that the Earth induces the same acceleration for all objects, independent of their mass. This fact—that acceleration is independent of the mass of the accelerated object—is unique to the gravitational force, because the strength of no force other than gravity depends on mass. And because the gravitational force law and Newton's law of motion depend on mass in the same way, the mass cancels out when you calculate acceleration. Acceleration therefore doesn't depend on mass.

This relatively straightforward deduction has profound implications. Since all objects have the same acceleration in a uniform gravitational field, if this *single* acceleration could be canceled, the evidence of gravity would be canceled as well. And that is exactly what happens to a freely falling body: it is accelerated precisely so as to cancel the evidence of gravity.

The equivalence principle says that if you and everything around you were freely falling, you would not be aware of a gravitational field. Your acceleration would cancel the acceleration that the gravitational field would otherwise have produced. This state of weightlessness is now familiar from pictures from orbiting spacecraft, where the astronauts and the objects that surround them don't experience any gravity.

Textbooks often illustrate the absence of gravity's effects (from the vantage point of the freely falling observer) with a picture of someone dropping a ball in a free-falling elevator. You see the person and the ball falling together in the picture. The person in the elevator would always see the ball at the same height above the elevator's floor. He wouldn't see the ball drop (see Figure 36).

Physics texts always present the freely falling elevator as if it were the most natural thing in the world that the observer inside would

---

*He did the experiment by timing objects rolling down an inclined plane.

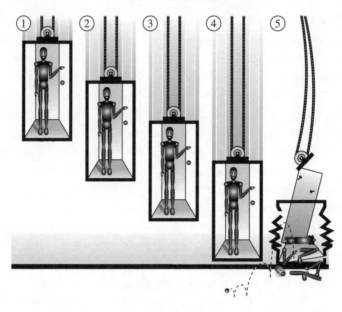

Figure 36. *An observer in a falling elevator who releases a ball will not see it drop. However, when a freely falling elevator meets the stationary Earth, the observer will not be very happy.*

calmly watch a ball not drop with complete equanimity, with no concern at all for his personal well-being. This is in sharp contrast to the terrified faces in movies in which the cables of an elevator are cut and the actors hurtle towards the ground. Why such different responses? If everything were freely falling, there would be no cause for alarm. The situation would be indistinguishable from everything being at rest, albeit in a zero-gravity environment. But if, as in the movies, someone is falling but the ground below him stays put, he has good reason to be petrified. If someone is on a freely falling elevator, but solid ground awaits his descent, you can be sure that he will notice the consequences of gravity when his free fall is ended (as is illustrated in the last frame of Figure 36).

The reason that Einstein's conclusion seems so surprising and strange is that our upbringing here on Earth, with a stationary planet beneath our feet, biases our intuition. When the force of the Earth keeps you stationary on the ground, you notice the effects of gravity because you are not following the path towards the center of the Earth that gravity would have you follow. On Earth, we're accustomed to gravity making things fall. But "falling" really means "falling relative to us." If we were falling along with a dropped ball, as we would be in a free-falling elevator, the ball would not go down any faster than we would. We therefore would not see it drop.

In your freely falling reference frame, all the laws of physics would coincide with the laws of physics that would be obeyed if you and everything near you were at rest. A freely falling observer would observe that motion is described by the same equations, consistent with special relativity, that apply for observer in an inertial, non-accelerating reference frame. In the review paper he wrote in 1907 about relativity, Einstein explains how the gravitational field has only a relative existence, "because for an observer falling freely from the roof of a house there exists—at least in his immediate surroundings— no gravitational field."*

This was Einstein's major insight. The equations of motion for a freely falling observer are the equations of motion for an observer in an inertial reference frame. A freely falling observer does not feel the force of gravity—only objects that are not in free fall experience a gravitational force.

In our lives we don't generally encounter things or people in free fall. When free fall happens, it looks scary and dangerous. But, as an Irishman said to the physicist Raphael Bousso when he was visiting Ireland's Cliffs of Moher, "It's not the fall that kills you, but the %&!# crash when you stop." And when I broke several bones in a rock-climbing accident and had to miss a conference I had organized, there were quite a few jokes about my testing the theory of gravity. I can state with complete confidence that gravitational acceleration agrees with predictions.

---

*Albert Einstein, "Über das Relativitätsprinzip und die aus demselben gezogene Folgerungen" ["On the relativity principle and the conclusions drawn from it"], *Jahrbuch der Radioaktivität und Electronik*, vol. 4, pp. 411–62 (1907); see also Abraham Pais, *Subtle is the Lord* (Philadelphia: American Philological Association, 1982).

Tests of General Relativity

There's more to general relativity; soon we'll get to the rest, which took considerably longer to develop. But the equivalence principle alone explains many results from general relativity. Once Einstein had recognized that gravity could be canceled in an accelerating reference frame, he could calculate gravitational influence by imagining an accelerating system equivalent to the one with gravity. This allowed him to calculate the gravitational effects for some interesting systems which others could use to check his conclusions. We'll now consider a few of the most significant experimental tests.

First is the *gravitational redshift* of light. A redshift causes us to detect light waves at a lower frequency than the frequency at which they were emitted. (You've probably encountered the analogous effect in sounds waves when a motorcycle roared past you and the sound waves rose and fell in pitch.)

There are several ways to understand the origin of the gravitational redshift, but probably the simplest is through an analogy. Imagine that you throw a ball up into the air. The rising ball slows down as it moves against the force of gravity. But the ball's energy is not lost, even though the ball is slowing down. It is converted into potential energy, which is then released as kinetic energy, or energy of motion, when the ball falls back down.

The same reasoning applies to the particle of light, the *photon*. Just as a ball loses momentum when it is thrown up into the air, a photon loses momentum as it escapes from a gravitational field. As with the ball, this means that the photon loses kinetic energy but gains potential energy as it fights its way out of the gravitational field. But a photon cannot slow down as a ball would, since it always travels at the constant speed of light. To jump the gun a bit, we will see in the next chapter that one consequence of quantum mechanics is that a photon lowers its energy when it lowers its frequency. And that is exactly what happens to the photon that is going through the changing gravitational potential. In order to lower its energy, the photon decreases its frequency, and this lowered frequency is the gravitational redshift.

Conversely, a photon that was moving towards a gravitational source would increase its frequency. In 1965, the Canadian-born physicist Robert Pound and one of his students, Glen Rebka, measured this effect by studying gamma rays emitted from radioactive iron that was placed at the top of the "tower" of Harvard's Jefferson Lab, the building where I now work. (Though it's part of the building, an elevated attic area in Jefferson Lab and the floors beneath it are known as "the tower."). The gravitational fields at the top and bottom of the tower were slightly different, since the top is slightly further from the center of the Earth. A high tower would be best for this measurement, since it would maximize the difference in height between where the gamma rays were emitted (the top of the tower) and where they were detected (the basement). But even though the tower consists of just three floors, an attic, and some windows that peer out above the attic—it's all of 74 feet high—Pound and Rebka managed to measure the difference in frequency between the emitted and absorbed photons with incredible precision, five parts in a million billion. They thereby established that the general relativity predictions for the gravitational redshift were correct to at least 1% accuracy.

A second experimentally observable consequence of the equivalence principle is the bending of light. Gravity can attract energy as well as mass. After all, the famous relation $E = mc^2$ means that energy and mass are closely connected. If mass experiences gravity, then so should energy. The Sun's gravity influences mass, and likewise affects the trajectory of light. Einstein's theory predicted exactly the amount light should bend under the Sun's influence. These predictions were first confirmed during the solar eclipse of 1919.

The English scientist Arthur Eddington organized expeditions to the island of Principe off the coast of West Africa and to Sobral in Brazil, where the eclipse could best be seen. Their purpose was to photograph the stars in the neighborhood of the eclipsed Sun and check whether stars that appeared near the Sun moved relative to their usual positions. If the stars did appear to be shifted, that would mean that their light was traveling along a bent trajectory. (The scientists needed to make their measurements during an eclipse so that the sunlight wouldn't overwhelm the much dimmer light of the stars.) Sure enough, the stars appeared in just the right "wrong" places. The measurement of the correct bending angle provided strong evidence supporting Einstein's theory of general relativity.

Incredibly, the bending of light is now so well established and understood that it is one of the tools that was used to probe the distribution of mass in the universe and look for dark matter in the form of small, burnt-out stars that no longer emit light. Like black cats on a moonless night, such objects are very hard to see. The only way to observe them is through their gravitational effects.

Gravitational lensing is one way that astronomers can learn about dark objects; dark objects, like everything else, interact via gravity. Although the burnt-out stars do not themselves emit light, there can be bright objects behind them (from our perspective) whose light we can see. Without any dark star near its path, the light would travel in straight lines. But light emitted by a bright star will bend when it passes by the dark star. Light passing on the left will bend in the opposite direction than light passing on the right and light passing on the top will bend in the opposite direction than light passing on the bottom. This will create multiple images of a bright object behind a dark star and the effect is called *gravitational lensing*. Figure 37 shows an example of a multiple image of a star that appeared when an intervening massive object bent the star's light rays in different directions.

The Graceful Curves of the Universe

The equivalence principle says that the force of gravity is indistinguishable from constant acceleration. I'm glad you made it to this point, because I need to confess that I simplified, and the two aren't entirely indistinguishable after all. How could they be? If gravity were equivalent to acceleration, it would not be possible for people in opposite hemispheres to simultaneously fall to Earth. After all, the Earth cannot accelerate in two directions at once. Gravitational pull in the different directions felt in America and China, for example, cannot possibly be accounted for by a single acceleration.

The resolution of this paradox is that the equivalence principle asserts only that gravity can be replaced by acceleration *locally*. At

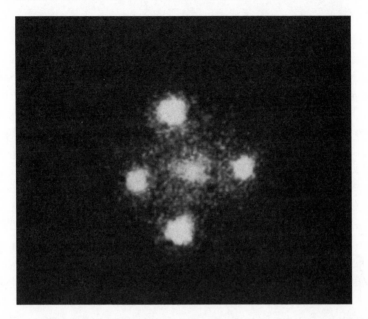

Figure 37. *The "Einstein Cross" is formed when multiple images of a bright, distant quasar are formed by light bending in different directions as it passes by a massive foreground galaxy.*

different places in space, the acceleration that would replace gravity according to the principle would generally be in different directions. The answer to our problem with Chinese/American relations is that American gravity is equivalent to an acceleration in a different direction from the acceleration that would reproduce Chinese gravity.

This critical insight led Einstein to a complete reformulation of the theory of gravity. He no longer saw gravity as a force that acts directly on an object. Instead, he described it as a distortion of the geometry of spacetime that reflects the different accelerations required to cancel gravity in different places. Spacetime is no longer a parenthetical background to an event—it is an active player. With Einstein's theory of general relativity, the force of gravity is understood in terms of the curvature of spacetime, which in turn is determined by the matter and energy that are present. Let's now consider the notion of the curvature of spacetime, on which Einstein's revolutionary theory rests.

Curved Space and Curved Spacetime

A mathematical theory must be internally consistent but, unlike a scientific theory, it has no obligation to correspond to an external physical reality. True, mathematicians have often drawn inspiration from what they see in the world around them. Mathematical objects such as cubes and natural numbers do have real-world counterparts. But mathematicians extend their assumptions about these familiar concepts to objects whose physical reality is less certain, such as tesseracts (hypercubes in four-dimensional space) and quaternions (an exotic number system).

Euclid wrote his five fundamental postulates of geometry in the third century BC. From these assumptions a beautiful logical structure developed, one that you might have had a taste of in high school. But later mathematicians found themselves having trouble with the fifth postulate, the one known as the parallel postulate. This postulate states that, given a line and a point outside that line, there is one and only one line that can be drawn through the point that is parallel to the initial line.

For two millennia after Euclid formulated his postulates, mathematicians argued about whether this fifth postulate was actually independent or merely a logical consequence of the other four. Could there be a system of geometry for which all but the last postulate was true? If no such system of geometry existed, the fifth postulate would not be independent, and would therefore be disposable.

Only in the nineteenth century did mathematicians put the fifth postulate in its proper place. The great German mathematician Carl Friedrich Gauss discovered that Euclid's fifth assumption was exactly what Euclid had claimed: a postulate that could be replaced by another. He went ahead and replaced it, discovering other systems of geometry and thereby demonstrating that the fifth postulate was independent. With that, non-Euclidean geometry was born.

A Russian mathematician, Nikolai Ivanovich Lobachevsky, also developed non-Euclidean geometry, but when he sent his work to Gauss he was disappointed to learn that the older mathematician had come up with the same idea fifty years before. But neither Lobachevsky nor anyone else had known about Gauss's results, which the German had hidden for fear that his colleagues would ridicule him.

Gauss shouldn't have worried. It is obvious that Euclid's fifth postulate is not always true, because we all know about alternatives. For example, lines of longitude meet at the North Pole and at the South Pole, even though they are parallel at the equator. Geometry on a sphere is an example of non-Euclidean geometry. Had the ancients written on spheres rather than scrolls, this might have been obvious to them, too.

But there are many examples of non-Euclidean geometry which, unlike the sphere, cannot be realized physically in a three-dimensional world. The original non-Euclidean geometries of Gauss, Lobachevsky, and the Hungarian mathematician János Bolyai* dealt with such undrawable theories, which makes it less surprising that they took so long to discover them.

A few examples illustrate what makes curved geometries different from the flat geometry of this page. Figure 38 shows three two-dimensional surfaces. The first, the surface of a sphere, has constant positive curvature. The second, a section of a flat plane, has zero

Figure 38. *Surfaces of positive, zero, and negative curvature.*

---

*János Bolyai was a genius, but although his father, Farkas Bolyai, wanted him to be a mathematician, János was poor and joined the military and not the academy. Others initially discouraged János about his work on non-Euclidean geometry, and he eventually published it only because his father insisted on putting it in a book he was writing. Farkas, who was friends with Gauss, sent him the appendix that János wrote. But once again, János was in for disappointment. Although Gauss recognized János Bolyai's genius, he replied only, "To praise it would amount to praising myself. For the entire content of the work . . . coincides almost exactly with my own meditations which have occupied my mind for the past thirty or thirty-five years." (Letter from Gauss to Fartas Bolyair, 1832.) So once again, János's mathematical career was thwarted.

curvature. And the third, a hyperbolic paraboloid, has constant negative curvature. Examples of negatively curved surfaces are the shape of a horse's saddle, the terrain between two mountain peaks, and a Pringles potato chip.

There are many litmus tests that will tell us which of the three possible types of curvature any particular geometric space possesses. For example, you can draw a triangle on each of the three surfaces. On the flat surface the sum of the angles of a triangle is always precisely 180 degrees. But what about a triangle on the surface of the sphere, with one vertex on the North Pole and the remaining two vertices on the equator, a quarter of the way around the equator from each other? Each of the angles of this triangle is a right angle of 90 degrees. Therefore the sum of the angles on the triangle is 270 degrees. This could never happen on a flat surface, but on a surface of positive curvature the sum of the angles of a triangle must exceed 180 degrees because the surface bulges out.

Similarly, the sum of the angles of a triangle drawn on a hyperbolic paraboloid is always less than 180 degrees, a reflection of its negative curvature. This is a bit harder to see. Draw two vertices near the top of the saddle and one down low, along one of the lower parts of the hyperbolic paraboloid, where one of your feet would go if you were sitting on a horse. This last angle is less than it would be if the surface were flat. The angles add up to less than 180 degrees.

Once it was established that non-Euclidean geometries were internally consistent—that is, their premises didn't result in paradoxes or contradictions—the German mathematician Georg Friedrich Bernhard Riemann developed a rich mathematical structure to describe them. A piece of paper cannot be rolled into a sphere, but it can be rolled into a cylinder. You can't flatten a saddle without having it crumple or fold back on itself. Building on Gauss's work, Riemann created a mathematical formalism that encompassed such facts. In 1854 he found a general solution to the problem of how to characterize all geometries through their intrinsic properties. His studies laid the groundwork for the modern mathematical field of differential geometry, a branch of mathematics that studies surfaces and geometry.

Because I will almost always consider space and time together from now on, we will generally find the notion of *spacetime* more useful than the notion of space. Spacetime has one more dimension than space: in addition to "up-down," "left-right," and "forwards-backwards," it includes time. In 1908 the mathematician Hermann Minkowski used geometric notions to develop this idea of an absolute spacetime fabric. Whereas Einstein studied spacetime using time and space coordinates that depended on a frame of reference, Minkowski identified the observer-independent spacetime fabric that can be used to characterize a given physical situation.

In the rest of the book, when I refer to dimensionality I will be giving the number of spacetime dimensions, except where I explicitly state otherwise. For example, when we look around us we see what I will from now on refer to as a four-dimensional universe. Occasionally I will single out time and talk about a "three-plus-one"-dimensional universe, or three spatial dimensions. Bear in mind that all these terms refer to the same setting—one that has three dimensions of space and one of time.

The spacetime fabric is a very important notion. It concisely characterizes the geometry that corresponds to the gravitational field produced by a particular distribution of energy and matter. But Einstein initially disliked the idea, which had seemed to him like an overly fancy way to reformulate the physics that he had already explained. However, he eventually recognized that the spacetime fabric was essential for completely describing general relativity and calculating gravitational fields. (For the record, Minkowski wasn't overly impressed with Einstein on first acquaintance, either. Based on Einstein's performance in Minkowski's calculus class back when Einstein was a student, Minkowski had concluded then that Einstein was a "lazy dog.")

Einstein wasn't alone in resisting non-Euclidean geometry. His friend Marcel Grossmann, a Swiss mathematician, also considered it unduly complicated and tried to talk Einstein out of using it. However, they eventually agreed that the only tractable way of explaining gravity was by using non-Euclidean geometry to represent the spacetime fabric. Only then could Einstein interpret and calculate the warping of spacetime that was equivalent to gravity, which turned out to be the key to completing general relativity. After Grossmann conceded defeat, both he and Einstein struggled through the intricacies of differential geometry to simplify their highly complicated earlier attempts to arrive at a formulation of the theory of gravity. In the end, they completed the theory of general relativity and reached a deeper understanding of gravity itself.

Einstein's Theory of General Relativity

General relativity presented a radical revision of the concept of gravity. We now understand gravity—the force that keeps your feet on the ground and binds together our galaxy and the universe—not as a force acting directly on objects, but as a consequence of the geometry of spacetime, an idea that took Einstein's view of the union of space and time to its logical conclusion. General relativity exploits the deep connection between inertial and gravitational mass to formulate the effect of gravity *solely* in terms of the geometry of spacetime. Any distribution of matter or energy curves or warps spacetime. Curved pathways in spacetime determine gravitational motion, and the matter and energy of the universe cause spacetime itself to expand, undulate, or contract.

In flat space the shortest distance between two points, the *geodesic*, is a straight line. In curved space we still can define a geodesic as the shortest path between two points, but that path won't necessarily look straight. For example, routes of airplanes that follow great circles on the Earth are geodesics. (A great circle is any circle, such as the equator or a line of longitude, that goes around the fattest part of a sphere.) Although these paths are not straight, they are the shortest routes that don't tunnel through the Earth.

In curved four-dimensional spacetime, we can also define a geodesic. For two events separated in time, a geodesic is the natural path things would take in spacetime to connect one event to the other. Einstein realized that free fall, which is the path of least resistance, is motion along the spacetime geodesic. He concluded that, in the absence of external forces, dropped objects will fall along a geodesic, as with the path of the person on the falling elevator who doesn't feel his weight or see a ball drop.

However, even when things are following their geodesics through spacetime and there are no external forces, gravity has noticeable effects. We've already seen that the local equivalence between gravity and acceleration was one of the critical insights that led Einstein to develop an entirely new way of thinking about gravity. He deduced that, because the acceleration induced by a gravitational force is locally the same for all masses, gravity must be a property of spacetime itself. That's because "freely falling" means different things in different places, and it is only *locally* that gravity can be replaced by a single acceleration. My Chinese counterpart and I fall in different directions, even if we are both in our local version of Einstein's elevator. The fact that the direction of free fall is not the same everywhere is a reflection of the curvature of spacetime. There isn't a *single* acceleration that can cancel the effects of gravity everywhere. In curved spacetime, the geodesics of different observers will in general be different. So globally, gravity has observable consequences.

General relativity goes much further than Newtonian gravity because it allows us to calculate the relativistic gravitational field of any distribution of energy and matter. Moreover, the revelation that the geometry of spacetime encodes the effects of gravity permitted Einstein to close a major gap in his original formulation of gravity. Although physicists at the time knew how objects would respond to a gravitational field, they did not know what gravity was. Now they understood that the gravitational field is the distortion of the spacetime fabric caused by matter and energy. This distortion extends throughout the cosmos itself, or, as we will see shortly, throughout a higher-dimensional spacetime that might include branes. All of the gravitational effects of these more complicated situations can be embedded in the ripples and curves of a spacetime surface.

A picture gives perhaps the best description of how matter and energy distort the spacetime fabric to create a gravitational field. Figure 39 shows a sphere of matter sitting in space. The space surrounding the sphere is distorted: the ball makes a depression in the spatial surface whose depth reflects the ball's mass or energy. A ball

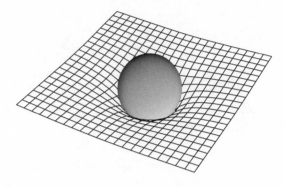

Figure 39. *A massive object distorts the surrounding space, thereby creating a gravitational field.*

passing nearby will roll towards the central depression, where the mass is located. According to general relativity, the spacetime fabric warps in an analogous fashion. Another ball passing through would be accelerated towards the center of the sphere. In this case, the result would agree with what Newton's law would predict, but the interpretation and calculation of the motion would be very different. According to general relativity, a ball follows the undulations of the spacetime surface, and thereby implements the motion induced by the gravitational field.

Figure 39 is a bit misleading, so you should keep in mind several caveats. First of all, I've shown the space surrounding the ball as two-dimensional. But really, the full three-dimensional space and the full four-dimensional spacetime are warped. Time is warped because it too is a dimension from the vantage point of special and general relativity. Warped time is how special relativity tells us that clocks run at different rates in different places, for example. A further caveat is that a second ball rolling in the curved geometry around the first ball would also affect the geometry of spacetime; we have assumed that its mass is much smaller than the larger ball's and neglected this small effect. The third thing that's important to keep in mind is that the object distorting spacetime can have any number of dimensions. Later on, a brane will play the role of the sphere in this picture.

Nonetheless, in all cases matter tells spacetime how to curve, and spacetime tells matter how to move. Curved spacetime sets up the geodesic paths along which, in the absence of other forces, things will travel. Gravity is encoded into the geometry of spacetime. It took Einstein the better part of a decade to deduce this precise connection between spacetime and gravity, and to incorporate the effects of the gravitational field itself—after all, the gravitational field carries energy, and is therefore bending spacetime.* It was a heroic effort.

In his famous equations, Einstein specified how to find the universe's gravitational field, given the contents of the universe. Although his best-known equation is $E = mc^2$, physicists use the term "Einstein's equations" to refer to the equations that determine the gravitational field. The equations accomplish this formidable task by showing how to determine the metric of spacetime from a known distribution of matter.[9] The metric you calculate determines the spacetime geometry by telling you how to translate numbers associated with arbitrary scale units into physical distances and shapes that determine the geometry.

With the final formulation of general relativity, physicists could determine the gravitational field and calculate its influence. As with previous formulations of gravity, physicists use these equations to figure out how matter moves in a given gravitational field. For example, they can plug in the mass and position of a big spherical body, such as the Sun or the Earth, and calculate the well-known Newtonian gravitational attraction. In this particular example, the results wouldn't be new—but their meaning would be. Matter and energy bend spacetime, and that bending gives rise to gravity. But general relativity has the further advantage that it incorporates any type of energy—including that of the gravitational field itself—into the distribution of matter and energy. This makes the theory useful even in situations where gravity itself contributes a significant amount of energy.

Because they apply to any distribution of energy, Einstein's equations changed the outlook for cosmologists—historians of the cosmos. Now, if scientists knew the matter and energy content of the universe, they could calculate its evolution. In an empty universe, space would

---

*Because the gravitational field carries energy, the energy of the field must be taken into account when using Einstein's equations. This makes solving for the gravitational field more subtle than it would be in Newtonian gravity.

be completely flat, with no ripples or undulations—no curvature at all. But when energy and matter fill the universe, they distort spacetime, producing interesting possibilities for the universe's structure and behavior over time.

We most definitely do not live in a static universe: as we will soon see, we just might live in a warped, five-dimensional one. Fortunately, general relativity tells us how to calculate their consequences. Just as there are examples of two-dimensional geometries with positive, zero, and negative curvature, there are four-dimensional geometrical configurations of spacetime with positive, zero, and negative curvature, which could arise from appropriate distributions of matter and energy. Later on, when we discuss cosmology and branes in extra dimensions, the distortions of spacetime arising from matter and energy—both in our visible universe and on the branes and in the bulk—will be of critical importance. We'll see that the three types of spacetime curvature (positive, negative, and zero) might be realized in higher dimensions as well.

General relativity has lots of consequences that you can't calculate with Newtonian gravity. Among its many merits, general relativity eliminated the annoying action-at-a-distance of Newtonian gravity, which asserted that an object's gravitational effects would be felt everywhere as soon as it appeared or moved. With general relativity, we know that before gravity can act, spacetime has to deform. This process does not happen instantaneously. It takes time. Gravity waves travel at the speed of light. Gravitational effects can kick in at a given position only after the time it takes for a signal to travel there and distort spacetime. That can never happen more quickly than the time it would take light, which travels as fast as anything we know, to get there. For example, you will never receive a radio signal or a cell phone call sooner than the time it would take for a light beam to travel to you.

Furthermore, physicists were able to use Einstein's equations to explore other types of gravitational field. With general relativity, scientists could describe and study black holes. These fascinating, enigmatic objects form when matter is highly concentrated within a very small volume. In black holes the geometry of spacetime is extremely distorted, so much so that anything entering a black hole gets trapped inside. Even light cannot escape. Although the German astronomer Karl Schwarzschild discovered that black holes were a consequence of Einstein's equations almost immediately after general relativity's development,* it was not until the 1960s that physicists took seriously the idea that they could be real things in our universe. Today, black holes are well accepted in the astrophysical community. In fact, it looks as though there is a supermassive black hole at the center of every galaxy, including our own. Moreover, if there are hidden dimensions then there exist higher-dimensional black holes which, when big, look like the four-dimensional black holes that astronomers have observed.

Coda

To conclude the story of the GPS system, it turns out that to calculate position to within a meter, we must measure time to better than one part in 10^{13}. The only possible way to get this accuracy is with atomic clocks.

But even if we had perfect clocks, time dilation would slow them down by about one part in 10^{10}. This error, if not corrected, would be a thousand times too big for our desired GPS system. We also have to account for the gravitational blueshift, a general relativity effect associated with a photon traveling in a changing gravitational field, which gives an error at least this great. This and other general relativity deviations would give errors that, if ignored, would build up at a rate greater than 10 km per day.† Ike (and current GPS systems) must correct for these relativistic effects.

Although by now relativity has been well tested and even gives rise to effects that need to be accounted for in practical devices, I do find it fairly remarkable that anyone listened to Einstein at first. He was completely unknown, working in the Bern patent office because he

---

*He did this on the Russian front while serving with the German army during World War I.

†Neil Ashby, "Relativity and the Global Positioning System," *Physics Today*, May 2002, p. 41.

couldn't get a better job. From this unlikely location he proposed a theory that went against the beliefs of all other physicists of his time.

Gerald Holton, a Harvard historian of science, tells me that the German physicist Max Planck was Einstein's first champion. Without Planck, who immediately recognized the brilliance of Einstein's work, it might have taken much longer for it to be recognized and accepted. Following Planck, a few other notable physicists knew enough to listen and pay attention. And shortly afterwards, so did the world.

What to Remember

- The speed of light is constant. It is independent of the speed of an observer.

- *Relativity* modifies our notions of space and time and tells us that we can treat them together as a single *spacetime* fabric.

- Special relativity relates the values of energy, momentum (which tells how an object responds to a force), and mass. For example, $E = mc^2$, where E is energy, m is mass, and c is the speed of light.

- Mass and energy make spacetime curve, and you can think of that curved spacetime as the origin of the gravitational field.

6

Quantum Mechanics: Principled Uncertainty, the Principal Uncertainties, and the Uncertainty Principle

> And you may ask yourself,
> Am I right? . . . Am I wrong?
>
> Talking Heads

Ike wondered whether Athena was making him watch too many movies or Dieter was talking too much about physics. But whatever the reason, the previous night Ike dreamed he met a quantum detective. Dressed in a fedora, a trench coat, and with a stone-faced expression, the dream detective spoke:

"I knew nothing about her except her name, and that she was standing there before me. But from the moment I set eyes on her I knew Electra* would be trouble. When I asked her where she came from, she refused to say. The room had two entrances, and she must have come through one. But Electra whispered hoarsely, 'Mister, forget it. I'll never tell you which.'

"Although I saw that she was shaking, I tried to pin this lady down. But Electra paced frenetically when I started to approach. She begged me to come no closer. Seeing she was agitated, I kept away. I was no stranger to uncertainty, but this time it had me beat. It looked like uncertainty was going to stick around here for a while."

Quantum mechanics, counterintuitive as it is, fundamentally altered the way scientists view the world. Much of modern science evolved

---

*The name refers to the electron, not to the character in Greek mythology.

from quantum mechanics: statistical mechanics, particle physics, chemistry, cosmology, molecular biology, evolutionary biology, and geology (through radioactive dating) were all either invented or revised as a result of its development. Many conveniences of the modern world, such as computers, DVD players, and digital cameras, wouldn't be possible without the transistor and modern electronics, whose development relied on quantum phenomena.

I'm not sure I fully appreciated how weird quantum mechanics is when I first studied it in college. I learned the basic principles and could apply them in various contexts. But it wasn't until I taught quantum mechanics many years later and carefully worked through quantum mechanical logic that I came to see just how fascinating it is. Although we can now teach quantum mechanics as part of the physics curriculum, it is nonetheless truly shocking.

The story of quantum mechanics beautifully exemplifies how science is supposed to evolve. Early quantum mechanics was done with a model building spirit—it addressed confusing observations even before anyone had formulated an underlying theory. Both experimental and theoretical advances happened fast and furiously. Physicists developed quantum theory to interpret experimental results that classical physics could not explain. And quantum theory, in turn, suggested further experiments with which to test hypotheses.

It took time for scientists to sort out the full implications of these experimental observations. The import of quantum mechanics was too radical for most scientists to immediately absorb. Scientists had to suspend their disbelief before they could accept the quantum mechanical premises, which were so different from familiar classical concepts. Even several of the theoretical pioneers, such as Max Planck, Erwin Schrödinger, and Albert Einstein, never really converted to the quantum mechanical way of thinking. Einstein voiced his objection in his famous remark, "God does not play dice with the universe." Most scientists did eventually accept the truth (as we currently understand it), but not immediately.

The radical nature of the scientific advances in the early twentieth century reverberated in modern culture. The fundamentals of art and literature and our understanding of psychology all changed radically at the time. Although some attribute these developments to the upheaval and havoc of World War I, artists such as Wassily Kandinsky used the fact that the atom was penetrable to justify the idea that everything can change, and that in art, therefore, everything is allowed. Kandinsky described his reaction to the nuclear atom: "The collapse of the atom model was equivalent, in my soul, to the collapse of the whole world. Suddenly the thickest walls fell. I would not have been amazed if a stone appeared before my eye in the air, melted, and became invisible."*

Kandinsky's reaction was a bit extreme. Radical as the fundamentals of quantum mechanics were, it's easy to overreach when applying them in nonscientific contexts. I find the most bothersome example to be the frequently abused uncertainty principle, which is often misappropriated to speciously justify inaccuracy. We will see in this chapter that the uncertainty principle is, in fact, a very precise statement about measurable quantities. Nonetheless, it is a statement with surprising implications.

We'll now introduce quantum mechanics and the underlying principles that make it so different from older, *classical* physics that came before. The strange and new concepts we'll encounter include quantization, the wavefunction, wave-particle duality, and the uncertainty principle. This chapter outlines these key ideas and gives a flavor of the history of how it was all worked out.

Shock and Awe

The particle physicist Sidney Coleman has said that if thousands of philosophers spent thousands of years searching for the strangest possible thing, they would never find anything as weird as quantum mechanics. Quantum mechanics is difficult to understand because its consequences are so counterintuitive and surprising. Its fundamental principles run counter to the premises underlying all previously known physics—and counter to our own experiences.

*Quoted in Gerald Holton and Stephen J. Brush, *Physics, the Human Adventure, from Copernicus to Einstein and Beyond* (Piscataway, NJ: Rutgers University Press, 2001).

One reason that quantum mechanics seems so bizarre is that we are not physiologically equipped to perceive the quantum nature of matter and light. Quantum effects generally become significant at distances of about an angstrom, the size of an atom. Without special instruments, we can see only sizes that are much larger. Even the pixels of a high-resolution television or computer monitor are generally too small for us to see.

Furthermore, we see only huge aggregates of atoms, so many that classical physics overwhelms quantum effects. We generally also perceive only many quanta of light. Although a photoreceptor in an eye is sufficiently sensitive to perceive the smallest possible units of light —individual quanta—an eye typically processes so many quanta that any would-be quantum effects are overwhelmed by more readily apparent classical behavior.

If quantum mechanics is difficult to explain, there is a very good reason. Quantum mechanics is sufficiently far-reaching to incorporate classical predictions, but not the other way round. Under many circumstances—for example, when large objects are involved— quantum mechanical predictions agree with those from classical Newtonian mechanics. But there is no range of size for which classical mechanics will generate quantum predictions. So when we try to understand quantum mechanics using familiar classical terminology and concepts, we are bound to run into trouble. Trying to use classical notions to describe quantum effects is something like trying to translate French into a restricted English vocabulary of only a hundred words. You would frequently encounter concepts or words that could be interpreted only vaguely, or which would be impossible to express at all with such a limited English vocabulary.

The Danish physicist Niels Bohr, one of the pioneers of quantum mechanics, was aware of the inadequacy of human language for describing the inner workings of the atom. Reflecting on the subject, he related how his models "had come to him intuitively . . . as pictures."* As the physicist Werner Heisenberg explained, "We simply have to remember that our usual language does not work any more,

*Gerald Holton, *The Advancement of Science, and Its Burdens* (Cambridge, MA: Harvard University Press, 1998).

that we are in the realm of physics where our words don't mean much."*

I will therefore not attempt to describe quantum phenomena with classical models. Instead, I will describe the key fundamental assumptions and phenomena that made quantum mechanics so different from the classical theories that came before. We'll reflect individually on several of the key observations and insights that contributed to quantum mechanics and its development. Although this discussion follows a roughly historical outline, my real purpose is to introduce the many new ideas and concepts intrinsic to quantum mechanics one at a time.

The Beginning of Quantum Mechanics

Quantum physics developed in stages. It began as a series of random assumptions that matched observations, although no one understood why they matched. These inspired guesses, which had no underlying physical justification but did have the virtue of giving the right answers, were embodied in what is now known as the *old quantum theory*. This theory was defined by the assumption that quantities such as energy and momentum couldn't have just any arbitrary values. Instead, the possibilities were confined to a discrete, *quantized* set of numbers.

Quantum mechanics, which developed from the humble antecedent of the old quantum theory, justifies the mysterious quantization assumptions that we'll shortly encounter. Furthermore, quantum mechanics provides a definite procedure for predicting how quantum mechanical systems evolve with time, greatly increasing the theory's power. But at the outset quantum mechanics developed only in fits and starts, since no one at the time really understood what was going on. At first, the quantization assumptions were all there were.

The old quantum theory began in 1900, when the German physicist

---

*Quoted in Gerald Holton and Stephen J. Brush, *Physics, the Human Adventure, from Copernicus to Einstein and Beyond* (Piscataway, NJ: Rutgers University Press, 2001).

Max Planck suggested that light could be delivered only in quantized units, just as bricks can only be sold in discrete chunks. According to Planck's hypothesis, the amount of energy contained in light of any specific frequency could only be a multiple of the fundamental energy unit for that particular frequency. That fundamental unit is equal to a quantity, now known as Planck's constant, h, multiplied by the frequency, f. The energy of light with a definite frequency f could be hf, $2hf$, $3hf$, and so on, but according to Planck's assumption you could never find anything in between. Unlike bricks, whose quantization is arbitrary and nonfundamental—bricks can be split apart—there is a minimum energy unit of light of a given frequency which is indivisible. Intermediate values of energy could never occur.

This remarkably prescient suggestion was made to address a theoretical puzzle known as the blackbody *ultraviolet** *catastrophe*. A blackbody is an object, such as a piece of coal, that absorbs all incoming radiation and then radiates it back.† The amount of light and other energy it emits depends on its temperature; temperature completely characterizes a blackbody's physical properties.

However, the classical predictions for the light radiated from a blackbody were problematic: classical calculations predicted that far greater energy would be emitted in high-frequency radiation than physicists had seen and recorded. Measurements showed that different frequencies do not contribute democratically to blackbody radiation; the very high frequencies contribute less than the lower ones. Only the lower frequencies emit significant energy. This is why radiating objects are "red-hot" and not "blue-hot." But classical physics predicted a large amount of high-frequency radiation. In fact, the total emitted energy predicted by classical reasoning was infinite. Classical physics faced an ultraviolet catastrophe.

An ad hoc way out of this dilemma would have been to assume that only frequencies below some specific upper limit could contribute to radiation from a blackbody. Planck disregarded this possibility in

---

*"Ultraviolet" means "high-frequency."
†A blackbody is actually an idealization; real objects like coal aren't perfect black-bodies.

favor of another, apparently equally arbitrary, assumption: that light is quantized.

Planck reasoned that if radiation at each frequency consisted of whole-unit multiples of a fundamental quantum of radiation, then no high-frequency radiation could be emitted because the fundamental unit of energy would be too large. Because the energy contained in a quantum unit of light was proportional to frequency, even a single unit of high-frequency radiation would contain a large amount of energy. When the frequency was high enough, the minimum energy a quantum would contain would be too large for it to be radiated. The blackbody could radiate only the lower-frequency quanta. Planck's hypothesis thereby forbade excessive high-frequency radiation.

An analogy might help elucidate Planck's logic. You've probably eaten dinner with people who protest when it is time to order dessert. They're afraid of eating too much fattening food, so they rarely order their own tasty treats. If the waiter promises that the desserts are small, they might order one. But they quail at the usual large, quantized portions of cake or ice cream or pudding.

There are two types of such people. Ike belongs to the first category. He has true discipline, and really doesn't eat dessert. When a dessert is too big, Ike simply refrains from eating it. I'm more like the second type of person—Athena is also one—who thinks that the desserts are too big, and therefore doesn't order any for herself, but, unlike Ike, has no compunction about taking bites from the desserts on everyone else's plate. So even when Athena refuses to order her own portion, she still ends up eating quite a lot. If Athena were eating dinner with a large number of people, and hence could pick off a large number of plates, she would suffer from an unfortunate "calorie catastrophe."

According to the classical theory, a blackbody is more like Athena. It would emit small amounts of light at any frequency, and theorists using classical reasoning would therefore predict an "ultraviolet catastrophe." To avoid this predicament, Planck suggested that a blackbody was analogous to the truly abstemious type. Like Ike, who never eats a fraction of a dessert, a blackbody behaves according to Planck's quantization rule and emits light of a given frequency only in quantized energy units, equal to the constant h times the frequency *f*. If the frequency were high, the quantum of energy would be simply too big for light to be emitted at that frequency. A blackbody would therefore emit most of its radiation at low frequencies, and high frequencies would be automatically cut out. In quantum theory, a blackbody doesn't emit a substantial amount of high-frequency radiation and therefore emits far less radiation than is predicted by the classical theory.

When an object emits radiation, we call the radiation pattern—that is, how much energy the object emits at each frequency at a given temperature—its *spectrum*[10] (see Figure 40). The spectra of certain objects such as stars can approximate that of a blackbody. Such blackbody spectra have been measured at many different temperatures, and they all agree with Planck's assumption. Figure 40 shows that the emission is all at lower frequency; at high frequency, emission shuts off.

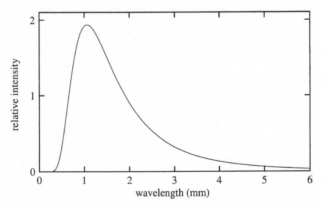

Figure 40. *The blackbody spectrum of the cosmic microwave background of the universe. A blackbody spectrum gives the amount of light that is emitted at all frequencies when the temperature of the radiating object is fixed. Notice that the spectrum cuts off at high frequency.*

One of the great achievements of experimental cosmology since the 1980s has been the increasingly accurate measurement of the blackbody spectrum that the radiation in our universe produces. Originally, the universe was a hot, dense fireball containing high-temperature radiation, but since then the universe has expanded and the radiation has cooled tremendously. That is because as the universe expanded, the wavelengths of the radiation did too. And longer wavelength corresponds to lower frequency, which corresponds to lower energy, which also corresponds to lower temperature. The universe now contains radiation that looks as if it has been produced by a blackbody with a temperature of only 2.7 degrees above absolute zero—considerably cooler than when it started.

Satellites have recently measured the spectrum of this cosmic microwave background radiation (which is what Figure 40 shows). It looks almost precisely like the spectrum of a blackbody with a temperature of 2.7 degrees K. The measurements tell us that deviations are smaller than one part in ten thousand. In fact, this relic radiation is the most accurately measured blackbody spectrum to date.

When asked in 1931 how he had come up with his outrageous assumption that light is quantized, Planck responded, "It was an act of desperation. For six years I had struggled with the blackbody theory. I knew the problem was fundamental and I knew the answer. I had to find a theoretical explanation at any cost . . ."* For Planck, light quantization was a device, a kludge that gave the correct blackbody spectrum. In his view, quantization was not necessarily a property of light itself, but could instead have been a consequence of some property of the atoms that were radiating the light. Although Planck's conjecture was the first step in understanding light quantization, Planck himself did not fully comprehend it.

Five years later, in 1905, Einstein made a major contribution to quantum theory when he established that light quanta were real things, not merely mathematical abstractions. Einstein was a very busy man that year, developing special relativity, helping to prove that atoms and molecules exist by studying the statistical properties of matter, and providing a validation of quantum theory—all while he was working at the Swiss patent office in Bern.

The particular observation that Einstein interpreted using the hypothesis of light quanta, thereby enhancing its credibility, is known

---

*". . . at any cost, that is, except for the inviolability of the two laws of thermodynamics." Quoted in David Cassidy, *Einstein and Our World*, 2nd edn (Atlantic Highlands, NJ: Humanities Press, 2004).

as the *photoelectric effect*. Experimenters shone a single frequency of radiation onto matter, and that incoming radiation propelled electrons out. Experiments had shown that bombarding material with more light, which carries more total energy, did not change the maximum kinetic energy (energy of motion) of the emitted electrons. This is contrary to what intuition might suggest: larger incident energy should surely produce electrons with larger kinetic energy. The limit on the electron's kinetic energy was therefore a puzzle. Why didn't the electron absorb more energy?

Einstein's interpretation was that radiation consists of individual quanta of light, and only a single quantum will donate its energy to any particular electron. Light is delivered to an individual electron like a single missile, not like a blitzkrieg. Because only one quantum of light ejects the electron, more incident quanta would not change the energy of the emitted electron. Increasing the number of incident quanta makes the light eject more electrons, but it doesn't influence the maximum energy of any particular electron.

Once Einstein interpreted the results of the photoelectric effect in terms of these definite packets of energy—the quantized units of light—it made sense that the emitted electrons always had the same maximum kinetic energy. The most kinetic energy an electron can have is the fixed energy that it receives from the quantum of light minus the energy required to eject the electron from the atom.

Using this logic, Einstein could deduce the energy of the light quanta. He found that their energy depended on the frequency of the incident light exactly as Planck's hypothesis predicted. To Einstein, this was clear evidence that light quanta were real. His interpretation gave a very concrete picture of light quanta: a single quantum hit a single electron, which it thereby ejected. It was this observation and not relativity that earned Einstein the Nobel Prize for Physics in 1921.

Oddly enough, however, although Einstein acknowledged the existence of quantized units of light, he was reluctant to accept that these quanta were actually massless particles, which carried energy and momentum but had no mass. The first convincing evidence for the particle nature of the quanta of light came from the 1923 measurement of *Compton scattering*, in which a quantum of light hits an electron

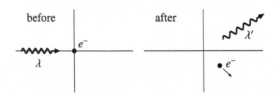

Figure 41. *In Compton scattering, a photon (γ) scatters off a stationary electron (e⁻)* and emerges with a different energy and momentum.

and is deflected (see Figure 41). In general, you can determine a particle's energy and momentum by measuring its deflection angle after a collision. If photons were massless particles, they would behave in a well-defined manner when they collided with other particles such as electrons. Measurements showed that the quanta of light behaved precisely as if the quanta were massless particles that interacted with the electrons. The inexorable conclusion was that light quanta were indeed particles, and we now call these particles *photons*.

It's perplexing that Einstein was so resistant to the quantum theory that he helped to develop. But his reaction is no more remarkable than Planck's response to Einstein's quantization proposal—which was disbelief. Planck and several others praised Einstein's many achievements, but qualified their enthusiasm.* Planck even said, somewhat disparagingly, "That he missed the target in his speculations, as, for example, in his hypothesis of light-quanta, cannot really be held too much against him, for it is not possible to introduce really new ideas even in the most exact sciences without sometimes taking a risk."† Make no mistake. Einstein's conjectured light-quanta were right on target. Planck's comment merely reflects the revolutionary nature of Einstein's insight and the initial reluctance of scientists to accept it.

---

*Abraham Pais, *Subtle Is the Lord: The Science and Life of Albert Einstein* (Philadelphia: American Philological Association, 1982).
†Gerald Holton, *Thematic Origins of Scientific Thought*, revised edn (Cambridge, MA: Harvard University Press, 1988).

Quantization and the Atom

The story of quantization and the old quantum theory didn't end with light. It turns out that *all* matter consists of fundamental quanta. Niels Bohr was next in line with a quantization hypothesis. In his case, he applied it to a well-established particle, the electron.

Bohr's interest in quantum mechanics developed, in part, from attempts at the time to clarify the atom's mysterious properties. During the nineteenth century, the notion of an atom was unbelievably vague: many scientists didn't believe that atoms existed other than as heuristic devices that were a useful tool but which had no grounding in reality. Even some of the scientists who did believe in atoms nonetheless confused them with molecules, which we now know to be composites of atoms.

The atom's true properties and composition were not accepted until the beginning of the last century. Part of the problem was that the Greek word "atom" meant a thing that could not be divided, and the original picture of the atom was indeed one of an unchanging, indivisible object. But as nineteenth-century physicists learned more about how atoms behave, they began to realize that this idea had to be incorrect. By the end of the century, radioactivity and *spectral lines*, the specific frequencies at which light is emitted and absorbed, were some of the best-measured properties of atoms. Yet both of these phenomena showed that atoms could change. On top of that, in 1897, J.J. Thomson identified electrons and proposed that the electron was an ingredient of the atom, which meant that atoms had to be divisible.

At the beginning of the twentieth century, Thomson synthesized the atomic observations of the time in his "plum pudding" model, named after the British dessert containing isolated pieces of fruit stuck in a bready blob. He suggested that there was a positively charged component spread throughout the atom (the bready part), with negatively charged electrons (the pieces of fruit) embedded inside.

The New Zealander Ernest Rutherford proved this model wrong in 1910, when Hans Geiger and a research student, Ernest Marsden, performed an experiment that Rutherford had suggested. They discovered a hard, compact atomic nucleus, much smaller than the atom itself. Radon-222, a gas produced in the radioactive decay of radium salts, emits alpha particles, which we now know to be helium nuclei. The physicists revealed the existence of the atom's nucleus by shooting alpha particles at atoms and recording the angles at which the alpha particles scattered. The dramatic scattering they recorded could arise only if there were a hard, compact atomic nucleus. A diffuse, positive charge spread throughout the extent of the atom could never have scattered the particles so widely. In Rutherford's words, "It was quite the most incredible event that has ever happened to me in my life. It was almost as incredible as if you fired a 15-inch shell at a piece of tissue paper and it came back and hit you."*

Rutherford's results disproved the plum pudding model of the atom. His discovery meant that the positive charge was not spread throughout the atom, but was instead confined to a much smaller inner core. There had to be a hard central component, the nucleus. An atom, according to this picture, consisted of electrons that orbited a small central nucleus.

In the summer of 2002 I attended the annual string theory conference, which happened to be held that year at the Cavendish Laboratory in Cambridge. Many important pioneers of quantum mechanics, including two of its heads, Rutherford and Thomson, did much of their important research there. The hallways are decorated with reminiscences of the exciting early years, and I learned some amusing facts while wandering the hallways.

For example, James Chadwick, the discoverer of the neutron, had studied physics only because he was too shy to point out that he had mistakenly waited in the wrong line when matriculating. And J.J. Thomson was so young when he became head of the lab (he was twenty-eight) that a congratulation read, "Forgive me if I have done wrong in not writing to wish you happiness and success as a professor. The news of your election was too great a surprise to permit me to do so." (Physicists aren't always the most gracious.)

Yet despite the coherent picture of the atom that had developed by the early twentieth century at the Cavendish and elsewhere, the

---

*Quoted in Abraham Pais, *Inward Bound: Of Matter and Forces in the Physical World* (Oxford: Oxford University Press, 1986).

behavior of its components was about to wreak havoc with physicists' most fundamental beliefs. Rutherford's experiments had suggested an atom consisting of electrons that traveled in orbits around a central atomic nucleus. This picture, simple as it was, had an unfortunate drawback: it had to be wrong. Classical electromagnetic theory predicted that when electrons orbited in a circle, they would radiate energy through photon emission (or, classically speaking, electromagnetic wave emission). The photons would thereby remove energy and leave behind a less energetic electron, which would orbit in ever smaller circles, spiraling in towards the center. In fact, classical electromagnetic theory predicted that atoms could not be stable, and would collapse in less than a nanosecond. The atom's stable electron orbits were a complete mystery. Why didn't electrons lose energy and spiral down into the atomic nucleus?

A radical departure from classical reasoning was required to explain the atom's electron orbits. Pursuing this logic to its inevitable conclusion exposed chinks in classical physics that could be filled only by the development of quantum mechanics. Niels Bohr made just such a revolutionary proposal when he extended Planck's notion of quantization to electrons. This, too, was an essential component of the old quantum theory.

Electron Quantization

Bohr decided that electrons couldn't move in just any old orbit: an electron's orbit had to have a radius that fit a formula he proposed. He found these orbits by making a lucky and ingenious guess. He decided that electrons must act as if they were waves, which meant that they oscillated up and down as they circulated about the nucleus.

In general, a wave with a particular wavelength oscillates up and down once over a fixed distance; that distance is the wavelength. A wave that goes around a circle also has an associated wavelength. In this case the wavelength sets the extent of the arc over which the wave will go up and down once as it winds around the nucleus.

An electron that orbits in a fixed radius cannot have any wavelength.

It can only have a wavelength that would permit the wave to go up and down a fixed number of times. That implied a rule for determining the allowed wavelengths: the wave has to oscillate an integer* number of times when going around the circle that defined the electron's orbit (see Figure 42).

Figure 42. *Possible wave patterns for an electron according to the Bohr quantization.*

Although Bohr's proposal was radical and its meaning obscure, his guess did the trick: if true, it would guarantee stable electron orbits. Only particular electron orbits would be allowed. Intermediate orbits would be forbidden. In the absence of an external agent that could make an electron jump from one orbit to another, there would be no way for the electron to move in towards the nucleus.

You can think of Bohr's atom with its fixed electron orbits as a multistory building in which you're restricted to the even-numbered floors, the second, fourth, sixth, and so on. Since you could never set foot on the in-between floors, such as the third and the fifth, you would be eternally stuck on the even-numbered floor you were on. There would be no way to reach the ground floor and exit.

Bohr's waves were an inspired assumption. He did not claim to know their meaning; he made his assumption simply to account for the stable electron orbits. Nevertheless, the quantitative nature of his proposal allowed it to be tested. In particular, Bohr's hypothesis correctly predicted atomic spectral lines. Spectral lines give the frequency of light that an *un-ionized* atom—a neutral atom with all its electrons that carries zero net charge—emits or absorbs.† Physicists had noticed that spectra show a barcode-like pattern of stripes rather

---

*Integers are the familiar whole numbers: 0, 1, 2, 3, and so on.
†We are focusing here on discrete spectra. When a free electron is absorbed by an ion, a continuous—not a discrete—spectrum of light is emitted.

than a continuous distribution (i.e., with all frequencies of light contributing). But no one understood why. Nor did they know the reason for the precise values of the frequencies they saw.

With his quantization hypothesis, Bohr could explain why photons were emitted or absorbed only at the measured frequencies. Although the electrons' orbits were stable for an isolated atom, they could change when a photon with the right frequency—and hence, according to Planck, the right energy—delivered or removed energy.

Using classical reasoning, Bohr calculated the energy of the electrons that obeyed his quantization assumption. From these energies he predicted the energies, and hence the frequencies, of the photons that the hydrogen atom, which contains a single electron, emitted or absorbed. Bohr's predictions were correct, and these correct predictions made his quantization assumption highly plausible. And this was what convinced Einstein, among others, that Bohr must be right.

The quantized packets of light, which could be emitted or absorbed and could thereby change electron orbits, can be compared to lengths of rope placed by the windows of the multistory building in our earlier analogy. If each piece of rope is precisely the length required to go from your floor to any of the other even-numbered floors, and only the windows to even-numbered floors are open, the rope would provide the means to change floors—but only between the even-numbered ones. In the same way, spectral lines could take only certain values, the values of the differences in energy between electrons that occupied permissible orbits.

Even though Bohr offered no explanation for his quantization condition, he certainly appeared to be correct. Many spectral lines had been measured, and his assumption could be used to reproduce them. If such agreement was a coincidence it would have been miraculous. Ultimately, quantum mechanics justified his assumption.

Particles' Commitment Phobia

Important as the quantization proposals were, the quantum mechanical connection between particles and waves began to gel only with the advances made by the French physicist Prince Louis de Broglie, the Austrian Erwin Schrödinger, and the German-born Max Born. The first key step off the random walk of the old quantum theory onto the road of a real theory of quantum mechanics was de Broglie's brilliant suggestion of turning Planck's quantization hypothesis on its head. Whereas Planck had associated quanta with the waves of radiation, de Broglie—like Bohr—postulated that particles could also act like waves. De Broglie's hypothesis meant that particles should exhibit wavelike properties and that those waves are determined by a particle's momentum. (For low speeds, momentum is mass multiplied by speed. For all speeds, momentum tells how something responds to an applied force. Although at relativistic speeds, momentum is a more complicated function of mass and speed, the generalization of momentum that applies at high speeds also indicates how something at relativistic speeds would respond to a force.)

De Broglie assumed that a particle with momentum p was associated with a wave whose wavelength was inversely proportional to momentum—that is, the smaller the momentum, the longer the wavelength. The wavelength was also proportional to Planck's constant, h.* The idea behind de Broglie's proposal was that a wave that oscillated frenetically (that is, one with small wavelength) carried more momentum than one that oscillated lethargically (with large wavelength). Smaller wavelengths mean more rapid oscillations, which de Broglie associated with larger momentum.

If you find the existence of this particle-wave perplexing, that's because it is. When de Broglie first suggested his waves, no one knew what they were supposed to be. Max Born proposed a surprising interpretation: that the wave was a function of position whose square gives the probability for finding a particle at any location in space.† He named this a *wavefunction*. Max Born's insight was that particles cannot be pinned down and can be described only in terms of probabilities. This is a big a departure from classical assumptions. It means that you cannot know the particle's exact location. You can only specify the *probability* of finding it somewhere.

---

*Wavelength is equal to Planck's constant, h, divided by momentum.
†Although we need three coordinates to specify a point in space, we sometimes simplify and pretend that the wavefunction depends only on a single coordinate. This makes it easier to draw pictures of wavefunctions on a piece of paper.

But even though a quantum mechanical wave describes only probabilities, quantum mechanics predicts this wave's precise evolution through time. Given the values at any one time, you can determine the values at any later time. Schrödinger developed the wave equation that shows the evolution of the wave associated with a quantum mechanical particle.

But what does this probability of finding a particle mean? It's a puzzling idea—after all, there's no such thing as a fraction of a particle. That a particle can be described by a wave was (and in some ways still is) one of the most surprising aspects of quantum mechanics, particularly as it is known that particles often behave like billiard balls, and not like waves. A particle interpretation and a wave interpretation seem incompatible.

The resolution to this apparent paradox hinges on the fact that you never detect the wave nature of a particle with just one particle. When you detect an individual electron, you see it in some definite location. In order to map out the entire wave, you need a set of identical electrons, or an experiment that is repeated many times. Even though each electron is associated with a wave, with a single electron you will measure only one number. But if you could prepare a large set of identical electrons, you would find that the fraction of electrons in each location is proportional to the probability wave assigned to an electron by quantum mechanics.

The wavefunction of an individual electron tells you about the likely behavior of many identical electrons with this same wavefunction. Any individual electron will be found only in a single place. But if there were many identical electrons, they would exhibit a wave-like distribution of locations. The wavefunction tells you the probability of the electron ending up in those locations.

This is analogous to the distribution of height in a population. Any individual has their own height, but the distribution tells us the likelihood that an individual will have any particular height. Similarly, even if one electron behaves like a particle, many electrons together will have a distribution of positions delineated by a wave. The distinction is that an individual electron is nonetheless associated with this wave.

In Figure 43 I've plotted an example of a probability function for

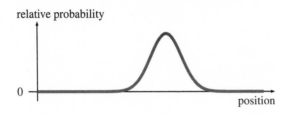

relative probability

0

position

Figure 43. *An example of a probability function for an electron.*

an electron. This wave gives the relative probability of finding the electron at a particular location. The curve I have drawn takes a definite value for every point in space (or rather, every point along a line, since the flatness of the paper forces me to draw only one dimension of space). If I could make many copies of this same electron, I could take a series of measurements of the electron's position. I would find that the number of times I measured the electron to be at any particular point was proportional to this probability function. A bigger value means that the electron would be more likely to be found there; a smaller value that it is less likely. The wave reflects the cumulative effect of many electrons.

Even though you map out the wave with many electrons, what makes quantum mechanics special is that an individual electron is nonetheless described by a wave. That means you can never predict everything about that electron with certainty. If you measure its location, you will find it in a definite spot. But until you make that measurement, you can predict only that the electron has a particular probability of winding up there. You can't say definitively where it will end up.

This particle-wave dichotomy is revealed in the famous double-slit experiment that Electra's unknown origin in the opening story referred to.[11] Until 1961, when the German physicist Claus Jonsson actually performed it in the lab, the electron double-slit experiment was merely a thought experiment that physicists used to elucidate the meaning and consequences of the electron wavefunction. The experiment consists of an electron-emitter that sends electrons through a barrier pierced by two parallel slits (see Figure 44). The electrons pass through the slits and hit a screen behind the barrier, where they are recorded.

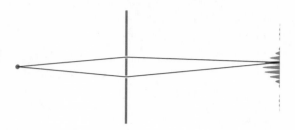

Figure 44. *Schematic arrangement of the double-slit electron interference experiment. Electrons can go through either of two slits before they hit a screen. The wave pattern that is recorded on the screen is a result of the interference of the two paths.*

This experiment was meant to mimic a similar experiment that demonstrated the wavelike nature of light in the early nineteenth century. At that time, Thomas Young, a British physician, physicist, and Egyptologist,* sent monochromatic light through two slits and observed the wavelike pattern that light made on a screen behind the slits. The experiment demonstrated that light behaved like a wave. The point of imagining the same experiment with electrons is to see how you might observe the electron's wavelike nature.

And indeed, if you were to perform the double-slit experiment with electrons, you would see what Young saw for light: a wavelike pattern on the screen behind the slits (see Figure 45). In the case of light, we understand that the wave is caused by interference. Some of the light goes through one slit and some of it goes though the other, and the wave pattern that is then recorded reflects the interference between the two. But what does a wavelike pattern mean for electrons?

The wavelike pattern on the screen tells us the very unintuitive fact that we should think of each electron as passing through both slits. You can't know everything about an individual electron. Any electron can pass through either slit. Even though each electron's location gets recorded when it reaches the screen, no one knows which of the two slits any individual electron passed through.

Quantum mechanics tells us that a particle can take any possible path from its starting point to its endpoint, and the wavefunction for

*He even helped to decipher the Rosetta Stone.

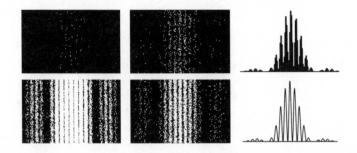

Figure 45. *The interference pattern that is recorded in the double slit experiment. The four panels on the left show, clockwise from the top left, the pattern seen after 50, 500, 5,000, and 50,000 electrons have been shot through. The curves on the right compare the distribution of the number of electrons (upper curve) to the pattern you would get for a wave that passes through the two slits. They are nearly identical, which shows that the electron wavefunction does in fact act like a wave.*

that particle reflects this fact. This is one of the many remarkable features of quantum mechanics. Unlike classical physics, quantum mechanics does not assign a particle a definite trajectory.

But how can the double-slit experiment indicate that an individual electron acts like a wave, when we already know that electrons are particles? After all, there is no such thing as half an electron. Any individual electron gets recorded in a definite location. What's really going on?

The answer is the one I gave earlier. You can see the wave pattern only when you record many electrons. Each individual electron is a particle. It hits the screen in a single location. However, the cumulative effect of many electrons being shot at the screen is a classical wave pattern, reflecting the fact that the two electron paths interfere. You can see this in Figure 45.

The wavefunction gives the probability that an electron will hit the screen in any particular location. The electron might go anywhere, but you would expect to find it only at some particular place with a definite probability given by the value of the wavefunction at that point. Many electrons together produce the wave that you could derive from the assumption that the electron passes through both slits.

In the 1970s, Akira Tonamura in Japan and Piergiorgio Merli,

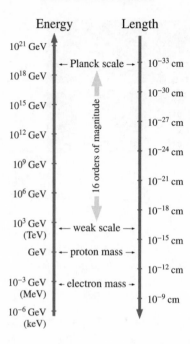

Figure 46. *Some important length scales and energy scales in particle physics. Larger energies correspond (via special relativity and the uncertainty principle) to smaller distances—a more energetic wave is sensitive to interactions that occur over shorter distance scales. The gravitational interaction is inversely proportional to the Planck scale energy. The large Planck scale energy means that gravitational interactions are weak. The weak scale energy is the energy which sets the scale (via E = mc²) for the weak gauge boson masses. The weak scale length is the distance over which the weak gauge bosons communicate the weak force.*

Giulio Pozzi, and Gianfranco Missiroli in Italy actually saw this explicitly in real experiments. They shot electrons through one at a time and saw the wave pattern develop as more and more electrons hit the screen.

You might wonder why it took until the twentieth century for anyone to notice something as dramatic as wave-particle duality. For example, why didn't people realize any earlier that light looks like a wave but is actually composed of discrete nuggets—namely, photons?

The answer is that none of us (with the possible exception of superheroes) sees individual photons,* so quantum mechanical effects cannot be easily detected. Ordinary light doesn't look as if it's made up of quanta. We see bunches of photons that constitute visible light. The large number of photons together act as a classical wave.

You need a very weak source of photons, or a very carefully prepared system, to observe the quantized nature of light. When there are too many photons, you can't distinguish the effect of any single one. Adding one more photon to classical light, which contains many photons, just doesn't make a big enough difference. If your lightbulb, which behaves classically, emitted one additional photon, you would never notice. You can observe detailed quantum phenomena only in carefully prepared systems.

If you don't believe that this one last photon is usually insignificant, think about how you feel when you go to the voting booth. Is it really worth the time and trouble to vote when you know that your vote can't possibly make a difference in the outcome, since millions of other people are voting? With the notable exception of Florida, the state of uncertainty, one vote generally gets lost in the crowd. Even though an election gets decided by the cumulative effect of individual votes, a single vote rarely, if ever, changes the result. (And, to take the comparison a step further, you might also observe that only in quantum systems—and in Florida, which acts like a quantum state—do repeated measurements produce different results.)

Heisenberg's Uncertainty

The wave nature of matter has many counterintuitive implications. We'll now turn from electoral uncertainty to Heisenberg's uncertainty principle, a favorite of physicists and after-dinner speakers.

The German physicist Werner Heisenberg was one of the major pioneers of quantum mechanics. In his autobiography, he told how

---

*People are actually capable of detecting individual photons, but only in carefully prepared experiments. Usually, you see more standard light composed of many photons.

his revolutionary ideas about atoms and quantum mechanics began to germinate when he was headquartered in the Theological Training College in Munich, where he was stationed in 1919 to fight off Bavarian communists. After the shooting had subsided, he sat on the college roof and read Plato's dialogues, the *Timaeus* in particular. Plato's text convinced Heisenberg that "in order to interpret the material world, we need to know something about its smallest parts."*

Heisenberg hated the external upheaval that surrounded him in his youth; he would have preferred a return to "the principles of Prussian life, the subordination of individual ambition to the common cause, modesty in private life, honesty and incorruptibility, gallantry and punctuality."† Nonetheless, with the uncertainty principle Heisenberg irrevocably changed people's worldview. Perhaps the tumultuous era in which Heisenberg lived gave him a revolutionary approach to science, if not to politics.‡ In any case, I find it a little ironic that the author of the uncertainty principle was a man of such conflicting dispositions.

The uncertainty principle says that certain pairs of quantities can never be measured accurately at the same time. This was a major departure from classical physics, which assumes that, at least in principle, you can measure all the characteristics of a physical system— position and momentum, for example—as accurately as you'd like.

The particular pairs are those for which it matters which one you measure first. For example, if you were to measure position and then momentum (the quantity which gives both speed and direction), you wouldn't get the same result as if you first measured momentum and then position. This would not be the case in classical physics, and is certainly not what we are used to. The order of measurements matters only in quantum mechanics. And the uncertainty principle says that for two quantities where the order of measurement matters, the product of the uncertainties of the two will always be greater than a

*Werner Heisenberg, *Physics and Beyond: Encounters and Conversations*, translated by Arnold Pomerans (New York: Harper & Row, 1971).
†Ibid. Owing to his German nationalism, he also participated in the German atomic bomb project.
‡Gerald Holton, *The Advancement of Science, and Its Burdens* (Cambridge, MA: Harvard University Press, 1998).

fundamental constant, namely Planck's constant, h, which is 6.582×10^{-25} GeV second for those who want to know.*[12] If you insist on knowing position very accurately, you cannot know momentum with a similar accuracy, and vice versa. No matter how precise your measuring instruments and no matter how many times you try, you can never simultaneously measure both quantities to very high accuracy.

The appearance of Planck's constant in the uncertainty principle makes a good deal of sense. Planck's constant is a quantity that arises only with quantum mechanics. Recall that, according to quantum mechanics, the quanta of energy of a particle with a particular frequency is Planck's constant times that frequency. If classical physics ruled the world, Planck's constant would be zero and there would be no fundamental quantum.

But in the true quantum mechanical description of the world, Planck's constant is a fixed, nonzero quantity. And that number tells us about uncertainty. In principle, any individual quantity can be accurately known. Sometimes physicists refer to the *collapse of the wavefunction* to specify that something has been accurately measured and therefore takes a precise value. The word "collapse" refers to the shape of the wavefunction, which is no longer spread out but takes a nonzero value at one particular place, since the probability of measuring any other value afterwards is zero. In this case—when one quantity is measured precisely—the uncertainty principle would tell you that after the measurement, you can know nothing at all about the other quantity that is paired with the measured quantity in the uncertainty principle. You would have infinite uncertainty in the value of that other quantity. Of course, had you first measured the second quantity, the first quantity would be the one you didn't know. Either way, the more accurately you know one of the quantities, the less precise the measurement of the other has to be.

I won't go into the detailed derivation of the uncertainty principle in this book, but I'll nonetheless try to give a flavor of its origin. Because this is not essential to what follows, feel free to skip ahead to the next section. But you might want to learn a little more about the reasoning that underlies uncertainty.

---

*The GeV is a unit of energy that I will soon explain.

In this derivation, we'll focus on time-energy uncertainty, which is a little easier to explain and understand. The time-energy uncertainty principle relates the uncertainty in energy (and hence, according to Planck's assumption, frequency) to the time interval that is characteristic of the rate of change of the system. That is, the product of the energy uncertainty and the characteristic time for the system to change will always be greater than Planck's constant, h.

A physical realization of the time-energy uncertainty principle happens, for example, when you turn on a light switch and hear static from a nearby radio. Turning on the light switch generates a large range of radio frequencies. That's because the amount of electricity going through the wire changed very rapidly, so the range of energy (and hence frequency) must be large. Your radio picks it up as static.

To understand the uncertainty principle's origin, let's now consider a very different example—a leaky faucet.* We will show that you need a long-lasting measurement to accurately determine the rate at which the faucet drips, which we will see is closely analogous to the uncertainty principle's claim. A faucet and the water passing through it, which involve many atoms, is too complex a system to exhibit observable quantum mechanical effects—they are overwhelmed by classical processes. It is nonetheless true that you need longer measurements to make more accurate frequency determinations—and that is the core of the uncertainty principle. A quantum mechanical system would take this interdependence a step further because for a carefully prepared quantum mechanical system, energy and frequency are related. So for a quantum mechanical system, a relation between frequency uncertainty and the length of time of a measurement (like the one we are about to see) therefore translates into the true uncertainty relation between energy and time.

Suppose that water is dripping at a rate of about once per second. How well could you measure the rate if your stopwatch had one-second accuracy—that is, it could be off by at most one second? If you were to wait one second, and saw a single drip, you might think that you could conclude that the faucet drips once per second.

---

*"Tap" for British readers. We are assuming in this example that the faucet drips nonuniformly, which is not always true of real faucets.

However, because your stopwatch could be off by as much as one second, your observation wouldn't tell you precisely how long it took for the faucet to drip. If your watch ticked once, the time might have been a little more than one second, or it might have been nearly two. At what time, between one and two seconds, should you say the faucet dripped? Without a better stopwatch or a longer measurement, there would be no preferred answer. With the watch you have, you can conclude only that the drops fall somewhere in the range between one per second and one per two seconds. If you said that the faucet drips once per second, you could have essentially 100% error in your measurement. That is, you could be off by as much as a factor of two.

But suppose that instead you waited 10 seconds while performing your measurement. Then, 10 drops of water would have fallen during the time it took your watch to click 10 times. With your crude stopwatch with only 1-second accuracy, all you could really infer is that the time it took for 10 drips was somewhere between 10 and 11 seconds. Your measurement, which would again say that the drips fall approximately once per second, would now have an error of only 10%. That's because by waiting 10 seconds, you could measure frequency to within $\frac{1}{10}$ of a second. Notice that the product of the time for your measurement (10 seconds) and the uncertainty in frequency (10%, or 0.1) is was roughly 1. Notice also that the product of uncertainty in frequency and time for the measurement in the first example, which had more error in the frequency measurement (100%) but took place over a shorter time (1 second), is also about 1.

You could continue along these lines. If you were to perform a measurement for 100 seconds, you could measure frequency to an accuracy of water dripping once per 100 seconds. If you were to measure water dripping for 1,000 seconds, you could measure frequency with an accuracy of once per 1,000 seconds. In all these cases the product of the time interval over which you performed your measurement and the accuracy with which you measure frequency is about 1.* The longer time required for a more accurate measurement of frequency is at the heart of the time-energy uncertainty principle.

---

*I will not derive the precise number here.

You can measure frequency more accurately, but to do so you would have to measure for longer. The product of time and the uncertainty in frequency is always about 1.*

To complete the derivation of our simple uncertainty principle, if you had a sufficiently simple quantum mechanical system—a single photon, for example—its energy would be equal to Planck's constant, h, times frequency. For such an object, the product of the time interval over which you measure energy and the error in energy would always exceed h. You could measure its energy as precisely as you like, but your experiment would then have to run for a correspondingly longer time. This is the same uncertainty principle we just derived; the added twist is only the quantization relation that relates energy to frequency.

Two Important Energy Values and What the Uncertainty Principle Tells Us About Them

That almost completes our introduction to the fundamentals of quantum mechanics. This and the following section review two remaining elements of quantum mechanics that we will use later on.

This section, which does not involve any new physical principles, presents one important application of the uncertainty principle and special relativity. It explores the relationships between two important energies and the smallest length scales of the physical processes to which particles with those energies could be sensitive—relationships that particle physicists use all the time. The following section will introduce spin, bosons, and fermions—notions that will appear in the next chapter, about the Standard Model of particle physics, and also later on, when we consider supersymmetry.

---

*The above reasoning is not entirely sufficient to fully explain the true uncertainty principle because you can never be sure that you are measuring true frequency if you measure for only a finite interval of time. Will the faucet leak for ever? Or did it leak only while you were making your measurement? Although it's somewhat more subtle to demonstrate, you will never do better than the true uncertainty principle, even if you have a more accurate stopwatch.

The position-momentum uncertainty principle says the product of uncertainties in position and in momentum must exceed Planck's constant. It tells us that anything—whether it is a light beam, a particle, or any other object or system you can think of that could be sensitive to physical processes occurring over short distances—must involve a large range of momenta (since the momentum must be very uncertain). In particular, any object that is sensitive to those physical processes must involve very high momenta. According to special relativity, when momenta are high, so are energies. Combining these two facts tells us that the only way to explore short distances is with high energies.

Another way of explaining this is to say that we need high energies to explore short distances because only particles whose wavefunctions vary over small scales will be affected by short-distance physical processes. Just as Vermeer could not have executed his paintings with a two-inch-wide brush, and just as you can't see fine detail with blurry vision, particles cannot be sensitive to short-distance physical processes unless their wavefunction varies over only small scales. But according to de Broglie, particles whose wavefunction involves short wavelengths also have high momenta. De Broglie said that the wavelength of a particle-wave is inversely proportional to its momentum. Therefore de Broglie would also have us conclude that you need high momenta, and hence high energies, to be sensitive to the physics of short distances.

This has important ramifications for particle physics. Only high-energy particles feel the effects of short-distance physical processes. We'll see in two specific cases just how high I mean.

Particle physicists often measure energy in multiples of an *electronvolt*, which is abbreviated as eV, and pronounced by saying the letters "e-V." An electronvolt is the energy required to move an electron against a potential difference, such as could be provided by a very weak battery, of 1 volt. I'll also use the related units *gigaelectronvolt*, or GeV (pronounced "G-e-V") and teraelectronvolt, or TeV; a GeV is 1 billion eV and a TeV is 1 trillion eV (or 1,000 GeV).

Particle physicists often find it convenient to use these units to measure not just energy, but also mass. We can do this because the special relativity relations between mass, momentum, and energy tell us that the three quantities are related through the speed of light, which is the constant $c = 299,792,458$ meters/second.[13] We can therefore use the speed of light to convert a given energy into mass or momentum. For example, Einstein's famous formula $E = mc^2$ means that there is a definite mass associated with any particular energy. Since everyone knows that the conversion factor is c^2, we can incorporate it and express masses in units of eV. The proton mass in these units is 1 billion eV—that is, 1 GeV.

Converting units in this way is analogous to what you do every day when you tell someone, for example, that "The train station is ten minutes away." You are assuming a particular conversion factor. The distance might be half a mile, corresponding to ten minutes at walking speed, or it might be ten miles, which is ten minutes at highway speeds. There is an agreed-upon conversion factor between you and your conversation partner.

These special relativity relationships, in conjunction with the uncertainty principle, determine the minimum spatial size of the physical processes that a wave or a particle of a particular energy or mass could experience or detect. We will now apply these relations to two very important energies for particle physics that will appear frequently in later chapters (see Figure 46).

The first energy, also known as the *weak scale energy*, is 250 GeV. Physical processes at this energy determine key properties of the weak force and of elementary particles, most notably how they acquire mass. Physicists (including myself) expect that when we explore this energy, we will see new effects predicted by as yet unknown physical theories and learn a good deal more about the underlying structure of matter. Fortunately, experiments are about to explore the weak scale energy and should soon be able to tell us what we want to know.

Sometimes I will also refer to the *weak scale mass*, which is related to the weak scale energy through the speed of light. In more conventional mass units, the weak scale mass is 10^{-21} grams. But as I just explained, particle physicists are content to talk about mass in units of GeV.

The associated *weak scale length* is 10^{-16} cm, or one ten thousand trillionth of a centimeter. It is the range of the weak force—the maximum distance over which particles can influence each other through this force.

Because uncertainty tells us that small distances are probed only with high energy, the weak scale length is also the minimum length that something with 250 GeV of energy can be sensitive to—that is, it is the smallest scale on which physical processes can affect it. If any smaller distances could be explored with that energy, the distance uncertainty would have to be less than 10^{-16} cm, and the distance-momentum uncertainty relation would be violated. The currently operating Fermilab accelerator and the future Large Hadron Collider (LHC), to be built at CERN in Geneva within the decade, will explore physical processes down to that scale, and many of the models I will discuss should have visible consequences at this energy.

The second important energy, known as the *Planck scale energy*, M_{Pl}, is 10^{19} GeV. This energy is very relevant to any theory of gravity. For example, the gravitational constant, which enters Newton's gravitational force law, is inversely proportional to the square of the Planck scale energy. Gravitational attraction between two masses is small because the Planck scale energy is large.

Moreover, the Planck scale energy is the largest energy for which a classical theory of gravity can apply; beyond the Planck scale energy, a quantum theory of gravity, which consistently describes both quantum mechanics and gravity, is essential. Later on, when we discuss string theory, we will also see that in the old string theory models the tension of a string is very likely determined by the Planck scale energy.

Quantum mechanics and the uncertainty principle tell us that when particles achieve this energy, they are sensitive to physical processes at distances as short as the *Planck scale length*,* which is 10^{-33} cm. This is an extremely small distance—far less than anything measurable. But to describe physical processes that occur over distances this small a theory of quantum gravity is required, and that theory might be string theory. For this reason, the Planck scale length, along with the Planck scale energy, are important scales that will reappear in later chapters.

---

*This is the same quantity that I referred to simply as the "Planck length" in earlier chapters.

Bosons and Fermions

Quantum mechanics makes an important distinction among particles, dividing the world of particles into *bosons* and *fermions*. Those particles could be fundamental particles such as the electron and quarks, or composite entities such as a proton or the atomic nucleus. Any object is either a boson or a fermion.

Whether such an object is a boson or a fermion depends on a property called *intrinsic spin*. The name is very suggestive, but the "spin" of particles does not correspond to any actual motion in space. But if a particle has intrinsic spin, it interacts as if it were rotating, even though in reality it is not.

For example, the interaction between an electron and a magnetic field depends on the electron's classical rotation—its actual rotation in space. But the electron's interaction with the magnetic field also depends on the electron's intrinsic spin. Unlike the classical spin that arises from actual motion in physical space,* intrinsic spin is a property of a particle. It is fixed and has a specific value now and for ever. For example, the photon is a boson and has spin-1. That is a property of the photon; it is as fundamental as the fact that the photon travels at the speed of light.

In quantum mechanics, spin is quantized. Quantum spin can take the value 0 (i.e., no spin at all), or 1, or 2, or any integer number units of spin. I'll call this spin-0 (pronounced "spin-zero"), spin-1, spin-2, and so on. Objects called bosons, named after the Indian physicist Satyendra Nath Bose, have intrinsic spin—the quantum mechanical spin that is independent of rotation—and that is also an integer: bosons can have intrinsic spin equal to 0, 1, 2, and so on.

Fermion spin is quantized in units that no one would have thought possible before the advent of quantum mechanics. Fermions, named after the Italian physicist Enrico Fermi, have half-integer values such as $\frac{1}{2}$ or $\frac{3}{2}$. Whereas a spin-1 object returns to its initial configuration after it is rotated a single time, a spin-$\frac{1}{2}$ particle would do so only after it were rotated twice. Despite the apparent weirdness of the

---

*For those who already know some physics, this is orbital angular momentum.

half-integer values of fermions' spins, protons, neutrons, and electrons are all fermions with spin-$\frac{1}{2}$. Essentially all familiar matter is composed of spin-$\frac{1}{2}$ particles.

The fermionic nature of most fundamental particles determines many properties of the matter around us. The *Pauli exclusion principle*, in particular, states that two fermions of the same type will never be found in the same place. The exclusion principle is what gives the atom the structure upon which chemistry is based. Because electrons with the same spin can't be in the same place, they have to be in different orbits.

That is why I could make the analogy with different floors of a tall building earlier on. The different floors represented the different possible quantized electron orbits that the Pauli exclusion principle tells us get occupied when a nucleus is surrounded by many electrons. The exclusion principle is also the reason you can't poke your hand through a table or fall into the center of the Earth. Tables and your hand take the solid structure they do only because the uncertainty principle gives rise to atomic, molecular, and crystalline structure in matter. The electrons in your hand, which are the same as the electrons in a table, have no place to go when you hit a table. No two identical fermions can be in the same place at the same time, so matter can't just collapse.

Bosons act in exactly the opposite fashion to fermions. They can and will be found in the same place. Bosons are like crocodiles—they prefer to pile up on top of one another. If you shine light where there is already light, it behaves very differently from your hand karate-chopping a table. Light, which is composed of bosonic photons, passes right through light. Two light beams can shine in exactly the same place. In fact, lasers are based on this fact: bosons occupying the same state allow lasers to produce their strong, coherent beams. Superfluids and superconductors are also made of bosons.

An extreme example of bosonic properties is the Bose-Einstein condensate, in which many identical particles act together as a single particle—something that fermions, which have to be in different places, could never do. Bose-Einstein condensates are possible only because the bosons of which they are composed, unlike fermions, can have identical properties. In 2001, Eric Cornell, Wolfgang Ketterle, and Carl Wieman received the Nobel Prize for Physics for their discovery of the Bose-Einstein condensate.

Later on I won't need these detailed properties of the way that fermions and bosons behave. The only facts I will use from this section are that fundamental particles have intrinsic spin and can act as if they were spinning in one direction or another, and that all particles can be characterized by whether they are bosons or fermions.

What to Remember

- Quantum mechanics tells us that both matter and light consist of discrete units known as *quanta*. For example, light, which seems continuous, is actually composed of discrete quanta called photons.

- Quanta are the basis of particle physics. The Standard Model of particle physics, which explains known matter and forces, tells us that all matter and forces can ultimately be interpreted in terms of particles and their interactions.

- Quantum mechanics also tells us that every particle has an associated wave, known as the particle's *wavefunction*. The square of this wave is the probability that the particle will be found in a particular location. For convenience, I will sometimes talk about a *probability wave*, the square of the more commonly used wavefunction. The values of this probability wave will give probabilities directly. Such a wave will appear later on when we discuss the *graviton*, the particle that communicates the force of gravity. The probability wave will also be important when discussing *Kaluza-Klein (KK) modes*, which are particles that have momentum along the extra dimensions—that is, directed perpendicular to the usual dimensions.

- Another major distinction between classical physics and quantum mechanics is that quantum mechanics tells us that you cannot precisely determine a particle's path—you can never know the precise path a particle took as it traveled from its starting point to its destination. This tells us that we have to consider all the paths that a particle can take when it communicates a force. Because quantum paths can involve any interacting particles, quantum mechanical effects can influence masses and interaction strengths.

- Quantum mechanics divides particles into *bosons* and *fermions*. The existence of two distinct categories of particles is critical to the structure of the Standard Model and also to a proposed extension of the Standard Model known as *supersymmetry*.

- The *uncertainty principle* of quantum mechanics, coupled with the relations of special relativity, tell us that, using physical constants, we can relate a particle's mass, energy, and momentum to the minimum size of the region in which a particle of that energy can experience forces or interactions.

- Two of our most frequent applications of these relations involve the two energies known as the *weak scale energy* and the *Planck scale energy*. The weak scale energy is 250 GeV (gigaelectronvolts) and the Planck energy is much bigger—ten million trillion GeV.

- Only forces with a range smaller than ten million billionths (10^{-17}) of a centimeter will produce measurable effects on a particle with weak scale energy. This is a very tiny distance, but it is relevant to the physical processes in a nucleus and to the mechanism by which particles acquire mass.

- Tiny as it is, the *weak scale length* is far greater than the *Planck scale length*, which is one million billion billion billionth (10^{-33}) of a centimeter. That is the size of the region where forces influence particles that have the Planck scale energy. The Planck scale energy determines the strength of gravity; it is the energy that particles would have to have for gravity to be a strong force.

7

The Standard Model of Particle Physics: Matter's Most Basic Known Structure

You're never alone,
You're never disconnected!
You're home with your own;
When company's expected, you're well protected!
. . . When you're a Jet, you stay a Jet!

Riff (*West Side Story*)

Of all the stories she had read, Athena was most thoroughly perplexed by Hans Christian Andersen's "The Princess and the Pea." The story tells of a Prince who searched unsuccessfully for a suitable princess to wed. After he had searched in vain for weeks, a potential princess arrived by chance at his palace, seeking shelter from a storm. This soggy visitor thereby became the unwitting subject of the Queen's litmus test for princesses.

The Queen prepared a bed, which she piled high with mattresses and eiderdown quilts. At the very bottom of the pile she placed a solitary pea. That night, she showed her visitor to the carefully prepared guest room. The next morning, the princess (as indeed she proved herself to be) complained that she had not been able to sleep at all. She had tossed and turned the whole night, and found she had actually turned black and blue—all because of the uncomfortable pea. The Queen and Prince were convinced that their visitor was truly of royal blood, for who else could be so delicate?

Athena turned the story round and round in her head. She thought it fairly ridiculous that anyone, even the most sensitive of princesses, would ever have discovered the pea by lying passively on top of the pile of mattresses. After many days' deliberation, Athena found a plausible interpretation, which she rushed to tell her brother.

She rejected the common interpretation that the princess proved her royal nature by demonstrating delicacy and refinement with her sensitivity to even something as minor as a pea under a pile of mattresses. She offered an alternative explanation.

Athena suggested that when the Queen went away and left the princess alone in the room, the princess threw decorum to the wind and gave vent to her boisterous youthful nature. The princess ran around and jumped up and down on her bed until she was exhausted, and only then lay down to try to sleep. Through her rambunctiousness, the princess compressed the mattresses so much that for a brief moment the pea stuck out like a sore thumb and gave her a small bruise. Athena thought this princess was still rather impressive, but found her revisionist interpretation much more satisfactory.

Finding substructure within the atom was as remarkable an accomplishment as the princess finding her pea. Particles called *quarks*, the building blocks of the proton, occupy about the same fractional volume of the proton as a pea does in a mattress. A 1 cubic centimeter pea in a 2 meters × 1 meter × $\frac{1}{2}$ meter mattress takes up one-millionth of the mattress's volume, which is not too different from the fraction of volume a quark occupies in a proton. And the way in which physicists discovered quarks bears some resemblance to the rambunctious princess's discovery. A passive princess would never discover a pea buried layers and layers down. Similarly, physicists didn't discover quarks until they slammed into the proton with energetic particles that could explore its innards.

In this chapter you will make a jump of your own, into the Standard Model of particle physics, the theory that describes the known elementary constituents of matter and the forces that act upon them.* The Standard Model, which represents the culmination of many surprising

---

*Despite the name "Standard Model," there is an ambiguity in convention. Some people also include the hypothetical Higgs particle as well. However, the name should refer to the known particles only, and that is the convention I use. We'll discuss the Higgs particle in Chapter 10.

and exciting developments, is a stupendous achievement. You don't need to remember all the details—I'll repeat the names of all the particles or the nature of their interactions when I refer to them later on. But the Standard Model underlies many of the exotic, extra-dimensional theories that I will describe shortly, and as you learn about the recent exciting developments, a feeling for the Standard Model and its key ideas will contribute to a deeper understanding of matter's fundamental structure and the way physicists think about the world today.

The Electron and Electromagnetism

When Vladimir I. Lenin used the electron as a metaphor in his philosophical book *Materialism and Empirio-Criticism*, he wrote that "the electron is inexhaustible," referring to the layers of theoretical ideas and interpretation through which we interpret it. Indeed, today we understand the electron very differently than we did in the early twentieth century, before quantum mechanics revised our ideas.

But in a physical sense the opposite of Lenin's quote is true: the electron *is* exhaustible. So far as has been determined, the electron is fundamental and indivisible. To a particle physicist, the electron, rather than having "inexhaustible" structure, is the simplest Standard Model particle to describe. The electron is stable and has no constituent parts, so we can characterize it completely by listing only a few properties, including mass and charge. (The Czech anti-Communist string theorist Luboš Motl quipped that this is not the only difference between his and Lenin's perspectives.)

An electron will move towards the positively charged anode of a battery. A moving electron also responds to a magnetic force: as an electron moves through a magnetic field, its path will bend. Both these phenomena are the result of the electron's negative charge, which makes the electron respond to electricity and magnetism.

Before the 1800s, everyone thought that electricity and magnetism were separate forces. But in 1819 the Danish physicist and philosopher Hans Oersted found that a current of moving charges generates a magnetic field. From this observation he deduced that there should be a single theory describing both electricity and magnetism: they must be two sides of the same coin. When a compass needle responds to a bolt of lightning, it confirms Oersted's conclusion.

The classical theory of *electromagnetism*, still in use today, was developed in the nineteenth century and used the observation that electricity and magnetism are related. The notion of a *field* was also critical to this theory. "Field" is the name physicists give to any quantity that permeates space. For example, the value of the gravitational field at any point tells how strong the effect of gravity is there. The same goes for any type of field: the value of the field at any location tells us how intense the field is there.

In the latter half of the nineteenth century, the English chemist and physicist Michael Faraday introduced the concepts of electric and magnetic fields, and these concepts persist in physics today. Given that he had to temporarily abandon his formal education at the age of fourteen to help support his family, it is quite remarkable that he managed to do physics research that had such a revolutionary impact. Fortunately for him (and for the history of physics), he was apprenticed to a bookbinder who encouraged him to read the books on which he was working, and educate himself.

Faraday's idea was that charges produce electric or magnetic fields everywhere in space, and these fields in turn act on other charged objects, no matter where those objects are. The magnitude of the effect of electric and magnetic fields on charged objects does depend on their location, however. The field exerts the most influence where its value is largest, and has a smaller effect where its value is less.

You can see evidence of a magnetic field by sprinkling iron filings in the vicinity of a magnet. The particles organize themselves in patterns according to the strength and direction of the field. You can also experience a field by holding two magnets close together. You'll feel the magnets' mutual attraction or repulsion well before they touch each other. Each is responding to the field that permeates the region between them.

The ubiquity of electric fields was brought home to me one day when I was finishing a climb on a ridge near Boulder, Colorado, with a partner who was new to climbing but had a lot of hiking experience.

An electrical storm was approaching rapidly, and I didn't want to make him nervous, so I encouraged him to move quickly without pointing out that the rope was crackling and his hair was standing on end. When we were safely down at the bottom happily reviewing our adventure, much of which had been a delightful climb, my partner told me that of course he had known we were in danger: my hair had been visibly standing on end too! The electric field wasn't only in one place—it was everywhere around us.

Before the nineteenth century, no one described electricity and magnetism in terms of fields. People conventionally used the term *action at a distance* to describe these forces. Action at a distance is the expression you might have learned in elementary school which describes how an electrically charged object instantly attracts or repels any other charge, no matter where it is. This might not seem mysterious, since it's what we're accustomed to. However, it would be extraordinary if something in one place could instantly affect another object some distance away. How would the effect be communicated?

Although it might sound like just a matter of semantics, there really is an enormous conceptual difference between a field and action at a distance. According to the field interpretation of electromagnetism, a charge doesn't affect other regions of space immediately. The field needs time to adjust. A moving charge creates a field in its immediate vicinity, which seeps (albeit very rapidly) throughout space. Objects learn of the motion of the distant charge only after light (which is composed of electromagnetic fields) has had time to reach them. The electric and magnetic fields therefore change no faster than the finite speed of light allows. At any given point in space, the field adjusts only after sufficient time has elapsed for the effect of the distant charge to reach that point.

However, despite the critical importance of Faraday's electromagnetic fields, they were more heuristic than mathematical. Perhaps because of his spotty education, math was not Faraday's strength. But another British physicist, James Clerk Maxwell, incorporated Faraday's field idea into classical electromagnetic theory. Maxwell was a brilliant scientist who counted among his many interests optics and color, the mathematics of ovals, thermodynamics, the rings of

Saturn, measuring latitude with a bowl of treacle, and the question of how cats land upright while conserving angular momentum when dropped upside down.*

Maxwell's most important contribution to physics was the set of equations that describe how to derive the values of electric and magnetic fields from a distribution of charges and currents.[14]† From these equations, he deduced the existence of electromagnetic waves—the waves in all forms of electromagnetic radiation, as in your computer, television, microwave oven, and the many other conveniences of the modern era.

However, Maxwell made one mistake. Like all other physicists of his day, he took the field idea too materially. He assumed that the field arose from the vibrations of an aether—an idea that Einstein, as we have seen, ultimately debunked. Nonetheless, Einstein credited Maxwell with the origin of the special theory of relativity: Maxwell's electromagnetic theory gave Einstein the insight about the constant speed of light that instigated his monumental work.

The Photon

Maxwell's classical electromagnetic theory made many successful predictions, but it predated quantum mechanics so it obviously didn't include quantum effects. Today, physicists study the electromagnetic force with particle physics. The particle physics theory of electromagnetism includes the predictions of Maxwell's well-studied and well-verified classical theory, but incorporates the predictions of quantum mechanics as well. It is therefore a more comprehensive and more accurate theory of electromagnetism than its classical predecessor. In fact, the quantum theory of electromagnetism has yielded incredibly

---

*They have a very flexible spine and no collarbone, so can twist their bodies while conserving angular momentum. Actually, this is still actively studied.

†Richard Feynman said, "From a long view of the history of mankind—seen from, say, ten thousand years from now—there can be little doubt that the most significant event of the nineteenth century will be judged as Maxwell's discovery of the laws of electrodynamics." (*The Feynman Lectures on Physics*, Vol. II (Reading, MA: Addison-Wesley Longman, 1970).)

precise predictions that have been tested with the unbelievable precision of one part in a billion.*

The quantum electromagnetic theory attributes the electromagnetic force to the exchange of the particle called the *photon*, the quantum of light that we considered in the previous chapter. The way it works is that an incoming electron emits a photon, which travels to another electron, communicates the electromagnetic force, and then disappears. Through their exchange, photons transmit, or *mediate*, a force. They act as confidential letters that convey information from one place to another, but are afterwards immediately destroyed.

We know that the electric force is sometimes attractive and sometimes repulsive: it's attractive when oppositely charged objects interact, and repulsive when the charges have the same sign, either both positive or both negative. You might think of the repulsive force communicated by the photon as an interaction between two ice skaters throwing a bowling ball back and forth; each time one of them catches the ball, he slides away from the other across the ice. Attractive forces, on the other hand, are more like two novices tossing a frisbee to each other; unlike the ice skaters, who slide further apart, these beginning frisbee players would approach each other with each successive throw.

The photon is the first example we will encounter of a *gauge boson*, a fundamental, elementary particle that is responsible for communicating a particular force. (The word "gauge" sounds more daunting than it really is; physicists first used it in the late 1800s because of a tangential analogy to railroad gauges that tell you the distance between the rails—a term that was far more familiar a hundred years ago.) Weak bosons and gluons are other examples of gauge bosons. These particles communicate the weak and strong forces respectively.

Between the late 1920s and the 1940s, the English physicist Paul Dirac and the Americans Richard Feynman and Julian Schwinger—as well as Sin-Itiro Tomonaga working independently in post-war Japan—developed the quantum mechanical theory of the photon.

---

*This is through the measurement of a quantity known as the electron anomalous magnetic moment.

They named the branch of quantum theory that they developed *quantum electrodynamics* (QED). Quantum electrodynamics includes all the predictions of the classical electromagnetic theory as well as particle (quantum) contributions to physical processes—that is, interactions that are generated by exchanging or producing quantum particles.

QED predicts how photon exchange produces the electromagnetic force. For example, in the process illustrated in Figure 47, two electrons enter the interaction region, exchange a photon, and then emerge with their resultant paths (speed and direction of motion, for example) influenced by the electromagnetic force that was communicated. Field theory associates numbers with each part of the diagram so that we can use it to make quantitative predictions. This picture is an example of a *Feynman diagram,* named after Richard Feynman, and is a pictorial way of describing interactions in quantum field theory. (Feynman was so proud of his invention that he had some diagrams painted on his van.)

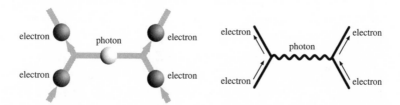

Figure 47. *The Feynman diagram, on the right, has several interpretations. One interpretation (reading bottom to top) is that two electrons enter an interaction region, exchange a photon, and two electrons leave, as illustrated schematically on the left. (This diagram can also be interpreted in terms of electron-positron annihilation.)*

Not all QED processes involve a photon that is destroyed, however. In addition to the ephemeral *intermediate* or *internal particles**—like the photons leading to electromagnetic interactions that are produced and almost immediately destroyed—there are also real, *external* photons, particles that enter or leave an interaction region.

---

*In Chapter 11 we will see that these are also called *virtual particles.*

Sometimes those particles are deflected and sometimes they turn into other particles. Either way, the particles that enter or leave are real physical particles.

Quantum Field Theory

Quantum field theory, the tool with which we study particles,* is based on eternal, omnipresent objects that can create and destroy those particles. These objects are the "fields" of quantum field theory. Like the classical electromagnetic fields that inspired their name, quantum fields are objects that permeate spacetime. But quantum fields play a different role. They create or absorb elementary particles. According to quantum field theory, particles can be produced or destroyed anywhere and at any time.

For example, an electron or a photon can appear or disappear anywhere in space. Quantum processes allow the number of charged particles in the universe to change through particle creation or destruction. Each particle is created or destroyed by its own particular field. In quantum field theory, not only electromagnetism but all forces and interactions are described in terms of fields, which can create new particles or eliminate particles that were already present.

According to quantum field theory, you can think of particles as excitations of the quantum field. Whereas the *vacuum*, a state with no particles, contains only constant fields, states with particles present contain fields with bumps and wiggles corresponding to the particles. When the field acquires a bump, a particle is created, and when it absorbs this bump to become constant once again, the particle is destroyed.

The fields that create electrons and photons must exist everywhere to guarantee that all interactions can occur at any point in spacetime. This is essential because interactions are *local*, which is to say that only particles in the same place can participate. Action at a distance would be more like magic. Particles don't have ESP—they have to be in contact to interact directly.

---

*QED is quantum field theory applied to electromagnetism.

Electromagnetic interactions do occur between distant charges that are not in direct contact, but only through the auspices of the photon or some other particle that has direct contact with both of the inter- acting charged particles. In that case, charges appear to affect each other instantaneously, but only because the speed of light is so fast. Really, the interaction only occurred through local processes; the photon first coincided with one of the charged particles and then the other. The field therefore had to create and destroy the photon at the precise locations of the charged particles.

Antiparticles and the Positron

Quantum field theory also tells us that for each particle, a counterpart must exist, known as an antiparticle. Tom Stoppard talks about anti- particles in his play *Hapgood*: "When a particle meets an anti-particle, they annihilate each other, they turn into energy-bang, you under- stand." Any science fiction fan knows about antiparticles—they are what you make guns from to destroy the universe and are also what powers *Star Trek*'s USS *Enterprise*.

Those last applications are fictitious, but antiparticles are not. Anti- particles are truly a part of the particle physics view of the world. In field theory and the Standard Model, they are as essential as particles. In fact, antiparticles are just like particles, except that all their charges are opposite.

Paul Dirac first encountered antiparticles when he developed the quantum field theory describing the electron. He found that a quantum field theory that is consistent with both quantum mechanics and special relativity necessarily includes antiparticles. He hadn't deliber- ately added them. When he incorporated special relativity, the theory spit them out. Antiparticles are a necessary consequence of relativistic quantum field theory.

Here's a rough argument for why antiparticles follow from special relativity. Charged particles can travel forwards and backwards in space. Naively, special relativity would therefore tell us that those particles should be able to travel forwards and backwards in time as well. But so far as we know, neither particles nor anything else we are aware of can actually travel backwards in time. What happens instead is that oppositely charged antiparticles replace the reverse-time-traveling particles. Antiparticles reproduce the effects the reverse-time-traveling particles would have so that even without them, quantum field theory's predictions are compatible with special relativity.

Imagine a movie of a current of negatively charged electrons traveling from one point to another. Now imagine running the movie in reverse. Negative charge would then travel backwards, or, equivalently (so far as charge is concerned), positive charge would travel forwards. A current of *positrons*, the positively charged antiparticles of electrons, produces this positively charged forward-traveling current and therefore acts like a time-reversed electron current.

Quantum field theory tells us that if any type of charged particle exists, such as an electron, so must a corresponding antiparticle with opposite charge. For example, since an electron carries charge -1, the positron has charge $+1$. The antiparticle is like the electron it in all respects aside from its charge. A proton also has charge $+1$, but it is 2,000 times heavier than an electron and therefore could not be its antiparticle.

As Stoppard said, antiparticles do indeed annihilate particles when the two come into contact. Because the charges of a particle and its antiparticle always add up to zero, when a particle meets an antiparticle, they can annihilate each other and be destroyed. The particle and antiparticle together carry no charge, so Einstein's relation $E = mc^2$ tells us all the mass can convert into energy.

On the other hand, energy can convert into a particle-antiparticle pair when the energy is sufficient to produce them. Both particle annihilation and particle creation occur in high-energy particle accelerators, where physicists conduct the experiments that study heavy particles, particles too massive to be found in ordinary matter. In these colliders, a particle and an antiparticle meet and annihilate each other, thereby creating a burst of energy from which new particle-antiparticle pairs emerge.

Because matter—and atoms in particular—are composed of particles and not antiparticles, antiparticles such as positrons are generally not found in nature. But they can be produced temporarily at particle colliders, in hot regions of the universe, and even in hospitals, where positron emission tomography (PET) is used to scan for signs of cancer.

Gerry Gabrielse, a colleague of mine in the Harvard physics department, makes antiparticles all the time in the basement of Jefferson Laboratories, where I work. Thanks to the work of Gerry and others, we know at a very high level of precision that antiparticles really are like their particle counterparts in mass and gravitational pull, despite their opposite charge. But there aren't enough of them to do any harm. I can assure science fiction fans that these antiparticles do far less damage to the building than the perpetual construction of new labs and offices, which is always preceded by a large amount of visible and audible destruction.

Electrons, positrons, and photons are the simplest and most accessible particles. It is no coincidence that electric forces and electrons were the first Standard Model ingredients that physicists understood. The electron, positron, and the photon are not the only particles, however, and electromagnetism is not the only force.

I listed the known particles and nongravitational forces* in Figures 32 and 33. I left gravity out of the picture because it is qualitatively different from the other forces and must be treated separately. Despite the prosaic names of two of the forces—the weak force and the strong force—they have many interesting properties. In the next two sections, we'll see what they are.

The Weak Force and the Neutrino

Even though you don't notice the weak force in your daily existence because it is indeed weak, it is essential to many nuclear processes. The weak force explains some forms of nuclear decay, such as that of potassium-40 (found here on Earth, with a decay that is sufficiently slow—about a billion years on average—to continue to heat the Earth's core) and, indeed, of the neutron itself. Nuclear processes

---

*In particle physics, this means the fundamental forces other than gravity (i.e., the weak force, the strong force, and the electromagnetic force).

change the structure of the nucleus, and through such processes the number of neutrons in a nucleus changes, releasing a large amount of energy. This energy can be harnessed for nuclear power or nuclear bombs, but has other purposes as well.

For example, the weak force plays a role in the creation of heavy elements, which are created during cataclysmic supernova explosions. The weak force is also essential for stars, including the Sun, to shine: it kicks off the chain of reactions that convert hydrogen to helium. The nuclear processes that are triggered by the weak force help make the composition of the universe continuously evolve. From our knowledge of nuclear physics, we can deduce that about 10% of the universe's primordial hydrogen has been used as nuclear fuel in stars. (Happily, the 90% that remains guarantees that the universe won't need to rely on foreign energy sources any time soon.)

Despite its importance, scientists identified the weak force only relatively recently. In 1862, William Thomson (later Lord Kelvin*), one of the most respected physicists of his day, grossly underestimated the age of the Sun and the Earth because he didn't know about nuclear processes originating from the weak force (which, in fairness to him, had not yet been discovered). J.J. Thomson based his estimate on the only known source of illumination, incandescence. He deduced that the energy that had been available could not have supported the Sun for more than about 30 million years.

Charles Darwin didn't like this result. He had come up with a minimum age far closer to the correct one by estimating the number of years required for erosion to wash away the Weald, a valley in the south of England. Darwin's estimate of 300 million years had the further appeal for him that it allowed enough time for natural selection to provide the large range of species found on Earth.

However, everyone—including Darwin himself—assumed that Thomson, the physicist of stellar reputation, was correct. Darwin was so persuaded by Thomson's calculation and reputation that he removed his own time estimates from later editions of his book *The Origin of Species*. Only after Rutherford's discovery of the significance

---

*This honor was given not only for his science, but for his opposition to Home Rule in Ireland.

of radiation* was Darwin's idea for an older age vindicated and the age of the Earth and the Sun established as about 4.5 billion years— far larger than Thomson's estimate, and Darwin's.

In the 1960s, the American physicists Sheldon Glashow and Steven Weinberg, and the Pakistani physicist Abdus Salam, all working independently (and not necessarily harmoniously), developed the *electroweak theory*, a theory that explains the weak force and provided insight into the force of electromagnetism.† According to the electroweak theory, the exchange of particles called *weak gauge bosons* produces the effects of the weak force, just as photon exchange communicates electromagnetism. There are three weak gauge bosons. Two are electrically charged, the W+ and W− (the W stands for weak force, and the + or − sign is the gauge boson's charge). The other one is neutral and is called the Z (because of its zero charge).

As with photon exchange, weak gauge boson exchange produces forces that can be attractive or repulsive, depending on the particles' *weak charges*. Weak charges are numbers that play the same role for the weak force that electric charge plays for the electromagnetic force. Only particles that carry weak charge experience the weak force, and their particular charge determines the strength and type of interactions they will experience.

However, there are several important distinctions between the electromagnetic force and the weak force. One of the most surprising is that the weak force distinguishes left from right, or, as physicists

---

*Rutherford presented his results, but knew that in doing so he was contradicting Kelvin. A.S. Eve's biography of Rutherford quotes him: "I came into the room, which was half dark, and presently spotted Lord Kelvin in the audience and realized that I was in for trouble at the last part of my speech dealing with the age of the earth, where my views conflicted with his. To my relief, Kelvin fell fast asleep, but as I came to the important point, I saw the old bird sit up, open an eye and cock a baleful glance at me! Then a sudden inspiration came, and I said, 'Lord Kelvin had limited the age of the earth, provided no new source was discovered. That prophetic utterance refers to what we are now considering tonight, radium!' Behold! the old boy beamed upon me." (Eve, *Rutherford: Being the Life and Letters of the Rt. Hon. Lord Rutherford,* O.M. (Cambridge: Cambridge University Press, 1939).)
†Weak interactions had, however, been observed earlier, and nuclear mechanisms inside the Sun were known to occur. But the connection to a weak force was understood only later.

would say, *violates parity symmetry*. Parity violation means that the mirror image of particles would behave differently to each other. The Chinese-American physicists C.N. Yang and T.D. Lee formulated the theory of parity violation in the 1950s, and another Chinese-American physicist, C.S. Wu, confirmed it experimentally in 1957. Yang and Lee received the Nobel Prize for Physics that year. Curiously, Wu, the only woman who played a role in the Standard Model developments I'm discussing, didn't receive a Nobel prize for her momentous discovery.

Some violations of parity invariance should be familiar. For example, your heart is on the left side of your body. But if evolution had proceeded differently, and people had ended up with the heart on the right, you would expect that all its properties would be the same as the ones we now see. That the heart is on one side and not the other shouldn't matter for any fundamental biological processes.

For many years prior to Wu's 1957 measurement, it had been "obvious" that physical laws (though not necessarily physical objects) couldn't have a preferred handedness. After all, why should they? Certainly gravity and electromagnetism and many other interactions make no such distinction. Nonetheless, the weak force, a fundamental force of nature, distinguishes left from right. Although it's very surprising, the weak force violates parity symmetry.

How could a force prefer one handedness over the other? The answer lies in fermionic intrinsic spin. Just as a screw is threaded so that you screw it in by twisting it clockwise, but not counterclockwise, particles can also have a handedness, which indicates the direction in which they spin (see Figure 48). Many particles, such as the electron and the proton, can spin in one of two directions: either to the left or the right. The word *chirality*, derived from the Greek word *cheir*, which means hand, refers to the two possible directions of spin. Particles can be left- or right-handed, just like the fingers of your hands, one set of which curls to the left and the other set to the right.

The weak force violates parity symmetry by acting differently on left-handed and right-handed particles. It turns out that only left-handed particles experience the weak force. For example, a left-handed electron would experience the weak force, whereas one spinning to the right would not. Experiments show this clearly—

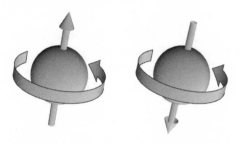

Figure 48. *Quarks and leptons can be either right- or left-handed.*

it's the way the world works—but there is no intuitive, mechanical explanation for why this should be so.

Imagine a force that could act on your left hand but not on your right! All I can say is that parity violation is a startling but well-measured property of weak interactions; it is one of the Standard Model's most intriguing features. For example, the electrons that emerge when a neutron decays are always left-handed. Weak interactions violate parity symmetry, so when I list the full set of elementary particles and the forces that act on them (in Figure 52, p. 168) I'll need to list separately the left- and right-handed particles.

The violation of parity symmetry, strange as it seems, is not the only novel property of the weak force. A second, equally important property is that the weak force can actually convert one particle type into another (while nonetheless preserving the total amount of electromagnetic charge). For example, when a neutron interacts with a weak gauge boson, a proton might emerge (see Figure 49). This is very different from a photon interaction, which would never change the net number of charged particles of any particular type (that is, the number of particles minus the number of antiparticles), such as the number of electrons minus the number of positrons. (For comparison,

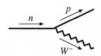

Figure 49. *The interaction with a* W⁻ *gauge boson changes a neutron into a proton (and a down quark contained in the neutron into an up quark contained in the proton).*

a photon interacting with an electron that enters and emerges is illustrated in Figure 50, along with the schematic figure type we used before.) The interaction of a charged weak gauge boson with the neutron and the proton is what allows an isolated neutron to decay and turn into an entirely different particle.

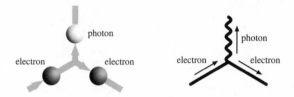

Figure 50. *The Feynman diagram (on the right) representation of a photon-electron interaction. The squiggly line is the photon. It interacts with the electron that comes in and leaves the interaction vertex, as illustrated schematically on the left.*

However, because the neutron and proton have different masses and carry different charges, the neutron must decay into a proton plus other particles, so as to conserve charge, energy, and momentum. And it turns out that when a neutron decays, it produces not only a proton, but also an electron and a particle called a *neutrino*.* This is the process known as *beta decay*, illustrated in Figure 51.

When beta decay was first observed, no one knew about the neutrino, which interacts only through the weak force and not through the electromagnetic force. Particle detectors can find only charged

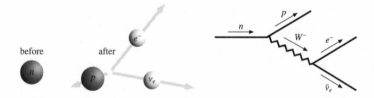

Figure 51. *In beta decay, a neutron decays via the weak force into a proton, an electron, and an antineutrino. A Feynman diagram representation of this process is shown on the right. A neutron turns into a proton and a virtual W⁻ gauge boson, which then turns into an electron and an electron antineutrino.*

*It is actually an antineutrino, but that is not important to us here.

particles or those that deliver energy. Because the neutrino has no electric charge and does not decay, it was invisible to detectors and no one knew it existed.

But without the neutrino, beta decay looked as if it wouldn't conserve energy. The conservation of energy is a fundamental principle in physics, and says that energy can be neither created nor destroyed— it can only be transferred from one place to another. The assumption that beta decay failed to conserve energy was outrageous, yet many respected physicists, unaware of the neutrino's existence, were willing to make this radical (and erroneous) claim.

In 1930, Wolfgang Pauli paved the way to the doubters' scientific salvation by proposing what he called "a desperate way out": a new electrically neutral particle.* His idea was that the neutrino spirits away some energy when a neutron decays. Three years later, Enrico Fermi gave the "little" neutral particle, which he named the neutrino, a firm theoretical foundation. Yet the neutrino seemed such a shaky proposition at the time that the leading scientific journal *Nature* rejected Fermi's paper because "it contained speculations too remote to be of interest to the reader."

But Pauli's and Fermi's ideas were correct, and physicists today universally agree on the existence of the neutrino.† In fact, we now know that neutrinos constantly stream through us, released along with photons from the nuclear processes in the Sun. Trillions of solar neutrinos pass through you each second, but interact so weakly that you never notice. The only neutrinos that we know for sure exist are left-handed; right-handed neutrinos either don't exist or are very heavy—too heavy to be produced—or interact very weakly. No matter which is true, right-handed neutrinos have never been produced at colliders, and we have never seen them. Because we are much more certain about left-handed neutrinos than right-handed ones, I've included only left-handed neutrinos in Figure 52, where I list left- and right-handed particles separately.

---

*The exact words are known because they were contained in a 1934 letter sent to participants of an important scientific meeting, which Pauli missed in order to attend a ball.
†Neutrinos were finally detected at a nuclear reactor by Clyde Cowan and Fred Reines in 1956, eliminating any residual doubts.

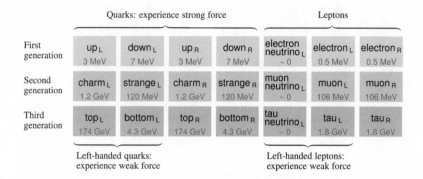

Figure 52. *The three generations of the Standard Model. Left- and right-handed quarks and leptons are listed separately. Each column contains particles with the same charge (different flavors of the particle type). The weak force can change elements of the first column into elements of the second, and elements of the fifth column into elements of the sixth. The quarks experience the strong force, whereas the leptons do not.*

So we now know that weak interactions act only on left-handed particles, and can change particle type. But to truly understand the weak force we need a theory that predicts the interactions of the weak gauge bosons that communicate the force. Physicists initially found that constructing that theory was not simple. They needed to make a major theoretical advance before they could truly understand the weak force and its consequences.

The problem was the final bizarre feature of the weak force: it falls away precipitously over a very short distance, one ten thousand trillionth (10^{-16}) of a centimeter. That makes it quite unlike gravity and electromagnetism, for both of which, as we saw in Chapter 2, strength decreases with distance in proportion to the inverse square of the separation. Although gravity and electromagnetism become weaker as you go further out, they don't drop off nearly as quickly as the weak force. The photon conveys the electromagnetic force to large distances. Why does the weak force behave so differently?

It was clear that physicists needed to find a new type of interaction to account for nuclear processes such as beta decay, but it was not clear what this new interaction could possibly be. Before Glashow,

Weinberg, and Salam developed their theory of the weak force, Fermi made a stab at it with a theory that included new types of interaction involving four particles, such as the proton, neutron, electron, and neutrino. This *Fermi interaction* directly produced beta decay without invoking an intermediate weak gauge boson. In other words, the interaction permitted a proton to turn directly into its decay products—the neutron, electron, and neutrino.

However, it was clear, even at the time, that the Fermi theory could not be the true theory that would work at all energies. Although its predictions were correct for low energies, they were obviously completely wrong for high energies, at which particle interactions became much too strong. In fact, if you incorrectly assumed that you could apply the Fermi theory when the particles were highly energetic, you would get nonsensical predictions, such as particles that should interact with a probability greater than one. That's impossible, since nothing can happen more often than always.

Although the theory based on the Fermi interaction was a fine effective theory for explaining interactions at low energies and between sufficiently distant particles, physicists saw that they needed a more fundamental explanation of processes such as beta decay if they were to know what happened at high energies. A theory based on forces communicated by weak gauge bosons looked as if it would work much better at high energies—but no one knew how to account for the weak force's short range.

That short range turns out to be a consequence of nonzero masses for the weak gauge bosons. In particle physics the relationships implied by the uncertainty principle and special relativity have noticeable consequences. At the end of Chapter 6 I discussed the smallest distances at which a particle of a particular energy, such as the weak scale energy or the Planck scale energy, can be affected by forces. Because of the special relativity relation between energy and mass ($E = mc^2$), massive particles, such as the weak gauge bosons, automatically incorporate similar relationships between mass and distance.

In particular, the force communicated by the exchange of a particle with a given mass dies away over a larger distance when the mass is smaller. (That distance is also proportional to Planck's constant and inversely proportional to the speed of light.*) The relationship between mass and distance given in Chapter 6 tells us that the weak gauge boson, whose mass is about 100 GeV, automatically transmits the weak force only to particles that lie within one ten thousand trillionth of a centimeter. Beyond this distance, the force conveyed by the particle becomes extremely small, too small to do anything we would ever detect.

The nonzero mass of the weak gauge boson is critical to the success of the weak force theory. The mass is the reason that the weak force acts only over very short distances and is so weak as to be almost nonexistent at longer distances. The weak gauge bosons are different in this respect from the photon and graviton, both of which are massless. Because the photon and the *graviton*, the particle that communicates the gravitational force, carry energy and momentum but have no mass, they can communicate forces across great distances.

The concept of massless particles might sound strange, but from the particle physics perspective it is nothing very remarkable. The masslessness of the particles tells us that they travel at the speed of light (after all, light is composed of massless photons), and also that energy and momentum always obey a particular relation: energy is proportional to momentum.

The carriers of the weak force, on the other hand, do have mass. And from the perspective of particle physics, a massive gauge boson—not a massless one—is the oddity. The key development that paved the way for the theory of the weak force was understanding the origin of the weak gauge boson masses, which make the distance dependence of the weak force so different from that of electromagnetism. The mechanism that gives rise to the weak gauge boson masses, known as the *Higgs mechanism*, is the subject of Chapter 10. As we will see in Chapter 12, the underlying theory—that is, the precise model that gives particles their mass—is one of the biggest puzzles facing particle

---

*One way to see that quantum mechanics and special relativity are relevant to this relation is that Planck's constant tells us that quantum mechanics is involved, and the speed of light tells us that special relativity is too. The distance would be zero if Planck's constant were zero (and classical physics applied) or if the speed of light were infinite.

physicists today. One of the attractions of extra dimensions is that they might help solve this mystery.

Quarks and the Strong Force

A physicist friend once explained to one of my sisters that he worked on "the strong force which is called the strong force because it is so strong." Although she did not find this particularly edifying, the strong force is in fact aptly named. It is an extremely powerful force. It binds together the constituents of the proton so powerfully that ordinarily they never separate. The strong force is only tangentially relevant to later parts of this book, but here I'll give some basic facts about it for completeness.

The strong force, described by the theory called *quantum chromo-dynamics* (QCD), is the last of the Standard Model forces that we can explain with gauge boson exchange. It too was discovered only in the last century. The strong gauge bosons are known as gluons because they communicate the force, the "glue," that binds strongly interacting particles together.

In the 1950s and 1960s, physicists discovered many particles in rapid succession. They gave the individual particles various Greek-letter names such as the π pronounced "pion"), the θ (pronounced "eta"), and the Δ (pronounced "Delta"—written with a capital "D" to reflect the case of the Greek letter). Collectively, these particles are called *hadrons*, after the Greek work *hadros*, which means "fat, heavy."

Indeed, hadrons were all much more massive than the electron. They were mostly comparable in mass to the proton, which has 2,000 times the electron's mass. The enormous multiplicity of hadrons was a mystery until the physicist Murray Gell-Mann* suggested in the 1960s that the many hadrons were not fundamental particles but were instead themselves composed of particles that he named *quarks*.

Gell-Mann got the word "quark" from a poem in James Joyce's

---

*And George Zweig, though his paper was never published.

Finnegans Wake: "Three quarks for Muster Mark! Sure he hasn't got much of a bark. And sure any he has it's all beside the mark." This, so far as I can deduce, is pretty much unrelated to the physics of quarks except for two things: there were three of them, and they were difficult to understand.*

Gell-Mann proposed that there are three varieties of quark†— they're now called *up*, *down*, and *strange*—and that the numerous hadrons corresponded to the many possible combinations of quarks that could be bound together. If his proposal was correct, hadrons would have to fall neatly into predictable patterns. As was often the case when new physical principles are suggested, Gell-Mann did not actually believe in the existence of quarks when he first proposed them. Nonetheless, his proposal was quite daring since only some of the predicted hadrons had been discovered. It was therefore a major victory for him when the missing hadrons were found and the quark hypothesis was confirmed, paving the way for Gell-Mann's 1969 Nobel Prize for Physics.

Even though physicists agreed that hadrons were made of quarks, nine years elapsed after the suggestion of quarks before hadron physics was explained in terms of the strong force. Paradoxically, the strong force was the last force to be understood, in part because of its enormous strength. We now know that the strong force is so large that the fundamental particles, such as quarks, that experience the strong force are always bound together and are difficult to isolate and therefore to study. Particles that experience the strong force are not free to roam unchaperoned.

There are three types of every quark variety. Physicists playfully label the different types with colors and sometimes call them red, green, and blue. And these colored quarks are always found with other quarks and antiquarks, bound together into *color-neutral combinations*. These are the combinations in which the strong force "charges" of the quarks and antiquarks cancel each other, analogously

---

*Quark is also a type of German cheese. The name would be doubly appropriate if it contained curds, which would be floating in the cheese like quarks in a hadron. However, my German friends tell me it does not.
†We now know that there are six.

to the way colors cancel in white light.* There are two types of color-neutral combination. Stable hadronic configurations contain either a quark and an antiquark that team up with each other, or else three quarks (and no antiquarks) that bond among themselves. For example, a quark pairs with an antiquark in particles called pions, and three quarks bind together in the proton and the neutron.

The strong force "charge" cancels among the quarks in hadrons, much as the charge of the positively charged proton and the negatively charged electron cancel in an atom. But whereas you can readily ionize an atom, it is very difficult to pry apart the objects, such as the proton and neutron, that are bound extraordinarily tightly by the gluons of the strong force. Gluons would be more aptly named "crazygluons,"† since their bonds are so difficult to break.

We are now almost ready to return to the discovery of quarks that Athena's revisionist tale metaphorically described. The proton and neutron consist of combinations of three quarks in which the charge associated with the strong force cancels out. The proton contains two up quarks and one down quark—different types of quark with different electric charge. Because the up quark has electric charge $+\frac{2}{3}$ and the down quark has charge $-\frac{1}{3}$, the proton has electric charge $+1$. A neutron, on the other hand, contains one up and two down quarks, so it has zero (the sum of $-\frac{1}{3}$, $-\frac{1}{3}$, and $+\frac{2}{3}$) electric charge.

Quarks can be thought of as hard, pointlike objects in a big, mushy proton. Quarks are embedded in a proton or neutron, like a pea buried under a mattress. But as with our bouncing princess who bruises herself on the pea, an active experimenter can shoot in a high-energy electron that emits a photon, which bounces directly off the quark. This looks very different from a photon bouncing off a big fluffy object, just as Rutherford's alpha particle bouncing off a hard nucleus looked very different from one bouncing off more diffuse positive charge.

The Friedman-Kendall-Taylor *deep inelastic scattering* experiment,

---

*This is the origin of the name "quantum chromodynamics." *Chromos* is Greek for "color."
†Or "supergluons" in the UK.

conducted at the Stanford Linear Accelerator Center (SLAC), demon-strated the existence of quarks by registering this effect. The experi-ment showed how electrons behave when they scatter off protons, thereby providing the first experimental evidence that quarks really exist. For this discovery, Jerry Friedman and Henry Kendall (who were my colleagues at MIT) and Richard Taylor won the 1990 Nobel Prize for Physics.

When quarks are produced in high-energy collisions, they aren't yet bound into hadrons, but that doesn't mean they're isolated—they will always have a retinue of other quarks and gluons accompanying them which make the net combination neutral under the strong force. Quarks never appear as free, unaccompanied objects but are always shielded by many other, strongly interacting particles. Instead of a single, isolated quark, a particle experiment would register a set of particles composed of quarks and gluons, going in more or less the same direction.

Collectively, the groups of particles composed of quarks and gluons that move in unison in a particular direction are known as *jets*. Once an energetic jet is formed, it is like a rope in that it will never dis-appear. When you cut a rope, all you do is create two new pieces of rope. Similarly, when interactions divide jets, the pieces can only form new jets: they will never separate into individual, isolated quarks and gluons. Stephen Sondheim was presumably not think-ing about high-energy particle colliders when he wrote the lyrics to the Jet song from *West Side Story*, but his words apply admirably to jets of strongly interacting particles. Energetic, strongly interact-ing particles remain together. "They're never alone . . . they're well protected."

The Known Fundamental Particles

This chapter has described three of the four known forces: electromag-netism, the weak force, and the strong force. Gravity, the remaining force, is so weak that it would not change particle physics predictions in an experimentally observable way.

But we have not yet finished introducing the particles of the Standard Model. They are identified by their charges, and also by their handedness. As I described earlier, the left- and right-handed particles can (and do) have different weak charges.

Particle physicists categorize these particles as either quarks or *leptons*. Quarks are fundamental fermionic particles that experience the strong force. Leptons are fermionic particles that do not. Electrons and neutrinos are examples of leptons. The word "lepton" derives from the Greek *leptos*, which means "small" or "fine," referring to the tiny mass of the electron.

The bizarre thing is that in addition to the particles that are essential to the structure of the atom, such as the electron and the up and down quarks, there are additional particles that, though heavier, have the same charges as the particles we have already introduced. All of the lightest stable quarks and leptons have heavier replicas. No one knows why they are there, or what they are good for.

When physicists first realized that the muon, a particle first seen in cosmic rays, was nothing other than a heavier version of the electron (200 times heavier), the physicist I.I. Rabi asked, "Who ordered that?" Although the muon is negatively charged, like the electron, it is heavier than the electron, into which it can decay. That is, a muon is unstable (see Figure 53) and quickly converts into an electron (and two neu-

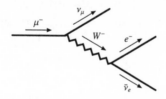

Figure 53. *In muon decay, the muon turns into a muon neutrino and a virtual W⁻ gauge boson, which then converts to an electron and an electron antineutrino.*

trinos). So far as we know, it serves no purpose to matter here on Earth. Why does it exist? This is one of the many mysteries of the Standard Model that we hope scientific progress will solve.

In fact, there are three copies of the full set of particles with the same Standard Model charges (see Figure 52). Each of these copies is called a *generation*, or sometimes a *family*. The first generation of particles contains a left- and a right-handed electron, a left- and a right-handed up quark, a left- and a right-handed down quark, and a left-handed neutrino. This first generation contains all the stable stuff of which atoms, and therefore all stable matter, is composed.

The second and third generations contain particles that decay and are not present in "normal" known matter. They are not exact copies of the first generation; they have charges identical to those of their first generation counterparts but are heavier. They were discovered only when they were produced at high-energy particle colliders, and their purpose remains obscure. The second generation consists of a left- and a right-handed muon, a left- and a right-handed *charm quark*, and a left- and a right-handed *strange quark*, as well as a stable left-handed *muon neutrino*.* The third generation contains a left- and a right-handed *tau*, a left- and a right-handed *top quark*, a left- and a right-handed *bottom quark*, and a left-handed *tau neutrino*. The identical copies of a particular particle with the same charge assignments, each a member of a different generation, are often called *flavors* of the particle type.

From Figure 52 you can see that although there were only three known flavors of quark when Gell-Mann first proposed them, we now know of six: three "up types" and three "down types"—one in each generation. In addition to the up quark itself, there are two identically charged up-type quarks—the charm and the top. Similarly, the down, strange, and bottom quarks are different flavors of down-type quark. And the muon and tau leptons are heavier versions of the electron.

Physicists are still trying to understand the reason for three generations and why particles have their particular masses. These are major questions about the Standard Model that fuel the research being conducted today. Along with many others, I've worked on these problems throughout my career, but we're still searching for the answers.

The heavier flavors are significantly heavier than the lighter ones. Although the next heaviest quark, the bottom, was discovered in 1977, the very heavy top quark eluded discovery until 1995. Two

---

*The neutrinos are named after the charged leptons with which they directly interact through the weak force.

particle experiments, including the remarkable one that discovered the top quark, are the subject of the following chapter.

What to Remember

- The Standard Model consists of the nongravitational forces and the particles that experience those forces. In addition to the well-known force of electromagnetism, there are two forces that act within a nucleus: the *strong force* and the *weak force.*

- The weak force poses the most important remaining mystery about the Standard Model. Whereas the other two forces are communicated by massless particles, the gauge bosons that communicate the weak force have mass.

- In addition to the particles that communicate forces, the Standard Model contains particles that experience those forces. These particles are divided into two categories: *quarks*, which experience the strong force, and *leptons*, which do not.

- The light quarks and leptons found in matter (the up quark, the down quark, and the electron) are not the only known particles. Heavier quarks and leptons also exist: the up quark, the down quark, and the electron each have two heavier versions.

- These heavy particles are unstable, which means that they decay to lighter quarks and leptons. But experiments at particle accelerators have produced them and shown that these heavier particles experience the same forces as the familiar light, stable particles.

- Each of the groups of particles that include a charged lepton, an up-type quark, and a down-type quark is known as a *genera-tion*. There are three generations, each of which contains successively heavier versions of each particle type. These particle varieties are known as *flavors*. There are three up-type quark flavors, three down-type quark flavors, three charged lepton flavors, and three neutrino flavors.

- I won't use the details or names of any particular quark or lepton later on. However, you will need to know about flavors and generations because of the strong constraints on the particles' properties, which give us vital clues and constraints on the physics that lies beyond the Standard Model.

- Chief among these constraints is that different flavors of quarks and leptons with the same charges rarely, if ever, turn into one another. Theories in which particles readily change flavor are ruled out. We will see later that this poses a big challenge for models of broken supersymmetry and other proposed extensions of the Standard Model.

8

Experimental Interlude:
Verifying the Standard Model

One way, or another
I'm gonna find you . . .   Blondie

Ike once again dreamed he met the quantum detective. This time, the sleuth knew what he was after—and he had a pretty good idea where it should be. All he had to do was wait—sooner or later, if he wasn't mistaken, his quarry would appear.

Finding heavy particles is not easy. Yet that's what we must do if we are to discover the structure underlying the Standard Model and, ultimately, the physical makeup of the universe. Most of what we know about particle physics comes from high-energy *particle accelerator* experiments, which first accelerate a rapidly moving beam of particles and then smash them into other matter.

In a *high-energy particle collider*, the accelerated beam of particles actually collides with an accelerated beam of antiparticles so that they meet in a small collision region containing a huge amount of energy. This energy is then sometimes converted into heavy particles not readily found in nature. High-energy particle colliders are the only place where the heaviest known particles have appeared since the time of the Big Bang, when the much hotter universe contained all particles in abundance. Colliders can create pairs of any kind of particle and antiparticle, in principle, as long as they have enough energy for that particular pair, the energy given by Einstein's $E = mc^2$.

But the goal of high-energy physics is not merely to find new particles. Experiments at high-energy colliders will tell us about fundamental laws of nature that cannot be observed in any other way—laws that operate at too close a range to be visible more directly. High-energy experiments are the only way to probe any short-distance interactions that operate at extremely tiny distance scales.

This chapter is about two of the collider experiments that were important in confirming the predictions of the Standard Model and constraining what physical theories might lie beyond. These experiments are both impressive in their own right. But they should also give you a sense of what physicists will be up against when they will search for new phenomena, such as extra dimensions, in the future.

The Top Quark Discovery

The search for the top quark beautifully illustrates the difficulties of finding a particle at a collider when the collider's energy is barely adequate to produce it, and the ingenuity with which experimenters can rise to this challenge. Although the top quark is not part of any atom or known matter, the Standard Model would be inconsistent without it, so most physicists had been confident of its existence since the 1970s. Yet as recently as 1995, no one had ever detected one.

At that time, experiments had been looking for the top quark in vain for many years. The bottom quark, the next-heaviest Standard Model particle, which weighs in at five times the mass of a proton, was discovered in 1977. But although physicists back then thought the top quark would soon show up, and experimenters raced to find it and claim the glory, to everyone's surprise experiment after experiment failed. It wasn't found at colliders that operated at 40 times, or 60 times, or even 100 times the energy required to produce a proton. The top quark was evidently heavy—remarkably heavy compared with the other quarks, all of which had been detected. When it finally made its appearance after twenty years of searching, it turned out to have a mass almost 200 times that of the proton.

Because the top quark is so heavy, the relations of special relativity tell us that only a collider that operated at extremely high energy could produce it. High energy always requires a very large accelerator, which is technically difficult to design and expensive to construct.

The collider that eventually produced the top quark was the Tevatron in Batavia, Illinois, thirty miles west of Chicago. The collider at Fermilab was initially designed with far too low an energy to produce a top quark, but engineers and physicists had made many changes that improved its potential enormously. By 1995 the Tevatron, the culmination of these improvements, operated at far higher energy and produced many more collisions than the original machine could have managed.

The Tevatron, which is still in operation, is located at Fermilab, an accelerator center that was commissioned in 1972 and named after the physicist Enrico Fermi. I was very amused when I first visited Fermilab and found there were wild corn, geese, and for some strange reason, buffalo on the site. Buffalo aside, the region is fairly flat and boring. The movie *Wayne's World* was set in Aurora, about five miles south of Fermilab, and if you are familiar with this movie, you might have some idea of the Fermilab surroundings. Fortunately, the physics there is exciting enough to keep people happy anyway.

The Tevatron gets its name because it accelerates both protons and antiprotons to an energy of a TeV (pronounced T-e-V, although the "Tev" in "Tevatron" rhymes with "Bev"), which is the same as 1,000 GeV, the highest energy that has been achieved so far at any accelerator. The energetic beams of protons and antiprotons that the Tevatron produces circulate in a ring and smash together every 3.5 microseconds at two collision points.

Two separate experimental collaborations set up detectors at each of the two collision points, where the beams of particles and antiparticles cross paths and the interesting physical processes can happen. One of these experiments was named CDF (Collider Detector at Fermilab) and the other was called D0, the designation of the collision point between protons and antiprotons at which the detector was located. The two experiments searched extensively for new physical particles and processes, but in the early 1990s the top quark was their Holy Grail. Each experimental collaboration wanted to be the first to find it.

Many heavy particles are unstable and decay almost immediately.

When that is the case, experiments search for visible evidence of a particle's decay products, rather than the particle itself. The top quark, for example, decays into a bottom quark and a W (the charged gauge boson that communicates the weak force). And the W also decays, either into leptons or quarks. So experiments seeking the top quark look for the bottom quark in conjunction with other quarks or leptons.

Particles do not come with nametags, however, so detectors have to identify them by their distinguishing properties, such as their electric charge or the interactions in which they participate, and separate components of the detectors are needed to record these properties. The two detectors at CDF and D0 are each segmented into several pieces, each of which records different characteristics. One piece is a *tracker*, which detects charged particles by the electrons from ionized atoms that they leave in their wake. Another piece, called a *calorimeter*, measures the energy that particles deliver as they pass through. The detectors have other components which can identify particles with other specific distinguishing properties, such as a bottom quark, which lasts longer than most other particles before it decays.

Once a detector registers a signal, it transmits the signal through an extensive array of wires and amplifiers, and records resulting data. However, not everything that is detected is worth recording. When a proton and an antiproton collide, the interesting particles such as the top and antitop quarks are only rarely produced. Much more often, collisions produce only lighter quarks and gluons, and more often still, nothing of real interest. In fact, for every top quark that was produced at Fermilab, there were ten trillion collision events that didn't contain a top quark.

No computer system is sufficiently powerful to find the one interesting event in such a crowd of useless data. For this reason, experiments always include *triggers*—devices in which hardware and software elements act like nightclub bouncers and permit only potentially interesting events to be recorded. Triggers in CDF and D0 reduced the number of events that experimenters had to sift through to about one in one hundred thousand—still an enormously challenging task, but far more tractable than one in ten trillion.

Once information is recorded, physicists try to interpret it and reconstruct the particles that emerged from any interesting collision.

Because there are always many collisions and many particles and only a limited number of pieces of information, reconstructing the result of a collision is a formidable task, one that has stretched people's ingenuity and is likely to lead to further data processing advances in the years to come.

By 1994, several of CDF's working groups had seen events that looked like the top quark (see Figure 54 for an example), but they

Figure 54. *A top quark event as recorded by Do, which detects the decay products of the top quark and top antiquark that are produced simultaneously. The line in the upper right is a muon, which reaches the outer portion of the detector. The four rectangular-like blocks are four jets that were produced. The line to the right is the missing energy of the neutrino.*

weren't really sure. Although CDF couldn't say with certainty that they found the top quark that year, both Do and CDF confirmed discovery in 1995. A friend of mine on Do, Darien Wood, described the intensity of the final editorial board meeting at which Do completed the data analysis and the paper that would report their results. The meeting went through the night and into the next day, with people occasionally napping on table tops.

Do and CDF received joint credit for discovering the top quark. A new particle was produced that had never been seen before. This newly discovered particle joined the ranks of other, established Standard Model particles. By now, so many top quarks have been seen that we know the top quark's mass and its other properties extremely precisely. In the future, we expect higher-energy colliders to produce so many top quarks that there is a danger that the top quarks themselves will become the *background* that mimics and interferes with the discovery of other particles.

New physics is almost certainly there to be seen. We will soon see why unresolved Standard Model issues are telling us that new particles and physical processes should appear when colliders reach only slightly higher energies than is possible at present. Experiments at the Large Hadron Collider (LHC) will look for evidence of structure beyond the Standard Model. If those experiments are successful, the reward will be fabulous—a better understanding of the underlying structure of all matter. High energy, many-particle collisions, and clever ideas will all contribute to accomplishing this difficult task.

Precision Tests of the Standard Model

We will now briefly move from the plains of Illinois to mountainous Switzerland—the location of CERN, the Conseil Européen pour la Recherche Nucléaire (now called the Organisation Européenne pour la Recherche Nucléaire or, in English, the European Organization for Nuclear Research, though the old acronym, CERN, has stuck). Many experiments have tested the Standard Model's predictions, but none were as spectacular as those performed between 1989 and 2000 at the Large Electron-Positron collider (LEP) located at the CERN accelerator facility.

The CERN site was chosen for its central location within Europe. CERN's main entrance is so close to the French border that the guard booth separating the two countries is almost directly outside. Many CERN employees live in France and cross the border twice daily. They are rarely bothered when crossing the border—unless their car isn't up to Helvetic standards, in which case the Swiss won't let them in. The only other danger is being an absent-minded professor, as one colleague can attest to. The guards stopped and searched him when he didn't stop at the border because he was distracted by thoughts about black holes.

The difference between the locations of Fermilab and CERN could not be more striking. CERN is adjacent to the beautiful Jura mountains (see Figure 55) and is only a short drive from Chamonix, a remarkable valley that runs between mountains covered with glaciers that descend practically to the road (though less so with global warming), and lies at the foot of Mont Blanc, the highest mountain in Europe. At CERN, many fortunate physicists pass the winter with tanned faces despite the persistent cloud cover in town because of the

Figure 55. *The CERN site with the Alps in the background. The Large Hadron Collider ring, in which two beams of protons will circulate underground, is indicated.*

time they spend in the mountains nearby skiing, snowboarding, or hiking.

CERN was created after World War II, in the nascent atmosphere of international collaboration. The original twelve member states were West Germany, Belgium, Denmark, France, Greece, Italy, Norway, the Netherlands, the United Kingdom, Sweden, Switzerland, and Yugoslavia (which left in 1961). Subsequently, Austria, Spain, Portugal, Finland, Poland, Hungary, the Czech and Slovak Republics, and Bulgaria have joined. Observer states involved in CERN activities include India, Israel, Japan, the Russian Federation, Turkey, and the United States. CERN is truly an international enterprise.

CERN, like the Tevatron, has many accomplishments to its credit. Carlo Rubbia and Simon van der Meer were awarded the 1984 Nobel Prize for Physics for designing the original CERN collider and discovering the weak gauge bosons, a success story that destroyed America's monopoly on particle discoveries. CERN was also where an employee, the Englishman Tim Berners-Lee, came up with the World Wide Web, HTML (hypertext markup language), and http (hypertext transfer protocol). He developed the Web so that many experimenters in scattered nations could be instantaneously linked to information and so that data could be shared among many computers. Of course, the repercussions of the Web have been felt far beyond CERN—it's often difficult to foresee the practical applications of scientific research.

In a few years, CERN will be the nexus of some of the most exciting physics results. The Large Hadron Collider, which will be able to reach seven times the present energy of the Tevatron, will be located there, and any discoveries made at the LHC will almost inevitably be something qualitatively new. Experiments at the LHC will seek—and very likely find—the as yet unknown physics that underlies the Standard Model, confirming or rejecting models such as the ones I describe in this book. Although the collider is in Switzerland, the LHC will truly be an international effort; experiments for the LHC are currently being developed all over the globe.

But back in the 1990s, physicists and engineers built the unbelievable LEP (Large Electron-Positron collider) at CERN, a Z boson "factory" that churned out millions of Zs. The Z gauge boson is one of the three gauge bosons that communicate the weak force. By studying millions of Zs, experimenters at LEP (and also at SLAC, the Stanford Linear Accelerator Center in Palo Alto, California) could do detailed measurements of the Z boson's properties, testing the predictions of the Standard Model to an unprecedented level of precision. It would take us too far off track to describe each of these measurements in detail, but in a moment I'll give you an idea of the stunning precision that was achieved.

The basic premise behind the Standard Model tests was very simple. The Standard Model makes predictions for the masses of the weak gauge bosons and the decays and interactions of the fundamental particles. We can test the consistency of the theory of weak interactions by checking whether the relationships among all these many quantities fit the theory's predictions. If there were a new theory with new particles and new interactions that became important at energies near the weak scale, there would be new ingredients that could change the weak interaction predictions from their Standard Model values.

Models that go beyond the Standard Model therefore make slightly different predictions for the Z boson's properties than those predicted by the Standard Model itself. In the early 1990s everyone used an incredibly cumbersome method for predicting the Z's properties in these alternative models so that the predictions could be tested. The method was very hard to penetrate and was outlined in a document with more pages than I cared to carry. At the time, I was a postdoctoral fellow at the University of California, Berkeley. In the summer of 1992, while I was attending a summer workshop at Fermilab, I decided that the relationships among different physical quantities could not possibly be as cumbersome as the method in the multipage document implied.

With Mitch Golden, then a postdoc at Fermilab, I developed a more concise way to interpret experimental results about the weak interactions. Mitch and I showed how to systematically incorporate the effects of new heavy (as yet unseen) particles by adding only three new quantities to the Standard Model that would summarize all possible non-Standard Model contributions. I spent a few weeks trying to get it all straight, and the answers finally came together during one intense weekend of work. It was extremely rewarding to discover how all the processes that the Z-factories would measure could be elegantly related. Mitch and I felt we had developed a much clearer picture of how theory and measurements were related, and it was very satisfying. We were not alone in our discovery, however. Michael Peskin at SLAC and his postdoc Takeo Takeuchi did similar work concurrently, and others followed rapidly in our footsteps.

But the real success story concerns LEP tests of the Standard Model, which were incredibly precise. I won't go into the details, but I will tell you two anecdotes that demonstrate their impressive sensitivity. The first is about finding the exact energy at which the positrons and electrons collided. The experimenters needed to know this energy to determine the precise value of the Z boson's mass. They had to take into account everything that might affect the value of its energy. But even after they had accounted for everything they could think of, they noticed that the energy seemed to rise and fall slightly when they measured it at particular times. What was causing the variation?

Incredibly, it turned out to be tides in Lake Geneva. The level of the lake rose and fell with the tides and with the heavy rain that year. This in turn affected the nearby terrain, which slightly altered the distance over which the electrons and positrons traveled inside the collider. Once the tidal effect was factored in, the spurious time-dependent measurement of the mass of the Z went away.

The second anecdote is also quite impressive. Electrons and positrons in the collider are kept in place by strong magnetic fields, which in turn require a large amount of power. It seemed that, periodically, the electrons and positrons would become slightly misaligned, indicating some variation in the collider's magnetic fields. A worker on the site observed that this variation correlated well with passages of the TGV, the express train that travels between Geneva and Paris. Apparently, there were power spikes associated with the French DC current that slightly disrupted the accelerator. Alain Blondel, a Parisian physicist working at CERN, told me the funniest part of this story. The experimenters had a real opportunity to absolutely confirm this hypothesis. Given that many of the TGV staff are French, there was inevitably a strike, so the experimenters were treated to a spike-free day!

What to Remember

- The most important experimental tool for studying particle physics is the high-energy *particle accelerator*. High-energy *colliders* are particle accelerators that smash together particles; if they have enough energy, colliders produce particles that are otherwise too massive to exist in the world around us.

- The *Tevatron* is the highest-energy collider currently in operation.

- The *Large Hadron Collider (LHC)* in Switzerland, which will have about seven times the Tevatron's energy and which will be built within a decade, will test many particle physics models.

9

Symmetry: The Essential Organizing Principle

La.
La la la la.
La la la la.
La la la la la la la la la.

Simple Minds

Athena uncaged three of her owls and let them fly around. Unfortunately for Ike, he had left the top of his convertible down that day and the curious owls flew right in. The most mischievous of the owls pecked at the car's interior and ended up tearing it a little.

When Ike saw the damage, he stormed into Athena's room and demanded that she watch her owls more carefully in the future. Athena protested that her owls were almost all well-behaved and she need only keep an eye on the bad one. But by that time the owls were back in their cages, and neither Ike nor Athena could identify which one was guilty.

The Standard Model works spectacularly well, but only because it is a theory in which quarks, leptons, and the weak gauge bosons—the charged Ws and the Z that communicate the weak force between weakly charged matter—all have mass. The mass of fundamental particles is, of course, critical to everything in the universe; if matter had been truly massless, it wouldn't form nice, solid objects, and structure and life in the universe as we know it would never have formed. Yet weak gauge bosons and other fundamental particles in the simplest theory of forces look as if they should be massless and should travel at the speed of light.

You might find it strange that a theory of forces should prefer zero masses. Why shouldn't any mass be allowed? But the most basic quantum field theory of forces is intolerant in this respect. It ostensibly forbids any nonzero values for the masses of fundamental Standard Model particles. One of the triumphs of the Standard Model is that it shows how to resolve this issue and fashion a theory in which particles have the masses that observations tell us they must have.

In the next chapter we will explore the mechanism by which particles acquire mass—the phenomenon known as the Higgs mechanism. But in this chapter we will discuss the important topic of *symmetry*. Symmetry and symmetry breaking help to determine how the universe goes from an undifferentiated point to the complex structure we now see. The Higgs mechanism is intimately connected with symmetry, and in particular with broken symmetry. Understanding how the elementary particles acquire mass requires some familiarity with these important ideas.

Things That Change but Remain the Same

Symmetry is a sacred word to most physicists. One might conjecture that other communities value symmetry highly as well, since the Christian cross, the Jewish menorah, the Dharma wheel of Buddhism, the crescent of Islam, and the Hindu mandala all exhibit symmetry (see Figure 56). Something has symmetry if you can manipulate it—for example, by rotating it, reflecting it in a mirror, or interchanging its

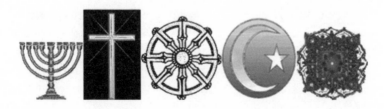

Figure 56. *A menorah, a cross, the Dharma wheel of Buddhism, the crescent of Islam, and the Hindu mandala all exhibit symmetry.*

parts—so that the new configuration is indistinguishable from the initial one. For example, if you were to interchange two identical candles on a menorah, you would see no visible difference. And the mirror image of a cross is identical to the cross itself.

Whether we are talking about mathematics, physics, or the world, we can make transformations that appear to do nothing when there is a symmetry. A system has symmetry if someone could exchange its components, reflect it in a mirror, or rotate it while your back was turned without your noticing any difference when you looked at it again.

Symmetry is often a static property: for example, the symmetry of a cross does not involve time. But physicists often prefer to describe symmetries in terms of imagined *symmetry transformations*—manipulations that one can apply to a system without changing any observable properties. For example, instead of saying that the candles of a menorah are equivalent, I might say instead that a menorah would look the same if I were to interchange two of the candles. I wouldn't actually have to exchange the candles in order to claim that there was a symmetry. But if, hypothetically, I did interchange the candles, I wouldn't be able to see any difference. Sometimes I will describe symmetry in this way for simplicity's sake.

We all are familiar with symmetries not only in science and sacred symbols, but in secular art as well. Symmetry can be found in most paintings, sculpture, architecture, music, dance, and poetry. Islamic art is perhaps the most spectacular in this respect, with its intricate and extensive use of symmetry in architecture and ornamental art, to which anyone who has seen the Taj Mahal can attest. Not only does the building look the same from any side, but when viewed from the edge of the long pool in front, it is perfectly reflected in the calm surface of the water. Even the trees have been planted to preserve the monument's symmetries. When I was there, I noticed a guide who was pointing out some symmetry points, so I asked him to show me the others. I ended up viewing the building from funny angles and scrambling up rubble on the edge of the site in order to see all the symmetries the monument presents.

In colloquial usage people often equate symmetry with beauty, and certainly some of the fascination with symmetry arises from the regularity and neatness that it guarantees. Symmetries also help us learn, since repetition, either in time or in space, can create indelible images in our mind. The brain's programmed response to symmetry and its sheer aesthetic appeal explain in large part why we surround ourselves with it.

But symmetries don't occur only in art and architecture, but also in nature, without any human intervention. For this reason you often encounter symmetries in physics. The goal of physics is to relate distinct quantities to one another so that we can make predictions based on observations. Symmetry is a natural player in this context. When a physical system has symmetry, you can describe the system on the basis of fewer observations than if the system had no symmetry. For example, if there are two objects with identical properties, I would know the physical laws that govern the behavior of one of the objects if I've already measured the behavior of the others. Because the two objects are equivalent, I know that they must behave the same way.

In physics, the existence of a symmetry transformation in a system means that there is some definite procedure for rearranging the system that leaves all its measurable physical properties unchanged.* For example, if a system possesses *rotational* and *translational symmetries*, two well-known examples of symmetries of space, physical laws apply the same way in all directions and in all places. Rotational and translational symmetries tell us, for example, that it doesn't matter which way you are facing or where you are standing when you swing a baseball bat at a ball: provided you apply the same force, the baseball will behave in exactly the same way. Any experiment should yield the same result if you rotate your setup or if you repeat your measurement in a different room or in a different place altogether.

It is difficult to overstate the importance of symmetry in physical laws. Many physical theories, such as Maxwell's laws of electrodynamics and Einstein's theory of relativity, are deeply rooted in symmetry. And by exploiting various symmetries we can usually simplify the task of using theories to make physical predictions. For

---

*I am describing the symmetry in terms of the consequences of a transformation, but, as always, symmetry is a property of the static system. That is, the system possesses symmetry, even if I don't actually make the transformation.

example, predictions of the orbital motion of the planets, the gravitational field of the universe (which is more or less rotationally symmetric), the behavior of particles in electromagnetic fields, and many other physical quantities are mathematically simpler once we take symmetry into account.

Symmetries in the physical world are not always completely obvious. But even when they are not readily apparent or when they are merely theoretical tools, symmetries usually greatly simplify the formulation of physical laws. The quantum theory of forces, which will soon be our focus, is no exception.

Internal Symmetries

Physicists generally classify symmetries into different categories. You are probably most familiar with symmetries of space—the symmetry transformations that move or rotate things in the external world. These symmetries, which include the rotational and translational symmetries I just mentioned, tell us that the laws of physics are the same for a system no matter which way the system points and no matter where it is located.

I now want to consider a different kind of symmetry, known as an *internal symmetry*. Whereas spatial symmetries tell us that physics treats all directions and all positions as the same, internal symmetries tell us that physical laws act the same way on distinct, but effectively indistinguishable, objects. In other words, internal symmetry transformations exchange or mix distinct things around in a way that can't be noticed. In fact, I have already given an example of an internal symmetry—the interchangeability of the candles on a menorah. The internal symmetry says that two candles are equivalent. It is a statement about the candles, not about space.

A traditional menorah, however, has both spatial and internal symmetries. While different candles are equivalent, which that means there is an internal symmetry, a menorah also looks the same if it is rotated 180 degrees about the central candle, which means that it has spatial symmetry as well. But an internal symmetry can exist even when there is no symmetry of space. For example, you can interchange identical green tiles in a mosaic even when the leaf they combine to portray has an irregular shape.

Another example of an internal symmetry is the interchangeability of two identical red marbles. If you hold one such marble in each hand, it wouldn't matter which was which. Even if you'd labeled them "1" and "2," you would never know whether I had somehow managed to interchange the two marbles. Notice that the example of the marbles is not tied to any spatial arrangement in the way that the examples of the menorah and the mosaic were; internal symmetries concern the objects themselves and not their locations in space.

Particle physics deals with somewhat abstract internal symmetries that relate different types of particle. These symmetries treat particles and the fields that create them as interchangeable. Just as two identical marbles behave in exactly the same way when you roll them or bang them against a wall, two particle types that have the same charges and mass obey identical physical laws. The symmetry that describes this is called *flavor symmetry*.

In Chapter 7 we saw that flavors are the three distinct particle types that have identical charges, one in each of the three generations. For example, electrons and muons are two flavors of charged leptons, which means that they have identical charges. Had we lived in a world in which the electron and the muon also had identical masses, the two would have been completely interchangeable. There would then have been a flavor symmetry, according to which the electron and muon would behave identically in the presence of any other particles or forces.

In our world the muon is heavier than the electron, so the flavor symmetry is not exact. But the difference in masses can be insignificant for some physical predictions, so flavor symmetries between light particles with identical charges, such as the muon and electron, are nonetheless often useful for calculations. Sometimes exploiting even slightly imperfect symmetries helps us to compute sufficiently accurate results. For example, the mass difference between particles is often so small (relative to energy or a large mass) that it doesn't make a measurable difference to predictions.

But the most important type of symmetry for us at this point is the symmetry that is relevant to the theory of forces, which is exact. This symmetry is also an internal symmetry among particles, but it's slightly more abstract than the flavor symmetry we just discussed. This particular type of internal symmetry is more analogous to the following example. As you might recall from high school physics, theater, or art class, three spotlights—generally one red, one green, and one blue—can shine together to produce white light. If we were to interchange the positions of three such lights, any of the new setups would still produce white light. It doesn't matter where any of the individual beams of light originate, so long as we see only the end result, white light. In that case, the internal symmetry transformation that exchanges the different lights would never produce any observable consequences.

We will now see that this symmetry is closely analogous to the symmetries associated with forces, because in both cases you are not able to observe everything. The light setup exhibits symmetry only because we are not allowed to look at everything, only the combined light. If you could see the lights themselves, you would know they had been interchanged. As mentioned earlier, this close analogy between colors and forces is the reason for the terms "color" and "quantum chromodynamics" (QCD) in the description of the strong force.

In 1927, the physicists Fritz London and Hermann Weyl demonstrated that the simplest quantum field theory description of forces involves internal symmetries similar to that in the spotlight example above. The connection between forces and symmetry is subtle, so you won't usually read about it outside textbooks. Because you don't really need to understand this connection to follow the discussion of the issues about masses—including the Higgs mechanism and the hierarchy problem of the next few chapters—you can skip ahead to the next chapter at this point if you want. But if you're interested in the role of internal symmetry in the theory of forces and the Higgs mechanism, read on.

Symmetry and Forces

Electromagnetism, the weak force, and the strong force all involve internal symmetries. (Gravity is related to symmetries of space and time, and must therefore be considered separately.) Without internal symmetries, the quantum field theory of forces would be an intractable mess. To understand these symmetries, we need first to consider the *gauge boson polarizations.*

You might be familiar with the notion of polarization of light; for example, polarizing sunglasses reduce glare by letting through only light that is vertically polarized and eliminating the horizontally polarized light. In this case, polarizations are the independent directions in which the electromagnetic waves associated with light can oscillate.

Quantum mechanics associates a wave with every photon. Each individual photon has different possible polarizations as well, but not all imaginable polarizations are allowed. It turns out that when a photon travels in any particular direction, the wave can oscillate only in directions that are perpendicular to its direction of motion. This wave acts like water waves on the ocean, which also oscillate perpendicularly. That is why you see a buoy or a boat bob up and down as a water wave passes by.

The wave associated with a photon can oscillate in any direction perpendicular to its direction of motion (see Figure 57). Really, there is an infinite number of such directions: imagine a circle perpendicular to the line of motion, and you can see that the wave is able to oscillate in any radial direction (from the center to the outside of the circle), and there are an infinite number of such directions.

But in the physical description of these oscillations, we need only

Figure 57. *A transverse wave oscillates perpendicularly to the direction of motion (in this case up and down, while the wave travels to the right).*

two independent perpendicular oscillations to account for them all. In physics terminology they are called *transverse polarizations*. It is as if you had labeled a circle with x and y axes. No matter what line you draw from the center of the circle, it will always intersect the circle at a particular position—a particular pair of x and y values— and can therefore be uniquely specified by only two coordinates. Similarly (without going into the details of how this works), although there are an infinite number of directions that are perpendicular to the direction in which a wave travels, all of those directions can be obtained from combinations of polarized light in any two perpendicular directions.

The important thing is that, in principle, there could have been a third polarization direction, one that oscillates along the direction in which the wave travels (had it existed, it would have been called the *longitudinal* polarization).[15] That is how sound waves travel, for example. But no such polarization of the photon exists. Only two of the three conceivable independent polarization directions exist in nature. A photon never oscillates along its direction of motion or in the time direction: it oscillates only along the directions perpendicular to its motion.

Even if we didn't already know from independent theoretical considerations that the longitudinal polarization was spurious, quantum field theory would have told us not to include it. If a physicist were to make calculations using a theory of forces that mistakenly included all three polarization directions, the theory's predictions of their properties wouldn't make sense. For example, she would predict ridiculously high gauge boson interactions rates. In fact, she would predict gauge bosons that interacted more often than always—that is, more than 100% of the time. Any theory that makes such nonsensical predictions is clearly wrong, and both nature and quantum field theory make it clear that this nonperpendicular polarization does not exist.

Unfortunately, the simplest theory of forces that physicists could formulate includes this spurious polarization direction. That is not so surprising because a theory that would work for any photon can't possibly contain information about one particular photon traveling in one particular direction. And without such information, special relativity would not distinguish any direction. In a theory that preserves the symmetries of special relativity (including rotational symmetry), you would need three directions—not two—to describe all the directions in which a photon could oscillate; in such a description, the photon could oscillate in any direction of space.

But we know that isn't true. For any particular photon, its direction of motion is singled out and oscillation in that direction is forbidden. But you wouldn't want to have to make a different theory for each and every photon, all with their own directions of travel. You would want a theory that works no matter which way a photon is travelling. Although you could try to make a theory that didn't include the spurious polarization direction at all, it is far simpler and cleaner to respect rotational symmetry and eliminate the bad polarization in some other way. Physicists, aiming for simplicity, have recognized that quantum field theory works best when they include the spurious longitudinal polarization in their theory but add an extra ingredient to filter out the good, physically relevant predictions from the bad.

This is where internal symmetries enter the picture. The role of internal symmetries in the theory of forces is to eliminate the contradictions that the unwanted polarization would create without making us forfeit the symmetries of special relativity. Internal symmetries are the simplest way to filter out the polarization along the direction of travel that independent theoretical considerations and experimental observations tell us does not exist. They classify polarizations into good and bad categories, those that are consistent with the symmetries and those that are not. The way it works is a bit too technical to explain here, but I can give you the general idea by using an analogy.

Suppose you have a shirt-making machine that can make left and right sleeves in two sizes, short and long, but for some reason the inventor of the machine neglected to include a control to ensure that the left and right sleeves are the same size. Half the time you will make useful shirts—ones with two long sleeves or two short sleeves—but half the time you will make useless, unbalanced shirts with one short and one long sleeve. Unfortunately for you, this is the only shirt-making machine you have.

You have a choice of throwing your shirt-making machine away and making no shirts at all, or keeping the machine and making some good shirts and some duds. All is not lost, however, because it will be pretty obvious which shirts to throw away: only the shirts that preserve a left-right symmetry are worth wearing. You will always be properly dressed if you let your machine make all kinds of shirts, but then keep only those shirts that have left-right symmetry.

The internal symmetry associated with forces accomplishes something analogous. It provides a useful marker to distinguish those quantities that we might in principle observe (the ones that involve the polarizations you want to keep) from those that should not be present (the ones involving the spurious polarization along the direction of motion). As with spam filters in computers that look for identifying features of unwanted e-mail to separate it from useful messages, the filter of internal symmetries distinguishes physical processes that preserve the symmetry from spurious ones that don't. Internal symmetries make it easy to eliminate the spamlike polarizations; if they were present, they would break the internal symmetry.

The way the symmetry works is very similar to the colored spotlight example we discussed earlier, in which we could observe only the white light produced by the three colors together, not the individual lights. Similarly, it turns out that only certain combinations of particles are consistent with the internal symmetries involved in the theory of forces, and those are the only combinations that appear in the physical world.

The internal symmetries associated with forces ban any process involving the bad polarizations—the ones oscillating along the direction of motion (the ones that don't really exist in nature). Just as the unbalanced shirts that were inconsistent with the left-right symmetry were readily distinguished and thrown away, the spurious polarizations that are inconsistent with the internal symmetry are automatically eliminated and never confuse calculations. A theory that stipulates the correct internal symmetry eliminates the bad polarizations that would otherwise be present.

Electromagnetism, the weak force, and the strong force are all communicated by gauge bosons: electromagnetism by the photon, the weak force by the weak gauge bosons, and the strong force by gluons. And each type of gauge boson is associated with waves that could in principle oscillate in any direction, but in reality oscillate only in the perpendicular directions. So each of the three forces requires its own particular symmetry to eliminate the bad polarizations of the gauge bosons that communicate that force. There is therefore a symmetry associated with electromagnetism, an independent symmetry associated with the weak force, and still another symmetry associated with the strong force.

Internal symmetries in the theory of forces might seem complicated, but they are the simplest way physicists know to formulate a useful quantum field theory of forces that allows us to make predictions. The internal symmetries are what discriminate between the true and the spurious polarizations.

The internal symmetries we have just explored are critical to the theory of forces. They also underlie the Higgs mechanism, which tells us how elementary particles in the Standard Model acquire their mass. For the next chapter we will not need the details of the internal symmetry, but we will see that symmetry (and symmetry breaking) are essential components of the Standard Model.

Gauge bosons, Particles, and Symmetry

So far we've considered the effect of symmetry only on gauge bosons. But the symmetry transformations associated with a force do not act only on the gauge bosons. A gauge boson interacts with the particles that experience the force associated with that gauge boson: the photon interacts with electromagnetically charged particles, the weak bosons interact with weakly charged particles, and the gluons interact with quarks.

Because of these interactions, each of the internal symmetries can be preserved only if it transforms both the gauge bosons and the particles with which they interact. We can see this by analogy. Rotations, for example, wouldn't be symmetry transformations if they acted on some objects but not others. If you rotate the top wafer of an Oreo cookie,* but not the rest of it, you would pull it apart. The

---

*A cookie consisting of a sandwich of two round wafers with "creme" in between.

Oreo cookie would look the same after a rotation only if you were to rotate the entire thing simultaneously.

For similar reasons, a transformation that transformed only the gauge bosons that communicate a force, but not the particles that experience that force, could never preserve a symmetry. The internal symmetry that eliminates the spurious polarizations of the gluons requires the quarks to be interchangeable as well as the gluons. In fact, the symmetry transformation that interchanges quarks is the same one that interchanges the gauge bosons. The only way to preserve the symmetry is to mix up both together, just as the only way to preserve the Oreo cookie is to rotate the whole thing at the same time.

The force that will interest us most in this book is the weak force. The internal symmetry associated with the weak force treats the three weak gauge bosons as equivalent. It also treats particle pairs such as the electron and the neutrino, or the up and down quarks, as equivalent. This weak force symmetry transformation interchanges the three weak gauge bosons and also these pairs of particles. As with gluons and quarks, the symmetry is preserved only when everything is interchanged at once.[16]

What to Remember

- *Symmetries* tell us when two different configurations behave the same way.

- In particle physics, symmetries are useful as a way of forbidding certain interactions: those that don't preserve the symmetries are not allowed.

- Symmetries are important to the theory of forces because the simplest workable theory of forces includes a symmetry associated with each force. Those symmetries eliminate unwanted particles. They also eliminate the false predictions that the simplest theory of forces would otherwise make about high-energy particles.

10

The Origin of Elementary Particle Masses: Spontaneous Symmetry Breaking and the Higgs Mechanism

One of these mornings the chain is gonna break.

Aretha Franklin

The stricter enforcement of speed limits made long-distance driving a nightmare for Icarus III. He longed to race as fast as he pleased, but police pulled him over nearly every half-mile. The cops never bothered with dull, neutral cars, but harassed only the lively, turbo-charged vehicles, like his own.

Ike resigned himself to driving only short distances, since that way he could avoid the police altogether. Within the half-mile-wide region around where he started, police never interfered and he could always drive impressively fast. Though the Porsche engine's force was unknown outside his neighborhood, closer to home it became legendary.

Symmetries are important, but the universe usually doesn't manifest perfect symmetry. Slightly imperfect symmetries are what makes the world interesting (but organized). For me, one of the most intriguing aspects of physics research is the quest for connections that make symmetry meaningful in an unsymmetric world.

When a symmetry is not exact, physicists say the symmetry is *broken*. Although broken symmetry is often interesting, it isn't always aesthetically appealing: the beauty and economy of the underlying system or theory can be lost (or lessened). Even the very symmetric Taj Mahal doesn't have perfect symmetry, since the builder's parsimonious heir decided not to build a planned second monument, adding instead an off-center tomb to the original. This second tomb destroys the Taj Mahal's otherwise perfect fourfold rotational symmetry, detracting slightly from its underlying beauty.

But fortunately for aesthetically minded physicists, broken symmetries can be even more beautiful and interesting than things that are perfectly symmetrical. Perfect symmetry is often boring. The *Mona Lisa* with a symmetric smile just wouldn't be the same.

In physics, as in art, simplicity alone is not necessarily the highest goal. Life and the universe are rarely perfect, and almost all symmetries you care to name are broken. Although we physicists value and admire symmetry, we still have to find a connection between a symmetric theory and an asymmetric world. The best theories respect the elegance of symmetric theories while incorporating the symmetry breaking necessary to make predictions that agree with phenomena in our world. The goal is to make theories that are richer and sometimes even more beautiful without compromising their elegance.

The concept of the *Higgs mechanism*, which relies on the phenomenon of *spontaneous symmetry breaking* (which we will consider in the following section), is an example of such a sophisticated, elegant theoretical idea. This mechanism, named after the Scottish physicist Peter Higgs, lets the Standard Model particles—quarks, leptons, and weak gauge bosons—acquire mass.

Without the Higgs mechanism, all elementary particles would have to be massless; the Standard Model with massive particles but without the Higgs mechanism would make nonsensical predictions at high energies. The magical property of the Higgs mechanism is that it lets you have your cake and eat it too: particles get mass, but they act as if they are massless when they have energies at which massive particles would otherwise cause problems. We will see that the Higgs mechanism allows particles to have mass but travel freely over a restricted range, in much the same way that Ike's car, which was stopped by policemen after half a mile, traveled undisturbed over limited distances, and that this suffices to solve high-energy problems.

Although the Higgs mechanism is one of the nicest ideas in quantum field theory and underlies all fundamental particle masses, it is also somewhat abstract. For this reason it is not well known by most people aside from specialists. While you can understand many features of ideas I discuss later in the book without knowing the details of the Higgs mechanism (and you can skip now to the summary bullets if you like), this chapter does provide an opportunity to delve a bit deeper into particle physics and into the ideas, such as spontaneously symmetry breaking, that buttress theoretical developments in particle physics today. As an added bonus, some familiarity with the Higgs mechanism will let you in on an amazing insight into electromagnetism that was discovered only in the 1960s, once the weak force and the Higgs mechanism were properly understood. And later on, when we come to explore extra-dimensional models, some understanding of the Higgs mechanism will make the potential merits of those recent ideas meaningful.

Spontaneously Broken Symmetry

Before describing the Higgs mechanism, we need first to investigate spontaneous symmetry breaking, a special type of symmetry breaking that is central to the Higgs mechanism. Spontaneous symmetry breaking plays a big role in many of the properties of the universe that we already understand and is likely to play a role in whatever we have yet to discover.

Spontaneous symmetry breaking is not only ubiquitous in physics, but is a prevalent feature of everyday life. A spontaneously broken symmetry is a symmetry that is preserved by physical laws but not by the way things are actually arranged in the world. Spontaneous symmetry breaking takes place when a system cannot preserve a symmetry that would otherwise be present. Perhaps the best way to explain how this works is to give a few examples.

Let's first consider a dinner arrangement in which a number of people are seated around a circular table with water glasses placed between them. Which glass should someone use, the one on the right or the one on the left? There is no good answer. I am told that Miss Manners says the one on the right, but aside from arbitrary rules of etiquette, left and right serve equally well.

However, as soon as someone chooses a glass, the symmetry is broken. The impetus to choose would not necessarily be part of the system; in this case it would be another factor—thirst. Nonetheless, if one person spontaneously drank from the glass on their left, so would that person's neighbors, and in the end everyone would have drunk from the glass on the left.

The symmetry exists until the moment someone picks up a glass. At that moment the left-right symmetry is spontaneously broken. No law of physics dictates that anyone has to choose left or right. But one has to be chosen, and after that, left and right are no longer the same in that there is no longer a symmetry that interchanges the two.

Here's another example. Imagine a pencil standing on end at the center of a circle. For the split second in which it rests on its tip and is exactly vertical, all directions are equivalent and a rotational symmetry exists. But a pencil standing on end won't just stay there: it will spontaneously fall in some direction. As soon as the pencil topples over, the original rotational symmetry is broken.

Notice that it would not be the physical laws themselves that determined the direction. The physics of the pencil falling over would be exactly the same no matter the direction in which it fell. What would break the symmetry would be the pencil itself, the state of the system. The pencil simply cannot fall in all directions at once. It has to fall in one particular direction.

A wall that is infinitely long and high would also look the same everywhere and in all directions along it. But because an actual wall has boundaries, if you are to see the symmetries you will have to get close enough to it that the boundaries are out of your field of vision. The wall's ends tell you that not everywhere along the wall is the same, but if you were to press your nose up against it so that you could see only a short distance away, the symmetry would appear to be preserved. You might want to briefly reflect on this example, which shows that a symmetry can appear to be preserved when viewed on one distance scale, even though it appears to be broken on another— a concept whose importance will become apparent very soon.

Almost any symmetry you care to name is not preserved in the world. For example, there are many symmetries that would be present in empty space, such as rotational or translational invariance, which tell us that all directions and positions are the same. But space is not empty: it is punctuated by structures such as stars and the solar system, which occupy particular positions and are oriented in particular ways that no longer preserve the underlying symmetry. They could have been anywhere, but they can't be everywhere. The underlying symmetries must be broken, although they remain implicit in the physical laws describing the world.

The symmetry associated with the weak force is also spontaneously broken. In the rest of this chapter I'll explain how we know this and discuss some of the consequences. We'll see that spontaneously breaking the weak force symmetry is the only way to explain massive particles while avoiding incorrect predictions for high-energy particles that cannot be avoided in any other candidate theory. The Higgs mechanism acknowledges both the requirement of an internal symmetry associated with the weak force and the necessity for it being broken.

The Problem

The weak force has one especially bizarre property. Unlike the electromagnetic force, which travels over large distances—which you benefit from each time you turn on the radio—the weak force affects only matter that is within extremely close range. Two particles must be within one ten thousand trillionth of a centimeter to influence each other via the weak force.

For the physicists who studied quantum field theory and quantum electrodynamics (QED, the quantum field theory of electromagnetism) in its earliest days, this restricted range was a mystery. QED made it look as if forces, such as the well-understood electromagnetic force, should be transmitted arbitrarily far away from a charged source. Why wasn't the weak force also communicated to particles at any distance and not just to those nearby?

Quantum field theory, which combines the principles of quantum mechanics and special relativity, dictates that if low-energy particles communicate forces only a short distance, they must have mass; and the heavier the particle, the shorter the particle's range. As explained in Chapter 6, this is a consequence of the uncertainty principle and special relativity. The uncertainty principle tells us that you need high-momentum particles to probe or influence physical processes at short distances, and special relativity relates that momentum to a mass. Although this is a qualitative statement, quantum field theory makes this relationship precise. It tells how far a massive particle will travel: the smaller the mass, the bigger the distance.

Therefore, according to quantum field theory, the short range of the weak force could mean only one thing: the weak gauge bosons communicating the force had to have nonzero mass. However, the theory of forces I described in the previous chapter works only for gauge bosons such as the photon, which communicates a force over large distances and has zero mass. According to the original theory of forces, the existence of nonzero masses was strange and problematic— the theory's high-energy predictions when gauge bosons have mass make no sense. For example, the theory would predict that very energetic, massive gauge bosons would interact much too strongly— so strongly in fact that particles would appear to be interacting more than 100% of the time. This naive theory is clearly wrong.

Furthermore, the masses for weak gauge bosons, quarks, and leptons (all of which we know to have nonzero mass) do not preserve the internal symmetry which, as we saw in the previous chapter, is a key ingredient in the theory of forces. Physicists who hoped to construct a theory with massive particles clearly needed a new idea.

Physicists have shown that the only way to make a theory that avoids nonsensical predictions about energetic, massive gauge bosons is to have the weak force symmetry break spontaneously through the process known as the Higgs mechanism. Here's why.

You might recall from the previous chapter that one of the reasons we wanted to include an internal symmetry that eliminates one of the three possible polarizations of a gauge boson was that a theory without the symmetry makes the same sort of nonsensical predictions I've just mentioned. The simplest theory of forces without an internal symmetry predicts that any energetic gauge boson, with or without a mass, interacts with other gauge bosons far too often.

The successful theory of forces eliminates this bad high-energy behavior by forbidding the polarization that is responsible for the incorrect predictions and doesn't actually exist in nature. Spurious polarizations are the source of the problematic predictions for high-energy scattering, so the symmetry allows only physical polarizations—the ones that really exist and are consistent with the symmetry—to remain. The symmetry, which rids the theory of nonexistent polarizations, also eliminates the incorrect predictions they would otherwise induce.

Although I didn't say so explicitly at the time, this idea works as stated only for massless gauge bosons. The weak gauge bosons, unlike the photon, have nonzero masses. Weak gauge bosons travel at less than the speed of light. And that puts a wrench in the works.

Whereas massless gauge bosons have only two polarizations that exist in nature, massive gauge bosons have three. One way to understand this distinction is that massless gauge bosons always travel at the speed of light, which tells us that they are never at rest. They therefore always single out their direction of motion, so you can always distinguish the perpendicular directions from the remaining polarization along the direction of travel. And it turns out that for massless gauge bosons, physical polarizations oscillate only in the two perpendicular directions

Massive gauge bosons, on the other hand, are different. Like all familiar objects, they can sit still. But when a massive gauge boson isn't moving, it doesn't single out a direction of motion. To a massive gauge boson sitting at rest, all three directions should be equivalent. But if all three directions are equivalent, then all three possible polarizations would have to exist in nature. And they do.

Even if you find the above logic mysterious, rest assured that experimenters have already seen the effects of a third polarization of a massive gauge boson and have confirmed its existence. The third polarization is called the *longitudinal polarization*. When a massive gauge boson is moving, the longitudinal polarization is the wave that oscillates along the direction of motion—the direction of sound wave oscillations, for example.

This polarization doesn't exist in the case of massless gauge bosons such as the photon. However, for massive gauge bosons, like the weak gauge bosons, the third polarization is truly a part of nature. This third polarization must be a part of the weak gauge boson theory.

Because this third polarization is the source of the weak gauge boson's overly large interaction rate at high energy, its existence poses a dilemma. We already know that we need a symmetry to eliminate the bad high-energy behavior. But this symmetry gets rid of the incorrect predictions by eliminating the third polarization as well, and that polarization is essential to a massive gauge boson and therefore to the theory that describes it. Although an internal symmetry would eliminate bad predictions for high-energy behavior, it would do so at too high a price: the symmetry would get rid of the mass as well! A symmetry in the theory of massive gauge bosons seems poised to throw away the baby with the bathwater.

The impasse at first glance looks insurmountable, since the requirements for a theory of massive gauge bosons appear to be entirely contradictory. On the one hand, an internal symmetry—the one described in the previous chapter—should not be preserved, since otherwise massive gauge bosons with three physical polarizations would be forbidden. On the other hand, without an internal symmetry to eliminate two of the polarizations, the theory of forces makes incorrect predictions when the gauge bosons have high energy. We still need a symmetry to eliminate the third polarization of each massive gauge boson if we are to have any hope of eliminating the bad high-energy behavior.

The key to resolving this apparent paradox and figuring out the correct quantum field theory description of a massive gauge boson was recognizing the difference between the ones with high energy and the ones with low energy. In the theory without an internal symmetry, only predictions about the high-energy gauge bosons looked as if they would be problematic. Predictions about low-energy massive gauge bosons were sensible (and true).

These two facts together imply something fairly profound: to avoid problematic high-energy predictions, an internal symmetry is essential—the lessons of the previous chapter still apply. But when the massive gauge bosons have low energy (low compared with the energy that Einstein's relation $E = mc^2$ associates with its mass), the symmetry should no longer be preserved. The symmetry must be eliminated so that gauge bosons can have mass and the third polarization can participate in the low-energy interactions where the mass makes a difference.

In 1964, Peter Higgs and others discovered how theories of forces could incorporate massive gauge bosons by doing exactly what we just said: keeping an internal symmetry at high energies, but eliminating it at low energies. The Higgs mechanism, based on spontaneous symmetry breaking, breaks the internal symmetry of the weak interactions, but only at low energy. That ensures that the extra polarization will be present at low energy, where the theory needs it. But the extra polarization will not participate in high-energy processes, and the nonsensical high-energy interactions will not appear.

Let's now consider a particular model that spontaneously breaks the weak force symmetry and implements the Higgs mechanism. With this exemplar of the Higgs mechanism, we'll see how the elementary particles of the Standard Model acquire mass.

The Higgs Mechanism

The Higgs mechanism involves a field that physicists call the *Higgs field*. As we have seen, the fields of quantum field theory are objects that can produce particles anywhere in space. Each type of field generates its own particular type of particle. An electron field is the source of electrons, for example. Similarly, a Higgs field is the source of Higgs particles.

As with heavy quarks and leptons, Higgs particles are so heavy that they aren't found in ordinary matter. But unlike heavy quarks and leptons, no one has ever observed the Higgs particles that the Higgs field would produce, even in experiments performed at high-energy accelerators. This doesn't mean that Higgs particles don't exist, just that Higgs particles are too heavy to have been produced with the energies that experiments have explored so far. Physicists expect that if Higgs particles exist, we'll create them in only a few years' time, when the higher-energy LHC collider comes into operation.

Nevertheless, we are fairly confident the Higgs mechanism applies to our world, since it is the only known way to give Standard Model particles their masses. It is the only known solution to the problems that were posed in the previous section. Unfortunately, because no one has yet discovered the Higgs particle, we still don't know precisely what the Higgs field (or fields) actually is.

The nature of the Higgs particle is one of the most hotly debated topics in particle physics. In this section, I will present the simplest of many candidate models—possible theories that contain different particles and forces—that demonstrates how the Higgs mechanism works. Whatever the true Higgs field theory turns out to be, it will implement the Higgs mechanism—spontaneously breaking the weak force symmetry and giving masses to elementary particles—in the same manner as the model I'm about to present.

In this model, a pair of fields experience the weak force. It will be useful later to think of these two Higgs fields, which are subject to the weak force, as carrying weak force charge. The Higgs mechanism terminology is sometimes sloppy, with "the Higgs" sometimes denoting the two fields together, and at other times one of the individual fields (and often the Higgs particles we hope to find). Here I will distinguish the possibilities and refer to the individual fields as $Higgs_1$ and $Higgs_2$.

Both $Higgs_1$ and $Higgs_2$ have the potential to produce particles. But they can also take nonzero values even when no particles are present. We haven't encountered such nonzero values for quantum fields up to this point. So far, aside from the electric and magnetic fields, we have considered only quantum fields that create or destroy particles but take zero value in the absence of particles. But quantum fields can also have nonzero values, just like the classical electric and magnetic fields. And according to the Higgs mechanism, one of the Higgs fields takes a nonzero value. We will now see that this nonzero value is ultimately the origin of particle masses.[17]

When a field takes a nonzero value, the best way to think about it is to imagine space manifesting the charge that the field carries, but not containing any actual particles. You should think of the charge that the field carries as being present everywhere. This is, alas, a rather abstract notion because the field itself is an abstract object. But when the field takes a nonzero value, its consequences are concrete: the charge that a nonzero field would carry exists in the real world.

A nonzero Higgs field, in particular, distributes weak charge throughout the universe. It is as if the nonzero weak-charge-carrying

Higgs field paints weak charge throughout space. A nonzero value for the Higgs fields means that the weak charge that $Higgs_1$ (or $Higgs_2$) carries is everywhere, even when there are no particles present. The vacuum—the state of the universe with no particles present—itself carries weak charge when one of the two Higgs fields takes a nonzero value.

Weak gauge bosons interact with this weak charge of the vacuum, just as they do with all weak charge. And the charge that pervades the vacuum blocks the weak gauge bosons as they try to communicate forces over long distances. The further they try to travel, the more "paint" they encounter. (Because the charge actually spreads throughout three dimensions, you might prefer to imagine a fog of paint.)

The Higgs field plays a role very similar to that of the traffic cop in the story, restricting the weak force's influence to very short distances. When attempting to communicate the weak force to distant particles, the force-carrying weak gauge bosons bump into the Higgs field, which gets in their way and cuts them off. Like Ike, who could travel freely only within a half-mile radius of his starting point, weak gauge bosons move unimpeded only for a very short distance, about one ten thousand trillionth of a centimeter. Both weak gauge bosons and Ike are free to travel short distances, but are intercepted at longer distances.

The weak charge in the vacuum is spread out so thinly that at short distance there is very little sign of the nonzero Higgs field and the associated charge. Quarks, leptons, and the weak gauge bosons travel freely over short distances, almost as if the charge in the vacuum didn't exist. The weak gauge bosons can therefore communicate forces over short distances, almost as if the two Higgs fields were both zero.

However, at longer distances, particles travel further and therefore encounter a more significant amount of weak charge. How much they encounter depends on the charge density, which depends in turn on the value of the nonzero Higgs field. Long-distance travel (and communication of the weak force) is not an option for low-energy weak gauge bosons, for on long-distance excursions the weak charge in the vacuum gets in the way.

This is exactly what we needed to make sense of weak gauge bosons. Quantum field theory says that particles that travel freely over short distances, but only extremely rarely travel over longer distances, have nonzero masses. The weak gauge bosons' interrupted travel tells us that they act as if they have mass, since massive gauge bosons just don't get very far. The weak charge that permeates space hinders the weak gauge bosons' travel, making them behave exactly as they should in order to agree with experiments.

The weak charges in the vacuum have a density that corresponds roughly to charges that are separated by one ten thousand trillionth of a centimeter. With this weak charge density, the masses of the weak gauge bosons—the charged Ws and the neutral Z—take their measured values of approximately 100 GeV.

And that's not all that the Higgs mechanism accomplishes. It is also responsible for the masses of quarks and leptons, the elementary particles constituting the matter of the Standard Model. Quarks and leptons acquire mass in a very similar fashion to the weak gauge bosons. Quarks and leptons interact with the Higgs field distributed throughout space and are therefore also hindered by the universe's weak charge. Like weak gauge bosons, quarks and leptons acquire mass by bouncing off the Higgs charge distributed everywhere throughout spacetime. Without the Higgs field, these particles would also have zero mass. But once again, the nonzero Higgs field and the vacuum's weak charge interfere with motion and make the particles have mass. The Higgs mechanism is also necessary for quarks and leptons to acquire their masses.

Although the Higgs mechanism is a more elaborate origin of mass than you might think necessary, it is the only sensible way for the weak gauge bosons to acquire mass according to quantum field theory. The beauty of the Higgs mechanism is that it gives the weak gauge bosons mass while accomplishing precisely the task I laid out at the beginning of this chapter. The Higgs mechanism makes it look as though the weak force symmetry is preserved at short distances (which, according to quantum mechanics and special relativity, is equivalent to high energy) but is broken at long distances (equivalent to low energy). It breaks the weak force symmetry spontaneously, and this spontaneous breaking lies at the root of the solution to the problem of massive gauge bosons. This more advanced topic is explained in the following section (but feel free to skip ahead to the following chapter if you wish).

Spontaneous Breaking of Weak Force Symmetry

We have seen that the internal symmetry transformation associated with the weak force will interchange anything that is charged under the weak force because the symmetry transformation acts on anything that interacts with weak gauge bosons. Therefore, the internal symmetry associated with the weak force must act on the $Higgs_1$ and $Higgs_2$ fields, or the $Higgs_1$ and $Higgs_2$ particles they would create, and treat them as equivalent, just as it treats up and down quarks, which also experience the weak force, as interchangeable particles.

If both of the Higgs fields were zero, they would be equivalent and interchangeable, and the full symmetry associated with the weak force would be preserved. However, when one of the two Higgs fields takes a nonzero value, the Higgs fields spontaneously break the symmetry of the weak force. If one field is zero and the other is not, the electroweak symmetry, by which $Higgs_1$ and $Higgs_2$ are interchangeable, is broken.

Just as the first person to choose his left or right glass breaks the left-right symmetry at a round table, one Higgs field taking a nonzero value breaks the weak force symmetry that interchanges the two Higgs fields. The symmetry is broken spontaneously because all that breaks it is the vacuum—the actual state of the system, the nonzero field in this case. Nonetheless, the physical laws, which are unchanged, still preserve the symmetry.

A picture might help convey how a nonzero field breaks the weak force symmetry. Figure 58 shows a graph with two axes, labeled x and y. The equivalence of the two Higgs fields is like the equivalence of the x and y axes with no points plotted. If I were to rotate the graph so that the axes switched places, the picture would still look the same. This is a consequence of ordinary rotational symmetry.[18]

Notice that if I plot a point at the position $x = 0$, $y = 0$, this

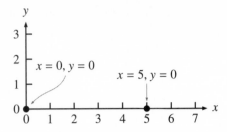

Figure 58. *When the point* x = 0, y = 0 *is singled out, rotational symmetry is preserved. But when* x = 5, y = 0 *is singled out, rotational symmetry is broken.*

rotational symmetry is completely preserved. But if I plot a point that has one nonzero coordinate value, for example where $x = 5$ and $y = 0$, the rotational symmetry is no longer preserved. The two axes are no longer equivalent because the x value, but not the y value, of this point is not zero.[19]

The Higgs mechanism spontaneously breaks weak force symmetry in a similar fashion. When the two Higgs fields are zero, the symmetry is preserved. But when one is zero and the other is not, the weak force symmetry is spontaneously broken.

The weak gauge boson masses tell us the precise value of the energy at which the weak force symmetry is spontaneously broken. That energy is 250 GeV, the weak scale energy, very close to the masses of the weak gauge bosons, the W⁻, the W⁺ and the Z. When particles have energy greater than 250 GeV, interactions occur as if the symmetry is preserved, but when their energy is less than 250 GeV, the symmetry is broken and weak gauge bosons act as if they have mass. With the correct value of the nonvanishing Higgs field, the weak force symmetry is spontaneously broken at the right energy, and the weak gauge bosons get precisely the right mass.

The symmetry transformations that act on the weak gauge bosons also act on quarks and leptons. And it turns out that these transformations won't leave things the same unless quarks and leptons are massless. That means that weak force symmetries would be preserved only if quarks and leptons didn't have mass. And because the weak force symmetry is essential at high energies, not only is spontaneous symmetry breaking required for the weak gauge boson masses, it's necessary for the quarks and lepton to acquire masses as well. The Higgs mechanism is the only way for all the massive fundamental particles of the Standard Model to acquire their masses.

The Higgs mechanism functions in exactly the way that is needed to ensure that any theory that incorporates it can have massive weak gauge bosons (as well as massive quarks and leptons) and nonetheless will make the correct predictions for high-energy behavior. Specifically, for high-energy weak gauge bosons—those with energy larger than 250 GeV—symmetry appears to be preserved, so there are no incorrect predictions. At high energies the internal symmetry associated with the weak force still filters out the problematic polarization of the weak gauge boson that would cause interactions at too high a rate. But at low energies, where the mass is essential to reproducing the measured short-range interactions of the weak force, the weak force symmetry is broken.

This is why the Higgs mechanism is so important. No other theory that assigns these masses has these properties. Other approaches fail either at low energies, where the mass will be wrong, or at high energies, where interactions will be predicted incorrectly.

Bonus

There is one more successful feature of the Standard Model that I have not yet explained. Although the Higgs field will be relevant to the next few chapters, this particular aspect of the Higgs mechanism will not. Yet it is so surprising and fascinating that it's worth mentioning.

The Higgs mechanism tells us about more than just the weak force. Surprisingly, it also gives new insight into why electromagnetism is special. Until the 1960s, no one would have believed that there was more to learn about the electromagnetic force, which had been so well studied and understood for over a century. In the 1960s, however, the electroweak theory proposed by Sheldon Glashow, Steven Weinberg, and Abdus Salam showed that when the universe began its evolution at high temperature and energy, there were three weak gauge bosons, plus a fourth, independent, neutral boson with a different interaction strength. The photon, ubiquitous and important as it is today, was not a member of this list. The authors of the electroweak theory deduced the nature of the four weak gauge bosons from both mathematical and physical clues, which I won't go into here.

The remarkable thing is that the photon was originally nothing special. In fact, the photon we talk about today is actually a mixture of two of the original four gauge bosons. The reason that the photon got singled out is that it is the only gauge boson involved in the electroweak force that is impervious to the weak charge of the vacuum. The chief distinguishing feature of the photon is that it travels unfettered through the weakly charged vacuum and therefore has no mass.

Photon travel, unlike that of the W and Z, is not obstructed by the nonzero value of a Higgs field. That's because although the vacuum carries weak charge, it does not carry electric charge. The photon, which communicates the electromagnetic force, interacts only with electrically charged objects. For this reason, the photon can communicate a long-range force without any interference from the vacuum. It is therefore the only gauge boson that remains massless even in the presence of the nonzero Higgs field.

The situation closely resembles the speed traps with which Ike had to contend (although this part of the analogy is admittedly a little more tenuous). The speed traps let dull cars pass through scot-free. Photons, like the dull neutral cars, always travel unimpeded.

Who would have thought it? The photon, the thing that physicists for years thought they understood completely, has an origin that can be understood only in terms of a more complex theory that combines the weak and electromagnetic forces into a single theory. This theory is therefore generally referred to as the *electroweak theory*, and the relevant symmetry is called *electroweak symmetry*. The electroweak theory and the Higgs mechanism are major successes of particle physics. Not only the weak gauge boson masses, but also the relevance of the photon are neatly explained within its framework. On top of that, it allows us to understand the origin of the quark and lepton masses. The rather abstract ideas we have just encountered nicely explain quite a wide range of features of the world.

Caveat

The Higgs mechanism works beautifully, and gives quarks, leptons, and weak gauge bosons their masses without making nonsensical high-energy predictions—and, furthermore, explains how the photon came to be. However, there is one essential property of the Higgs particle that physicists don't yet fully understand.

Electroweak symmetry must be broken at about 250 GeV to give particles their measured masses. Experiments show that particles with energy greater than 250 GeV look as if they are massless, whereas particles with energy below 250 GeV act as if they have mass. However, the electroweak symmetry will break at 250 GeV only if the Higgs particle (sometimes also called the Higgs boson)[20] itself has roughly this mass (again, using $E = mc^2$); the theory of the weak force wouldn't work if the Higgs mass were much greater. If the Higgs mass were greater, symmetry breaking would happen at a higher energy and the weak gauge bosons would be heavier—contradicting experimental results.

However, in Chapter 12, I will explain why a light Higgs particle poses a major theoretical problem. Calculations that take quantum mechanics into account point to a heavier Higgs particle, and physicists don't yet know why the Higgs particle mass should be so low. This quandary is critical to motivating new particle physics ideas and some of the extra-dimensional models that we'll consider later on.

Even without knowing the precise nature of the Higgs particle and the reason why it is so light, the mass requirement tells us that the Large Hadron Collider, which will start operating at CERN in Switzerland within the decade, should discover one or more crucial new particles. Whatever breaks electroweak symmetry must have a mass that is around the weak scale mass. And we expect that the LHC will find out what it is. When it does, this critically important discovery will greatly advance our knowledge of the underlying structure of matter. And it will also tell us which (if any) of the proposals for explaining the Higgs particle is correct.

But before we get to those proposals, we'll look at one possible extension of the Standard Model that was suggested purely in the interest of simplicity of nature. The next chapter explores virtual particles, the distance dependence of forces, and the fascinating topic of *grand unification*.

What to Remember

- Despite the importance of symmetries for making the right predictions about high-energy particles, the masses of quarks, leptons, and weak gauge bosons tell us that the *weak force symmetry* must be broken.

- Because we still have to guard against false predictions, the weak force symmetry must nonetheless be maintained at high energy. Therefore, the weak force symmetry must be broken only at low energy.

- *Spontaneous symmetry breaking* occurs when all physical laws preserve a symmetry but the actual physical system does not. Spontaneously broken symmetries are symmetries that are preserved at high energies but broken at low energies. The weak force symmetry is spontaneously broken.

- The process by which weak force symmetry is spontaneously broken is the *Higgs mechanism*. For the Higgs mechanism to spontaneously break the weak force symmetry, there has to be a particle with a mass of about the weak scale mass, which is 250 GeV (remember, special relativity relates energy and mass through $E = mc^2$).

I I

Scaling and Grand Unification: Relating Interactions at Different Lengths and Energies

> I hope someday you'll join us
> And the world will live as one.
>
> John Lennon

Athena often felt like she was the last to be told anything interesting. She didn't even hear about Ike's adventures with his car until after he had owned it for over a month. And she didn't learn them from him directly—she learned about them from a friend of hers who had heard about them from Dieter's cousin's brother, who had learned about them from Dieter's cousin, who had heard about them from Dieter.

Through this indirect route, Athena was told Ike's remark, "The influence of forces depends on where you are." Ike's uncharacteristic pronouncement completely mystified Athena until she realized that the message must have been distorted along the way. After thinking about it awhile, she decided that Ike's real remark must have been, "The performance of Porsches depends on the model of car."

We'll see that the statement Athena originally heard is true. This chapter is about how physical processes that take place between particles at one separation can be related to those that take place at another separation and why physical quantities, such as a particle's mass or interaction strength, depend on the particle's energy. This dependence on energy and distance is over and above the classical separation dependence of forces. For example, classically, the strength of electromagnetism, like that of gravity, decreases in proportion to the square of the interacting objects' separation (the inverse square law). But quantum mechanics changes this distance dependence by influencing the strength of the interaction itself so that particles at different separations (and energies) seem to interact with different charges.

Forces become weaker or stronger with increasing distance as a result of *virtual particles*—short-lived particles that exist as a consequence of quantum mechanics and the uncertainty principle. Virtual particles interact with gauge bosons and alter forces so that their effect depends on distance, much as Athena's friends distorted Ike's message as they passed it along.

Quantum field theory tells us how to compute the effect of virtual particles on the distance and energy dependence of forces. One triumph of such calculations was that they explained why the strong force is so strong. Another interesting fallout was the potential existence of a *Grand Unified Theory*, in which the three nongravitational forces, which are so different at low energies, merge into a single unified force at high energies. We'll explore both of these results and the quantum field theory ideas and calculations that underlie them.

When you are reading the next few chapters, bear in mind the very disparate energy scales we are discussing. The unification energy is about one thousand trillion GeV, and the Planck scale energy, where gravity gets strong, is about a thousand times greater than that. The weak scale energy, which is the energy where experiments currently operate, is a whole lot smaller: it is only about a hundred to a thousand GeV. The weak scale energy is about as small relative to the unification energy as a marble's size is compared with the distance between the Earth and the Sun. I'll therefore sometimes call the weak scale energy low even though it's a high energy from the perspective of experiments,* as it's so much smaller than both the unification and the Planck scale energies.

---

*This runs counter to American marketing terminology, which calls small things big.

Zooming In and Out

Effective field theories apply the effective theory idea that we learned about in Chapter 1 to quantum field theory. They focus only on those energy and length scales you can hope to measure. The effective field theory that applies at a particular energy or distance scale "effectively" describes those energies or distances we need to take into account. It concentrates on those forces and interactions that can occur when particles have that particular energy (or lower)* and ignores any energies that are inaccessibly higher. It doesn't ask for the details of physical processes or particles that occur only with higher energy than you can achieve.

One advantage of an effective field theory is that even if you don't know what interactions take place over short distances, you can still study the quantities that matter at the scales that interest you. You really need only to think about the quantities that you can (in principle) detect. When you mix paint, you don't need to know its detailed molecular structure. But you probably want to know the properties that you readily perceive, like color and texture. With this information, even without knowing the microstructure of your paint, you could categorize the paints' relevant properties and predict what mixtures of the paints would look like when you applied them to your canvas.

However, if you did know your paint's chemical composition, the rules of physics would allow you to deduce some of those properties. You don't need this information when you're painting (using the effective theory) but you would find it useful if you were making paint (deriving the effective theory's parameters from a more fundamental theory).

Similarly, if you don't know the short-distance (high-energy) theory, you won't be able to derive measurable quantities. However, when you do know the short-distance details, quantum field theory tells you

---

*Recall that quantum mechanics and special relativity make energies and distances interchangeable. For readability, I'll talk in terms of energies now, but processes involving high energies are the same as processes involving short distances.

precisely how to relate the different effective theories that apply to different energies. It lets you derive the quantities of one effective theory, such as masses or interaction strength, from the quantities of another.

The method for calculating how quantities depend on energy or distance, which was first developed by Kenneth Wilson in 1974, has a fancy name: the *renormalization group*. Along with symmetries, two of the most powerful tools in physics are the effective theory concept and the renormalization group, both of which involve physical processes with very different lengths or energy scales. The word "group" is a mathematical term that stuck, although its mathematical origin is largely irrelevant.

Renormalization is not such a bad word, though. It refers to the fact that at each distance scale of interest, you pause to get your bearings. You determine which particles and which interactions are relevant at the particular energies that interest you at the moment. You then apply a new normalization—that is, a new calibration—for any parameters in the theory.

The renormalization group uses ideas that are similar to those set out in Chapter 2, where we discussed the feasibility of interpreting a higher-dimensional theory in lower-dimensional language and treated a two-dimensional theory that had a small rolled-up dimension as if it were only one-dimensional. When we curled up dimensions, we ignored all the details of what happened inside the extra dimensions and assumed that everything could be described in lower-dimensional terms. Our new "normalization" was the four-dimensional description that could be used when focusing on large distances.

We can use a very similar procedure to derive a theory that applies to long distances from any theory appropriate to short distances: decide the minimum length you care about, and "wash out" the physics relevant to shorter scales. One way of doing this is to find the average value of those quantities whose details would make a difference only at the shorter distances you have chosen to ignore. If you had a grid filled with grayscale dots, you would literally average the shade density of the smaller dots to find the shade for bigger dots that would reproduce their effect. Your eyes do this automatically when you view something with fuzzy resolution.

If you can see things only with a given level of precision, you don't need to know what happens on smaller scales to make useful calculations that relate measurable quantities. Your most efficient course often involves choosing the "pixel size" in your theory to agree with your level of precision. That way, you can neglect heavy particles that you'll never produce and short-distance interactions that will never occur. Instead, you can focus your calculations on particles and interactions that are relevant at the energy you can achieve.

However, if you do know the more precise theory that applies at smaller distances, you can use that information to calculate quantities in the effective theory that interests you—that is, the effective theory with lower resolution. Just as with the grayscale dots, when you go from an effective theory with short-distance resolution to another with less precise resolution, in essence you change the "pixel size" with which you choose to analyze your theory. The renormalization group tells you how to calculate the influence that such short-distance interactions could have on the particles in your long-distance theory. You extrapolate physical processes from one length or energy scale to another.

Virtual Particles

Renormalization group calculations make these extrapolations by taking into account the effect of quantum mechanical processes and *virtual particles*. Virtual particles, a consequence of quantum mechanics, are strange, ghostly twins of actual particles. They pop in and out of existence, lasting only the barest moment. Virtual particles have the same interactions and the same charges as physical particles, but they have energies that look wrong. For example, a particle moving very fast clearly carries a lot of energy. A virtual particle, on the other hand, can have enormous speed but no energy. In fact, virtual particles can have any energy that is different from the energy carried by the corresponding true physical particle. If it had the same energy, it would be a real particle, not a virtual one. Virtual particles are a strange feature of quantum field theory that you have to include to make the right predictions.

So how can these apparently impossible particles exist? A virtual particle with its borrowed energy could not exist were it not for the uncertainty principle, which allows particles to have the wrong energy so long as they do so for such a short time that it would never be measured.

The uncertainty principle tells us that it would take infinitely long to measure energy (or mass) with infinite precision, and that the longer a particle lasts, the more accurate our measurement of its energy can be. But if the particle is short-lived and its energy cannot possibly be determined with infinite precision, the energy can temporarily deviate from that of a true long-lived particle. In fact, because of the uncertainty principle, particles will do whatever they can get away with for as long as they can. Virtual particles have no scruples and misbehave whenever no one is watching. (A physicist from Amsterdam even suggested that they are Dutch.)

You can think of the vacuum as a reservoir of energy—virtual particles are particles that emerge from the vacuum, temporarily borrowing some of its energy. They exist only fleetingly and then disappear back into the vacuum, taking with them the energy they borrowed. That energy might return to its place of origin, or it might be transferred to particles in some other location.

The quantum mechanical vacuum is a busy place. Even though the vacuum is by definition empty, quantum effects give rise to a teeming sea of virtual particles and antiparticles that appear and disappear—even though no stable, long-lasting particles are present. All particle-antiparticle pairs can in principle be produced, albeit only for very short visits, too short to be seen directly. But however brief their existence, we care about virtual particles because they nonetheless leave their imprint on the interactions of long-lived particles.

Virtual particles have measurable consequences because they influence the interactions of the real physical particles that enter and leave an interaction region. During its brief span of its existence, a virtual particle can travel between real particles before disappearing and repaying its energy debt to the vacuum. Virtual particles thereby act as intermediaries that influence the interactions of long-lived stable particles.

For example, the photon in Figure 47 (p. 157), which was

exchanged to generate the classical electromagnetic force, was in fact a virtual photon. It didn't have the energy of a true photon, but it didn't have to. It only needed to last long enough to communicate the electromagnetic force and make the real charged particles interact.

Another example of virtual particles is shown in Figure 59. Here, a photon enters an interaction region, a virtual electron-positron pair is produced, and then the pair is absorbed at another location. At the place where the particles are absorbed, another photon emerges from the vacuum that carries off the energy that the intermediate electron-positron pair temporarily borrowed. We'll now investigate one remarkable consequence of this type of interaction.

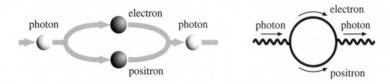

Figure 59. *A real physical photon can turn into a virtual electron and a virtual positron, which can then turn back into a photon. This is illustrated with a Feynman diagram on the right and schematically on the left.*

Why Interaction Strength Depends on Distance

The strengths of the forces we know about depend on the energies and distances involved in particle interactions, and virtual particles play a part in that dependence. For example, the strength of the electromagnetic force is smaller when two electrons are further apart. (Remember, this quantum mechanical decrease is over and above the classical distance dependence of electromagnetism.) The consequences of virtual particles and the distance dependence of forces is real; theoretical predictions and experiments match extremely well.

The reason that the quantities of an effective theory—the strength of forces or interactions, for example—depend on the energies and the separations of the particles involved follows from a feature of quantum field theory that the physicist Jonathan Flynn jokingly called the *anarchic principle*.* The anarchic principle follows from quantum mechanics, which tells us that all particle interactions that can happen will happen. In quantum field theory, everything that is not forbidden will occur.

I'll call each separate process by which a particular group of physical particles interacts a *path*. A path may or may not involve virtual particles. When it does, I'll call that path a *quantum contribution*. Quantum mechanics tells us that all possible paths contribute to the net strength of an interaction. For example, physical particles can turn into virtual particles, which can interact with each other and then turn back into other physical particles. In such a process, the original physical particles might reemerge or they might turn into different physical particles. Even though the virtual particles wouldn't last long enough for us to observe them directly, they would nonetheless affect the way real observable particles interacted with one another.

Trying to prevent virtual particles from facilitating an interaction would be like telling certain of your friends a secret and hoping it won't reach another friend. You know that sooner or later, some of the "intermediate virtual" friends will betray your confidence and relay the message to that other friend. Even if you already told that friend your secret, the fact that your virtual friends will discuss it with him as well will affect his opinion on the subject, too. In fact, his opinion will be the net result of everyone he has talked to.

Not only direct interactions between physical particles, but also *indirect interactions*—those that involve virtual particles—play a role in communicating forces. Just as your friend's opinion is affected by everyone who talks to him, the net interaction between particles is the sum of all possible contributions, including those from virtual particles. And because the importance of virtual particles depends on the distances involved, the strengths of forces depend on distance.

The renormalization group tells us precisely how to calculate the

---

*This is a modified version of Murray Gell-Mann's term, the "totalitarian principle," but I think that "anarchic principle" is a closer approximation to the physics to which it's applied.

impact of virtual particles in any interaction. All of the effects of intermediate virtual particles are added together, and this either strengthens or impedes the strength of a gauge boson's interactions.

Indirect interactions play a more important role when interacting particles are further apart. A greater distance is analogous to telling your secret to more "virtual" friends. Although you can't be sure that any single friend will betray your confidence, the more friends you tell, the more likely it is that at least one of them will. Whenever a path exists by which virtual particles can contribute to the net strength of an interaction, quantum mechanics ensures that they will. And the amount by which virtual particles affect that strength depends on the distance over which the force is communicated.

But actual renormalization group calculations are even more clever, since they add up the contributions of friends talking to one another as well. A better analogy for the contributions due to virtual particles resembles the paths of a message as it goes through a big bureaucracy. If a person at the top of the hierarchy sends a message, it will go through directly. But someone lower down in the hierarchy might have to have his messages vetted by his bosses. If someone at an even lower level sends a message, it might first have to circulate through even more layers of red tape before ultimately reaching its destination. In that case, at each level bureaucrats would send the message around before sending it on to successively higher levels. Only after the message finally reached the upper echelons would it be released. The message that emerged in this case would generally not be the original; instead, it would be the one that was filtered through this many-layered bureaucracy.

If you think of virtual particles as bureaucrats, and a higher-level bureaucrat as corresponding to a virtual particle with higher energy, a high-level message would get directly communicated, whereas the lower-level ones would pass through many stages. The quantum mechanical vacuum is the "bureaucracy" a photon encounters. Each interaction is vetted through intermediate virtual particles with less and less energy. As in a bureaucracy, there can be diversions at all levels (or distances). Some of the paths will bypass the "bureaucratic" detours imposed by virtual particles, and some will involve virtual particles that travel over ever-increasing distances. The shorter-distance (higher-energy) communication encounters fewer virtual processes than those that occur at larger distances.

However, there is a notable difference between virtual processes and a bureaucracy. In a bureaucracy, any one particular message takes one particular path, no matter how complicated that path. Quantum mechanics, on the other hand, says there can be many paths. And it insists that the net strength of an interaction is the sum of the contribution from all the possible paths that could occur.

Consider a photon traveling from one charged particle to another. Because it can turn into virtual electron-positron pairs en route (see Figure 60), quantum mechanics tells us that sometimes it will. And the paths with virtual electrons and positrons influence the efficacy with which the photon communicates the electromagnetic force.

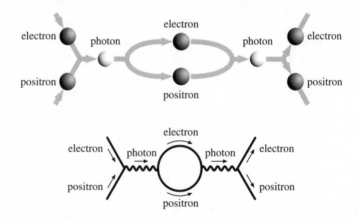

Figure 60. *Virtual correction to electron-positron scattering. Reading from left to right: an electron and a positron annihilate into a photon, which in turn splits into a virtual electron-positron pair, which then annihilate back into a photon, which in turn converts to an electron and positron. The intermediate virtual electron and positron thereby affect the strength of the electromagnetic force.*

And this is not the only quantum mechanical process that can occur. Virtual electrons and positrons can themselves emit photons, which can turn into other virtual particles, and so on. The distance between the two charged particles that exchange the photon determines how many such interactions the messenger photon will have with particles in the vacuum, and how large an impact the interactions will have. The strength of the electromagnetic force is the net result of the many paths the photon takes when all possible bureaucratic detours—quantum mechanical processes in which virtual particles might participate over long or short distances—are taken into account. Because the number of virtual particles that a photon will encounter depends on the distance it travels, the photon's interaction strength depends on the distance between the charged objects with which it interacts.

When all the contributions from all possible paths are added together, the calculation shows that the vacuum dilutes the message that the photon carries from the electron. The intuitive explanation for the dilution of the electromagnetic interaction is that opposite charges attract and like charges repel, and therefore, on average, virtual positrons are closer to an electron than are virtual electrons. The charges from the virtual particles therefore weaken the full impact of the initial electron's electric force. Quantum mechanical effects *screen* the electric charge. Electric charge screening means that the strength of the interaction between a photon and an electron decreases with distance.

The true electric force at long distances appears to be smaller than the classical short-distance electric force because a photon that communicates a force over short distances more frequently takes a path that doesn't involve virtual particles. A photon that travels a short distance wouldn't have to travel through a big, weakening cloud of virtual particles, as the photon that was communicating a force far away would have to do.

Not just the photon, but all force-carrying gauge bosons interact with virtual particles en route to their destination. Virtual particle pairs, the particle and its antiparticle, spontaneously erupt from and get absorbed by the vacuum, affecting the net strength of an interaction. These virtual particles temporarily waylay the gauge boson transmitting the force and change its overall interaction strength. Calculations show that the weak force's strength, like electromagnetism's, decreases with distance.

However, virtual particles don't always put the brakes on interactions. Surprisingly, they can sometimes help them along. In the early 1970s, David Politzer, who was then a Harvard graduate student of

Sidney Coleman (who suggested the problem), and separately David Gross and his then student, Frank Wilczek, who were both at Princeton, as well as Gerard 't Hooft in Holland, did calculations that demonstrated that the strong force behaves in precisely the opposite way from the electric force. Rather than screening the strong force at long distances and thereby making it weaker, virtual particles actually enhance the interactions of the gluons (the particles that communicate the strong force)—so much so that the strong force at long distances deserves its name. Gross, Politzer, and Wilczek won the 2004 Nobel Prize for Physics for their critical insight into the strong force.

The key to this phenomenon is the gluons themselves. One big difference between gluons and photons is that gluons interact with one another. A gluon can enter an interaction region and turn into a pair of virtual gluons which then influence the force's strength. These virtual gluons, like all virtual particles, exist only momentarily. But their effects pile up as you increase the distance, until the strong force is indeed extraordinarily strong. And the result of a calculation is that virtual gluons dramatically enhance the strong force's strength at larger particle separations. The strong force is much stronger when particles are well separated than when they are close together.

Compared with electric charge screening, the increase of the strong force with distance is a very counterintuitive result. How can it be that an interaction gets stronger when particles are further apart? Most interactions subside over distance. We would really need a calculation to show this, but there are examples in the world of such behavior.

For example, if someone sends a message through a bureaucracy whose importance some middle manager doesn't understand, the middle manager might blow up what should have been an ordinary memo into a critically important directive. Once the middle manager modified the message, it would have a far greater impact than if the original author had communicated it directly.

The Trojan War is another example in which long-distance forces were more powerful than those at short distances. According to the *Iliad*, the Trojan War began when the Trojan prince Paris decided to run off with Helen, the wife of the Spartan king Menelaus. Had Menelaus and Paris fought over Helen mano a mano before Paris and

Helen absconded to Troy, the war between the Greeks and the Trojans would have ended before it mushroomed into an epic. Once Menelaus and Paris were far apart, they interacted with many others and created the strong forces that participated in the very powerful Greek-Trojan interactions.

Though surprising, the growth of strong interactions with distance is sufficient to explain all the distinctive properties of the strong force. It explains why the strong force is powerful enough to keep quarks bound up into protons and neutrons, and quarks trapped inside jets—the strong force grows at long distance to the point where a particle that experiences the strong force cannot be separated overly far from other strongly interacting particles. Fundamental strongly interacting particles such as quarks are never found in isolation.

A well-separated quark and an antiquark would store an enormous amount of energy, so much so that it would be more energy-efficient to create additional physical quarks and antiquarks in between than to let them remain isolated. If you were to try to pull the quark and antiquark further apart, new quarks and antiquarks would be created from the vacuum. Just as in Boston traffic, where you can never be more than a car-length behind the car in front of you without a car coming in from the next lane, those new quarks and antiquarks would hover near the original ones so that no single quark or antiquark would become any more isolated than when it started—some other quark or antiquark is always nearby.

Because the strong force at large distances is so strong that it doesn't allow strongly interacting particles to be isolated from one another, particles that are charged under the strong force are always sur-rounded by other charged particles in strong-force-neutral combi-nations. The consequence is that we never see isolated quarks. We only see strongly bound hadrons and jets.

Grand Unification

The results of the previous section tell us about the distance depen-dence of the strong, weak, and electromagnetic forces.[21] In 1974, Georgi and Glashow made the bold suggestion that these three forces change with distance and energy in such a way that they unify into a single force at high energy. They called their theory a GUT, short for *Grand Unified Theory*. Whereas the strong force symmetry interchanges three colors of quarks (as discussed in Chapter 7) and the weak force symmetry interchanges different particle pairs, the GUT force symmetry acts on and interchanges all types of Standard Model particles, quarks and leptons.[22]

According to Georgi and Glashow's Grand Unified Theory, early in the evolution of the universe, when the temperature and energy were extremely high—the temperature was higher than one hundred trillion trillion degrees kelvin, and the energy was higher than one thousand trillion GeV—the strength of each of the three forces was the same as that of the others and the three nongravitational forces fused into a single one, "The Force."

As the universe evolved, the temperature dropped and the unified force split into three distinct forces, each with its own distinct energy dependence, through which they evolved into the three nongravitational forces we know today. Although the forces began as a single force, they ended up with very different interaction strengths at low energies because of the different influences that virtual particles had on each of them.

The three forces would be like identical triplets who developed from a single fertilized egg, but matured into three rather different individuals. One triplet might now be a punk rocker with dyed, spiked hair, one a marine with a crewcut, and one an artist with a long ponytail. They would nonetheless share the same DNA, and when they were babies would have been pretty much indistinguishable.

In the early universe, the three forces would also have been indistinguishable. But they would have split apart through spontaneous symmetry breaking. Just as the Higgs mechanism broke electroweak symmetry and left only electromagnetism unbroken, it would also break the GUT symmetry and leave the three separate forces that we witness today.

A single interaction strength at high energy is a prerequisite for a Grand Unified Theory. That means that the three lines representing interaction strength as a function of energy must all intersect at a single energy. But we already know how the strengths of the three nongravitational forces change with energy. And because quantum mechanics tells us that large distance is interchangeable with low energy and that short distance is interchangeable with high energy,* the results of the previous section can be interpreted equally well in terms of energy. At low energies the electromagnetic and weak forces are less powerful than the strong force, but they strengthen at higher energies, whereas the strong force weakens.

In other words, the strengths of the three nongravitational forces are becoming more comparable at high energies. They might even be converging to a single strength. This would mean that the three lines representing interaction strength as a function of energy intersect at high energies.

Two lines meeting at a single point is not such an exciting result— it is bound to happen when the lines approach each other. But three lines meeting at a point is either a strong coincidence or evidence of something more meaningful. If the forces do converge, their single interaction strength could be an indication that there is only a single type of force at high energy—in which case we would have a unified theory.

Although unification to this day remains a conjecture, the unification of forces, if true, would be a major leap towards a simpler description of nature. Because unifying principles are so intriguing, physicists studied the strength of the three forces at high energies to see whether or not they converge. Back in 1974, nobody had measured the interaction strengths of the three nongravitational forces with very great accuracy. Howard Georgi, Steven Weinberg, and Helen Quinn (who was then an unpaid Harvard postdoctoral fellow, now a physicist at the Stanford Linear Accelerator Center and president of the American Physical Society) used the imperfect measurements that were then available and did a renormalization group calculation to extrapolate the strength of the forces to high energies. They discovered that the three lines representing the strength of nongravitational forces did indeed appear to converge to a single point.

The famous 1974 Georgi-Glashow paper on their Grand Unified

---

*Remember that the uncertainty principle relates uncertainty in length to the inverse of the uncertainty in momentum.

Theory begins with these words: "We present a series of hypotheses and speculations leading inescapably to the conclusion . . . that all elementary particle forces (strong, weak, and electromagnetic) are different manifestations of the same fundamental interaction involving a single coupling strength. Our hypotheses may be wrong and our speculations idle, but the uniqueness and simplicity of our scheme are reasons enough that it be taken seriously."* Perhaps those were not the most modest of words. However, Georgi and Glashow did not really think that that uniqueness and simplicity were sufficient evidence that their theory was the correct description of nature. They also wanted experimental confirmation.

Although an enormous leap of faith was required to extrapolate the Standard Model to ten trillion times the energy anyone had directly explored, they realized that their extrapolation had a testable consequence. In their paper, Georgi and Glashow explained that their GUT "predicts that the proton decays," and that experimenters should try to test this prediction.

Georgi and Glashow's unified theory predicted that protons wouldn't last for ever. After a very long time, they would decay. Such a thing would never happen in the Standard Model. Quarks and leptons are ordinarily distinguished by the forces they experience. But in a Grand Unified Theory, the forces are all essentially the same. So, just as an up quark can change into a down quark via the weak force, a quark should be able to change into a lepton via the unified force. That means that if the GUT idea is correct, the net number of quarks in the universe would not remain the same, and a quark could change into a lepton, making the proton—a composite of three quarks— decay.

Because the proton can decay in a Grand Unified Theory that links quarks and leptons, all familiar matter would ultimately be unstable. However, the decay rate of the proton is very slow—the lifetime would far exceed the age of the universe. That means that even as dramatic a signal as a proton decaying would not stand much chance of being detected: it would happen much too rarely.

*Howard Georgi and S.L. Glashow, "Unity of all elementary-particle forces," *Physical Review Letters*, vol. 32, pp. 438–441 (1974).

To find evidence of proton decay, physicists had to build extremely large and long-lasting experiments that studied a huge number of protons. That way, even if any single proton is unlikely to decay, a large number of protons would greatly increase the odds that the experiment could detect the decay of one of them. Even though your likelihood of winning the lottery is small, it would be much greater if you bought millions of tickets.

Physicists did build such large, multi-proton experiments, including the Irvine/Michigan/Brookhaven (IMB) experiment located in the Homestake Mine in South Dakota, and the Kamiokande experiment, a vat of water and detectors buried a kilometer deep underground in Kamioka, Japan. Although proton decay is an extremely rare process, these experiments would already have found evidence of it if the Georgi-Glashow GUT were correct. Unfortunately for grand ambitions, no one has yet discovered such decay.

This doesn't necessarily rule out unification. In fact, thanks to more precise measurement of the forces, we now know that the original model proposed by Georgi and Glashow is almost certainly incorrect, and only an extended version of the Standard Model can unify forces. As it turns out, in such models the predictions for the proton lifetime are longer, and proton decay shouldn't have been detected yet.

Today, we don't actually know whether unification of forces is a true feature of nature or, if it is, what it signifies. Calculations show that unification could happen in several models I'll discuss later, including supersymmetric models, the Hořava-Witten extra-dimensional models, and the warped extra-dimensional models that Raman Sundrum and I developed. The extra-dimensional models are particularly intriguing because they could bring gravity into the unification fold and truly unify all four known forces. These models are also important because in the original unification models it was assumed there were no new particles to be found above the weak scale other than those with GUT scale masses.* These other models demonstrate that unification might happen even if there are many new particles that could be produced only at energies above the weak scale.

However, fascinating as unification of forces can be, physicists are

---

*This is known as the *desert hypothesis*.

currently divided about its theoretical merits according to whether they favor a top-down or a bottom-up approach to physics. The idea of a Grand Unified Theory embodies a top-down approach. Georgi and Glashow made a bold assumption about the absence of particles with mass between one thousand and one thousand trillion GeV and hypothesized a theory based on this assumption. Grand Unification was the first step in the particle physics debate that continues today with string theory. Both theories extrapolate physical laws from measured energies to energies at least ten trillion times higher. Georgi and Glashow later became skeptical about the top-down approach that string theory and the search for Grand Unification represent. They have since reversed their tracks and now concentrate on lower-energy physics.

Although unified theories have some appealing features, I'm not really sure whether studying them will lead to correct insights into nature. The gap in energy between what we know and what we extrapolate to is huge, and one can imagine many possibilities for what can happen in between. In any case, until proton decay (or some other prediction) is discovered—if it ever is—it will be impossible to establish with certainty whether forces truly unify at high energy. Until then, this theory remains in the realm of grand, but theoretical, speculation.

What to Remember

- *Virtual particles* are particles that have the same charges as true physical particles, but have energies that seem to be wrong.

- Virtual particles exist for only a very short time; they temporarily borrow energy from the *vacuum*—the state of the universe without any particles.

- *Quantum contributions* to physical processes arise from virtual particles that interact with real particles. These contributions from virtual particles influence the interactions of real particles by appearing and disappearing, and acting as intermediaries between the real particles.

- The *anarchic principle* tells us that quantum contributions always have to be taken into account when considering a particle's properties.

- In a *unified theory*, a single high-energy force turns into the three known nongravitational forces at low energies. For the three forces to unify, they must have the same strength at high energies.

12

The Hierarchy Problem: The Only Effective Trickle-Down Theory

The highway is for gamblers, better use your sense.
Take what you have gathered from coincidence.

Bob Dylan

Ike Rushmore III came to an ignominious end when he drove his resplendent new Porsche into a lamppost. He was nonetheless happy in Heaven, where he could play games all the time. He was a gambling man at heart.

One day, God Himself invited Ike to a rather strange game. God told him to write down a sixteen-digit number. God would roll the heavenly icosahedral die. Unlike a normal, cubic die with six sides, this die had twenty sides, with the digits 0 through 9 written twice. God explained that He would throw this die sixteen times and construct a sixteen-digit number by listing the results, one after the other. If God and Ike came up with the same enormous numbers—that is, if all the digits matched in the correct order—God would win. If the numbers weren't exactly the same—that is, if any of the digits failed to match—Ike would defeat God.

God began to roll. The first side that came up was the number 4. This agreed with the first digit of Ike's number, which was 4,715,031,495,526,312. Ike was surprised that God rolled correctly, since the odds were only one in ten. Nonetheless, he was pretty sure the second or third number would be wrong; the odds of God's rolling both numbers correctly in succession was only one in a hundred.

God threw the die for a second and then a third time. He rolled a

7 and then a 1, which were also correct. He kept rolling until, to Ike's astonishment, He had rolled all sixteen digits correctly. The chances of this happening randomly were only 1 in 10,000,000,000,000,000. How could God have won?

Ike was a bit angry (one can't get very angry in Heaven) and asked how something so ridiculously unlikely could have happened. God sagely replied, "I am the only one who could expect to win, since I am both omniscient and omnipotent. However, you must have heard, I do not like to play dice."

And with that, GAMBLING FORBIDDEN was posted on a cloud. Ike was furious (of course, only a little). Not only had he lost the game, he'd also lost the right to gamble.

By this point, you have hopefully learned quite a lot about particle physics and some of the beautiful theoretical elements with which physicists built the Standard Model. The Standard Model works exceptionally well in explaining many diverse experimental results. However, it rests uneasily on an unstable foundation that poses a deep and significant mystery, one whose solution is almost bound to lead to new insights into the underlying structure of matter. In this chapter we'll explore this mystery, known to particle physicists as the *hierarchy problem*.

The problem is not that the Standard Model's predictions disagree with experimental results. The masses and charges associated with the electromagnetic, weak, and strong forces have been tested to incredibly high accuracy. Experiments at the colliders at CERN, SLAC, and Fermilab have all confirmed with exquisite precision the Standard Model predictions for interactions and decay rates of the known particles. And the strengths of the forces in the Standard Model pose no significant mysteries either. Their relationship to one another is in fact highly suggestive, and underlies the idea of a Grand Unified Theory. Furthermore, the Higgs mechanism perfectly explains how the vacuum breaks electroweak symmetry and gives masses to the W and Z gauge bosons, as well as the quarks and the leptons.

However, even the most idyllic-seeming families can reveal undercurrents of tension when investigated more closely. Despite well-coordinated behavior and a happy, harmonious appearance, a devastating hidden family secret can be lurking underneath. The Standard Model has just such a skeleton in the closet. Everything agrees with predictions if you uncritically assume that the strength of the electromagnetic force, the strength of the weak force, and the gauge boson masses take the values that have been measured in experiments. But we'll soon see that the mass parameter (the weak scale mass that determines the elementary particle masses), though very well measured, is ten million billion times, or sixteen orders of magnitude, lower than the mass that physicists would expect from general theoretical considerations. Any physicist who would have guessed the value of the weak scale mass based on a high-energy theory would have gotten it (and therefore all particle masses) completely wrong. The mass seems to come out of thin air. This puzzle—the hierarchy problem—is a gaping hole in our understanding of particle physics.

In the Introduction I explained the hierarchy problem as the question of why gravity is so weak, but we will now see that this problem can be restated as the question of why the Higgs particle's mass, and hence the weak gauge boson masses, are so small. For those masses to take their measured values, the Standard Model has to incorporate a fudge that is as unlikely as someone winning the guessing game against Ike and randomly choosing a sixteen-digit number correctly. Despite its many successes, the Standard Model relies on this unconscionable fudge to accommodate the known elementary particle masses.

This chapter explains this problem, and why I, and most other particle theorists, think it's so important. The hierarchy problem tells us that whatever is responsible for electroweak symmetry breaking is bound to be more interesting than the two-field Higgs example presented in Chapter 10. Possible solutions all involve new physical principles, and the solution will very likely guide physicists to more fundamental particles and laws. Identifying what plays the role of the Higgs field and breaks electroweak symmetry will reveal some of the richest new physics we are likely to nail down in my lifetime. New physical phenomena will almost certainly appear at an energy of about a TeV. Experimental tests of competing hypotheses are near at hand, and within a decade there should be a dramatic revision in our understanding of fundamental physical laws that will incorporate whatever is discovered there.

The hierarchy problem tells us that before extrapolating physics to extremely high energies, we have at least one urgent low-energy problem to attend to. For the last thirty years or so, particle theorists have been searching for the structure that predicts and protects the weak scale energy, the relatively low energy at which electroweak symmetry breaks. I and others think that there must be a solution of the hierarchy problem, and that it will provide one of the best clues about what lies beyond the Standard Model. To understand the motivation for some of the theories I'll soon present, it is useful to know a little about this somewhat technical but very important problem. The search for its solution has already led us to investigate new physical concepts, such as the ones later chapters will explore, and the solution will almost certainly revise our current views.

Before we consider the most general version of the hierarchy problem, we'll first consider the hierarchy problem in the context of a Grand Unified Theory, where the problem was first identified and where it's a little simpler to understand. We'll then look at the problem in its larger (and more pervasive) context and see why it ultimately boils down to the weakness of gravity compared with all the other known forces.

The Hierarchy Problem in a GUT

Imagine that you visit a very tall friend, and discover that although he is 6' 5" tall, his fraternal twin brother is only 4' 11". That would be surprising. You'd expect both your friend and his brother, who should have similar genetic makeups, to be similar in height. Now imagine something even more bizarre: you walk into your friend's house and find that your friend's brother is ten times smaller or ten times bigger. That would be very strange indeed.

We don't think that particles should all have the same properties. But unless there is a good reason, we expect particles that experience similar forces to be somewhat similar. We expect them to have comparable masses, for example. Just as you have good reason to expect similar heights among family members, particle physicists have valid scientific reasons to expect similar masses among particles in a single theory, such as a Grand Unified Theory. But in a GUT the masses are not at all the same: even those particles that experience similar forces must have enormously different masses. And not by a mere factor of ten: the discrepancy between masses is more like a factor of ten trillion.

The problem in a Grand Unified Theory is that although the Higgs particle that breaks electroweak symmetry has to be "light"—with roughly the weak scale mass—a GUT partners the Higgs particle with another particle that interacts through the strong force. And that new particle in the GUT has to be extremely heavy—with a mass of roughly the GUT scale mass. In other words, two particles that are supposed to be related by a symmetry (the GUT force symmetry) have to have enormously different masses.

The two different but related particles must appear together in a GUT because the weak force and the strong force should be interchangeable at high energy. That's the whole idea behind a unified theory—all forces should ultimately be the same. So when the strong and the weak forces are unified, every particle that experiences the weak force, including the Higgs particle, must be partnered with another particle that experiences the strong force and has interactions similar to those of the original Higgs particle. However, there is a big problem with the new Higgs-related particle that experiences the strong force.

The strongly charged particle that is partnered with the Higgs particle can interact simultaneously with a quark and a lepton and thereby enable the proton to decay—even more rapidly than a GUT would otherwise predict. To avoid too rapid a decay, the strongly interacting particle—which must be exchanged between two quarks and two leptons for proton decay to take place—must be extremely heavy. The current limit on the proton lifetime tells us that the strongly charged Higgs partner, if it exists in nature, has to have a mass similar in size to the GUT scale mass, about one million billion GeV. If this particle existed but was *not* this heavy, you and this book would decay before you finished reading this sentence.

However, we already know that the weakly charged Higgs particle has to be light (around 250 GeV) to give the weak gauge boson masses that have been measured in experiments. So experimental constraints tell us that the Higgs particle's mass must be wildly different from the mass of the Higgs partner that experiences the strong force. The strongly charged Higgs particle, which is supposed to have very similar interactions to the weakly charged Higgs particle in a unified theory, must have a very different mass, or else the world would be nothing like what we see. This huge discrepancy between the two masses—one is ten trillion times the other—is very difficult to explain, especially in a unified theory in which both the weakly charged Higgs particle and the strongly charged Higgs particle are supposed to have similar interactions.

In most unified theories, the only way to make one particle heavy and the other one light is to introduce a huge fudge factor. No physical principle predicts that the masses should be so different; a very carefully chosen number is the only way to make things work. That number has to have thirteen digits of accuracy, otherwise either the proton would decay or the weak gauge boson masses would be too large.

Particle physicists call the necessary fudge *fine-tuning*. A fine-tune is when you adjust the parameter to get exactly the value you want. The word "tuning" is used because it is like tweaking a piano string to get precisely the right note. But if you wanted to get a frequency of a few hundred hertz correct to thirteen-digit accuracy, you would have to listen to it for ten billion seconds—a thousand years—to check that it was right. Thirteen-digit accuracy is hard to come by.

I could make other fine-tuning analogies, but I promise you they'll all sound contrived. For example, consider a huge corporation where one person is in charge of expenditure and another is in charge of receipts. Suppose that they never communicate with each other, but at the end of the year the corporation is supposed to have spent almost precisely the amount it took in, with less than a dollar remaining, or else the corporation will fail. Yep, that's a contrived example. And there's a good reason for that. No sensible situations depend on fine-tuning, no one wants their fate (or the fate of their business) to hang on an unlikely coincidence. Yet almost any Grand Unification Theory with a light Higgs particle has such a dependency problem. A theory in which the physical predictions depend so sensitively on a parameter is very unlikely to be the whole story.

But the only way to get a small enough Higgs particle mass in the simplest GUT is to fudge the theory. The GUT model offers no good alternative. This is a serious problem for most models that unify in four dimensions, and many physicists, including myself, are uncertain about unification of forces because of it.

And the hierarchy problem gets even worse. Even if you were willing to simply assume, without any underlying explanation, that one particle is light and the other extremely heavy, you would still run into problems with an effect called *quantum mechanical contributions*, or just *quantum contributions*. These quantum contributions must be added to the classical mass to determine the true, physical mass that the Higgs particle would have in the real world. And those contributions are generally far larger than the few hundred GeV mass that the Higgs particle requires.

Let me warn you that the discussion in the next section about quantum contributions, based as it is on virtual particles and quantum mechanics, is not going to be intuitive. Don't try to imagine a classical analog; what we are about to consider is a purely quantum mechanical effect.

Quantum Contributions to the Higgs Particle's Mass

The previous chapter explained how a particle generally will not travel through space unchallenged. Virtual particles can appear and disappear, and thereby influence the path of the original particle. Quantum mechanics tells us that we always have to add up the contributions to any physical quantity from all such possible paths.

We have already seen that such virtual particles make the strength of forces depend on distance in a way that has been measured and agrees quite well with predictions. The same types of quantum contribution that give energy dependence to the forces also influence the size of masses. But in the case of the mass of the Higgs particle—unlike the strengths of forces—the consequences of virtual particles don't look as if they'll coincide with what experiments require of the theory. They appear to be much too large.

Because the Higgs particle interacts with heavy particles whose mass is as high as the GUT scale mass, some of the paths that the Higgs particle takes involve the vacuum spitting out a virtual heavy particle and its virtual antiparticle, and the Higgs particle temporarily turning into those particles as it travels along (see Figure 61). The heavy particles that pop in and out of the vacuum influence the motion of the Higgs particle. They are the culprits responsible for large quantum contributions.

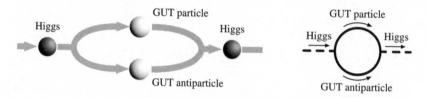

Figure 61. *Virtual contribution to the Higgs particle's mass from heavy particles in a GUT theory. The Higgs particle can convert into virtual heavy (GUT-mass) particles, which then turn back into a Higgs particle. This is illustrated schematically on the left, and with a Feynman diagram on the right.*

Quantum mechanics tells us that if we are to determine the mass that the Higgs particle actually possesses, we have to add such paths with virtual heavy particles to the single path without them. The problem is that the paths containing virtual heavy particles generate contributions to the Higgs particle's mass that are about the same size as the masses of the heavy particles in a GUT—thirteen orders of magnitude larger than the desired mass. All these enormous quantum mechanical contributions from virtual heavy particles must be added to the classical value for the Higgs particle's mass to yield the physical value that would appear in a measurement, which should be about 250 GeV if we want to get the weak gauge boson masses right. That means that, even though any individual GUT mass contribution is thirteen orders of magnitude too large, when we add together all the enormous contributions to the mass, some of which are positive and some of which turn out to be negative, the answer should be approximately 250 GeV. If even a single virtual heavy particle inter-acts with the Higgs particle, there is inevitably a problem.

If, as in the previous chapter, we think of virtual particles as members of a bureaucracy, it's as if the employees are U.S. Immi-gration and Naturalization Service (INS) officers whose job is to delay letters from certain suspect individuals, but they instead scrutin-ize all the letters that pass through. Instead of a two-tier system in which some letters quickly pass through and others are delayed, all the letters are treated the same way. Similarly, the Higgs mechanism requires that the "bureaucracy" of virtual particles should keep some particles heavy but let others, including the Higgs particle, be light. But instead, like the overzealous officers, quantum paths involving virtual particles give comparable contributions to all particle masses. So we would expect all particles, including the Higgs particle, to be as heavy as the GUT mass scale.

Without new physics, the only (and very unsatisfying) way around the problem of the overly large mass of the Higgs particle is to assume that its classical mass takes precisely the value (which could be nega-tive) that would cancel the large quantum contribution to its mass. The parameters in the theory that determine the masses would have to be such that all contributions add up to a very small number, even though each individual contribution is very large. This is the fine-tuning I mentioned in the previous section.

This is conceivable, but extremely unlikely to happen in reality. It is not simply a question of fudging a parameter a little bit to get the mass correct. This fudge is enormous, and enormously precise: anything less than thirteen digits of precision would give dramatic-ally incorrect results. Just to be clear, this bizarre fudge is not the same sort of thing as precisely measuring some quantity, say the speed of light. Ordinarily, qualitative predictions don't depend on a parameter taking any particular value. Only one value will match the precise quantity that is measured, but the world wouldn't be very different had that parameter taken a slightly different value. If Newton's constant of gravitation (which sets the strength of gravity) had a value that was 1% different, nothing would have changed dramatically.

With a Grand Unification Theory, on the other hand, a small change in a parameter is enough to completely ruin the theory's predictions, both quantitative and qualitative. The physical consequences of the value of the Higgs particle's mass that breaks electroweak symmetry depend extraordinarily sensitively on a parameter. For practically all the values of that parameter, the hierarchy between the GUT mass and the weak scale mass wouldn't exist, and structure and life, which rely on this hierarchy, would be impossible. If that parameter were off by as little as 1%, the Higgs particle's mass would be far too large. The weak gauge boson masses, and other particle masses as well, would then all be much larger, and the consequences of the Standard Model would be nothing like what we see.

The Hierarchy Problem of Particle Physics

The last section presented an enormous mystery, the hierarchy problem in a GUT. But the true hierarchy problem is even worse. Although GUTs first alerted physicists to the hierarchy problem, virtual particles will generate overly large contributions to the Higgs particle's mass, even in a theory without GUT-mass particles. Even the Standard Model is suspect.

The problem is that a theory consisting of the Standard Model combined with gravity contains two enormously different energy scales. One is the weak scale energy, the energy at which electroweak symmetry is broken, which is 250 GeV. When particles have energies below that scale, the effects of electroweak symmetry breaking are manifest, and weak gauge bosons and elementary particles have mass.

The other energy is the Planck scale energy, which is sixteen orders of magnitude, or ten million billion times, greater than the weak scale energy: a whopping 10^{19} GeV. The Planck scale energy determines the strength of gravitational interactions: Newton's law says that the strength is *inversely* proportional to the second power of that energy. And because the strength of gravity is small, the Planck scale mass (related to the Planck scale energy by $E = mc^2$) is big. A huge Planck scale mass is equivalent to extremely feeble gravity.

So far, the Planck scale mass hasn't come up in our particle physics discussions because gravity is so weak that for most particle physics calculations it can safely be ignored. But that is precisely the question particle physicists want answered: why is gravity so weak that it can be ignored in particle physics calculations? Another way of phrasing the hierarchy problem is to ask why the Planck scale mass is so huge—why is it ten million billion times higher than the masses relevant to particle physics scales, all of which are less than a few hundred GeV?

To give you a basis for comparison, consider the gravitational attraction between two low-mass particles, such as a pair of electrons. This gravitational attraction is about a hundred million trillion trillion trillion times weaker than the electric repulsion between the electron pair. The two kinds of forces would be comparable only if electrons were heavier than they really are by a factor of ten billion trillion. That's an enormous number—it's comparable to the number of times you could lay the island of Manhattan end to end in the extent of the observable universe.

The Planck scale mass is enormously bigger than the electron's mass and all other particle masses we know of, and that signifies that gravity is very much weaker than the other known forces. But why should there be such a huge discrepancy between the strengths of most forces—or, equivalently, why should the Planck scale mass be so enormous compared with known particle masses?

To particle physicists, the enormous ratio between the Planck scale mass and the weak scale mass, a factor of about ten million billion, is hard to stomach. This ratio is greater than the number of minutes that have passed since the Big Bang; it's about a thousand times the number of marbles you can line up from the Earth to the Sun. It's more than a hundred times the number of pennies in the U.S. budget deficit! Why should two masses that describe the same physical system be so enormously different?

If you are not a particle physicist, this might not sound like a very significant problem in itself, even though those numbers are dramatically big. After all, we can't necessarily explain everything, and two masses just might be different for no very good reason. But the situation is actually far worse than it appears. Not only is there the unexplained enormous mass ratio. In the following section, we'll see that in quantum field theory, any particle that interacts with the Higgs particle can participate in a virtual process that raises the Higgs particle mass to a value as high as the Planck scale mass, 10^{19} GeV .

In fact, if you asked any honest particle physicist who knew gravity's strength but knew nothing about the measured weak gauge boson masses to estimate the Higgs particle's mass using quantum field theory, he would predict a value for the Higgs particle—and hence the weak gauge boson masses—that is ten million billion times too large. That is, he would conclude from his calculation that the ratio between the Planck scale mass and the mass of the Higgs particle (or the weak scale mass, which is determined by the Higgs particle's mass) should be far closer to unity than to ten million billion! His estimate of the weak scale mass would be so close to the Planck scale mass that particles would all be black holes, and particle physics as we know it would not exist. Although he would have no a priori expectation for the value of either the weak scale mass or the Planck scale mass individually, he could use quantum field theory to estimate the ratio— and he would be totally wrong. Clearly, there is an enormous discrepancy here. The next section explains why.

Virtual Energetic Particles

The reason that the Planck scale mass enters quantum field theory calculations is a subtle one. As we have seen, the Planck scale mass determines the strength of the gravitational force. According to Newton's law, the gravitational force is inversely proportional to the value of the Planck scale mass, and the fact that gravity is so weak tells us that the Planck scale mass is huge.

Generally, we can ignore gravity when making predictions in particle physics because the gravitational effects on a particle with mass of about 250 GeV are completely negligible. If you really need to account for gravitational effects you can systematically incorporate them, but it's not usually worth the bother. Later chapters will explain the new and very different scenarios in which higher-dimensional gravity is strong and cannot be neglected. But for the conventional, four-dimensional Standard Model, neglecting gravity is a standard and justifiable practice.

However, the Planck scale mass has another role as well: it is the maximum mass that virtual particles can take in a reliable quantum field theory calculation. If particles carried more mass than the Planck scale mass, the calculation would be untrustworthy, and general relativity would not be reliable and would have to be replaced by a more comprehensive theory, such as string theory.

But when particles (including virtual particles) have mass that is less than the Planck scale mass, conventional quantum field theory should apply and calculations based on quantum field theory should be trustworthy. That means that a calculation involving a virtual top quark (or any other virtual particle) with mass almost as big as the Planck scale mass should be reliable.

The problem for the hierarchy is that the contribution to the Higgs particle's mass from virtual particles with extremely high mass will be about as big as the Planck scale mass, which is ten million billion times greater than the Higgs particle mass we want—the one that will give the right weak scale mass and elementary particle masses.

If we consider a path, such as the one shown in Figure 62, in which the Higgs particle turns into a virtual top quark-antitop quark pair, we can see that the contribution to the Higgs particle's mass will be far too large. In fact, any type of particle that can interact with the Higgs particle might appear as a virtual particle and have mass* up

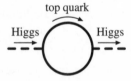

Figure 62. *A contribution to the Higgs particle's mass from a virtual top quark and a virtual antitop quark. The Higgs particle can convert to a virtual top quark and virtual antitop quark, and this gives an enormous contribution to the Higgs particle's mass.*

---

*Remember that virtual particles' masses are not the same as the masses of true physical particles.

to the Planck scale mass. And the result of all these possible paths is huge quantum contributions to the Higgs particle's mass. The Higgs particle has to be much less massive.

Particle physics in its present state is like a too effective "trickle-down" theory. In economics, a hierarchy of wealth is not difficult to achieve. The application of trickle-down economics has never raised poor people's financial well-being much at all, let alone to the level of the upper classes. In physics, though, the transfer of wealth is far too efficient. If one mass is large, then quantum contributions tell us that all masses of elementary particles are expected to be about as large. All particles end up rich in mass. But we know from measurements that high mass (the Planck scale mass) and low mass (particle masses) coexist in our world.

Without modifying or extending the Standard Model, particle physics theory can achieve a small mass for the Higgs particle only through a miraculous value for its classical mass. That value must be extremely large—and possibly negative—so that it can precisely cancel the large quantum contributions. All the mass contributions must add up to 250 GeV.

For this to happen, as in the Grand Unification Theory we considered earlier, the mass must be a fine-tuned parameter. And this fine-tuned parameter would have to be an enormous yet amazingly exact fudge specifically designed to give a small net mass to the Higgs particle. Either the quantum contributions from virtual particles or the classical contribution must be negative, and almost equal in magnitude to the other. The positive and negative terms, each of which is sixteen orders of magnitude too large, must add up to a much smaller value. The required fine-tuning, which must have sixteen-digit accuracy, is more extreme than the fine-tuning required to make your pencil stand on end. It's about as likely as someone randomly winning the guessing game with Ike.

Particle physicists would prefer a model that didn't involve the fine-tuning that is required in the Standard Model to ensure a light Higgs particle. Although we might fine-tune in an act of desperation, we hate it. Fine-tuning is almost certainly a badge of shame reflecting our ignorance. Unlikely things sometimes happen, but rarely when you want them to.

The hierarchy problem is the most urgent of the mysteries confronting the Standard Model. To put a positive spin on things, the hierarchy problem provides a clue to what plays the role of the Higgs particle and breaks the electroweak symmetry.

Any theory that replaces the two-field Higgs theory should naturally accommodate or predict a low electroweak mass scale—otherwise it is just not worth thinking about. Many underlying theories are compatible with the physical phenomena we see, but very few of them address the hierarchy problem and include a light Higgs particle in a compelling manner that avoids fine-tuning. While the task of unifying forces is a fascinating, if potentially tenuous, theoretical lure from high-energy physics, the task of solving the hierarchy is a concrete necessity urging progress at relatively low energies. What makes this challenge most exciting is that anything that addresses the hierarchy problem should have experimental consequences that will be measurable at the Large Hadron Collider, where experimenters expect to find particles with masses of about 250 to 1,000 GeV. Without such additional particles, there is no way to get around the problem. We'll soon see that the experimental consequences of solving the hierarchy problem might be the supersymmetric partners or the particles that travel in extra dimensions that we'll discuss later on.

What to Remember

- Although we know that the Higgs mechanism is responsible for particle masses, the simplest known example that implements the Higgs mechanism works only with a huge fudge. In the simplest theory you'd expect the masses of weak gauge bosons and quarks to be about ten million billion times greater than they are. The *hierarchy problem* is the question of why this is not the case.

- The hierarchy problem arises from the discrepancy between the low weak scale mass and the enormous Planck scale mass (see Figure 63). This latter mass is important for gravity—the large value of the Planck scale mass tells us that gravity is very

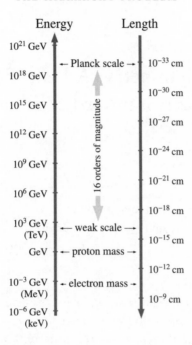

Figure 63. *The hierarchy problem is the question of why the Planck scale energy is so much bigger than the weak scale energy.*

weak. So another way of phrasing the hierarchy problem is to ask why gravity is so feeble, so much weaker than the other nongravitational forces.

- Any theory that solves the hierarchy problem will be experimentally testable because it will necessarily have experimental implications at colliders operating at energies above the weak scale energy. The Large Hadron Collider will explore such energies very soon.

13

Supersymmetry: A Leap Beyond the Standard Model

You were meant for me.
And I was meant for you.

Gene Kelly ("Singing in the Rain")

When Icarus first arrived in Heaven, he was directed to an orientation seminar where the authorities explained the local rules. To his surprise, he learned that right-wing religious groups were essentially correct, and family values were indeed a cornerstone of his new environment. The authorities had long ago established a traditional family structure premised on the separation of generations and the stability of marriages; a top would always marry a bottom, a charmer would always align with a strange bird, and an uptown girl would always marry a downtown cool cat. Everyone, including Ike, was satisfied with the arrangement.

But Ike later learned that the social structure in Heaven had not always been so secure. Originally, dangerous energetic infiltrators had threatened the hierarchical foundation of society. In Heaven, however, most problems can be solved. God had sent everyone a personal guardian angel, and the angels and their charges had heroically worked together to avert the threat to the hierarchy and preserve the ordered society that Ike could now enjoy.

Even so, Heaven was not entirely safe. The angels turned out to be free agents, with no contract binding them to a single generation. The fickle angels, who had so bravely rescued the hierarchy, now threatened to destroy Heaven's family values. Ike was appalled.

Despite Heaven's well-advertised attractions, he was finding it a surprisingly stressful place.

"Super" words abound in physics terminology. We have superconducting, supercooling, supersaturated, superfluid, the Superconducting Supercollider (the SSC)—which would have been the highest-energy collider today had Congress not canceled it in 1993—and the list goes on. So you can imagine the excitement when physicists discovered that spacetime symmetry itself has a bigger, "super" version.

The discovery of *supersymmetry* was truly surprising. At the time when supersymmetric theories were first developed, physicists thought they knew all the symmetries of space and time. Spacetime symmetries are the more familiar symmetries that we saw in Chapter 9, which declare that you can't tell where you are or which way you're facing or what time it is solely from physical laws. The trajectory of a basketball, for example, doesn't depend on which side of the court you're on if you play the game in California or New York.

In 1905, with the arrival of relativity theory, the list of spacetime symmetry transformations expanded to include those that change velocity (speed and direction of motion). But, physicists thought, that capped the list. No one believed that there could be other undiscovered symmetries involving space and time. Two physicists, Jeffrey Mandula and Sidney Coleman, codified this intuition in 1967 by proving that there could be no other such symmetries. However, they (and everyone else) had overlooked one possibility based on unconventional assumptions.

This chapter introduces *supersymmetry*, a strange new symmetry transformation that interchanges bosons and fermions. Physicists can now construct theories that incorporate supersymmetry. However, supersymmetry as a symmetry of nature is still hypothetical, since no one has yet discovered supersymmetry in the world around us. Nonetheless, physicists have two major reasons to think that it might exist in the world:

One reason is the superstring, which will be more thoroughly investigated in the chapter that follows. Superstring theory, which incorporates supersymmetry, is the only known version of string theory that has the potential to reproduce the particles of the Standard Model. String theory without supersymmetry doesn't look as if it could possibly describe our universe.

The second reason is that supersymmetric theories have the potential to solve the hierarchy problem. Supersymmetry doesn't necessarily explain the origin of the large ratio of the weak scale mass to the Planck scale mass, but it does eliminate the problematic enormous quantum contributions to the Higgs particle's mass. The hierarchy problem is a serious conundrum for which very few suggestions have survived experimental and theoretical scrutiny. Before extra-dimensional theories were introduced as potential alternatives, supersymmetry was the lone candidate solution.

Because no one yet knows whether or not supersymmetry exists in the external world, all we can do at this point is evaluate candidate theories and their consequences. This way, when experiments reach higher energy, we'll be prepared to figure out what the physical theory underlying the Standard Model really is. So let's take a look at what could lie in store.

Fermions and Bosons: An Unlikely Match

In a supersymmetric world every known particle is paired with another—its supersymmetric partner, also known as a *superpartner*—with which it is interchanged by a supersymmetry transformation. A supersymmetry transformation turns a fermion into its partner boson and a boson into its partner fermion. We saw in Chapter 6 that fermions and bosons are particle types that are distinguished in quantum mechanical theories by their spin. Fermionic particles have half-integer spin, while bosonic particles have integer spin. Integer spin values are those numbers that ordinary objects spinning in space could take, whereas half-integer values are a peculiar feature of quantum mechanics.

All fermions in a supersymmetric theory can be transformed into their partner bosons and the bosons can all be transformed into their partner fermions. Supersymmetry is a feature of the theoretical description of these particles. If you muck around with the equations that describe how particles behave by making a supersymmetry transformation that interchanges bosons and fermions, the equations will all end up looking the same. The predictions would all be identical to those you made before you did the transformation.

At first glance, such a symmetry defies logic. Symmetry transformations are supposed to leave systems unchanged. But supersymmetry transformations interchange particles that are manifestly different: fermions and bosons.

Although one would not expect a symmetry to mix things that are so different, several groups of physicists nevertheless proved that it could. In the 1970s, European and Russian physicists[*] showed that a symmetry could interchange such different particles, and that the laws of physics could be the same before and after bosons and fermions were interchanged.

This symmetry is a little different from previous symmetries we have considered because the objects that it interchanges clearly have different properties. But the symmetry can nonetheless exist if bosons and fermions are present in equal numbers. As an analogy, imagine an equal number of different-size red marbles and green marbles, with one marble of each color in each size. Suppose you are playing a game with a friend, and you get the red marbles and your friend gets the green ones. If the marbles were exactly paired, neither color choice would give you an advantage. However, if there weren't an equal number of red and green marbles of any given size, it wouldn't be an even playing field. It would matter if you chose red or green, and the game would proceed differently if you and your friend were to switch colors. For there to be a symmetry, every size of marble must come in both red and green, and there must be the same number of marbles of each color of any given size.

---

[*]Pierre Ramond, Julius Wess Bruno Zumino, Sergio Ferrara, and others in Europe; and, independently, Y.A. Gol'fand, E.P. Likhtman, D.V. Volkov, and V.P. Akulov in the Soviet Union.

Similarly, supersymmetry is possible only if bosons and fermions are exactly paired. You need the same number of boson and fermion particle types. And just as the marbles that were interchanged had to have identical sizes, the paired bosons and fermions must have the same mass and charges as each other, and their interactions must be controlled by the same parameters. In other words, each particle must have its own superpartner with similar properties. If a boson experiences strong interactions, so does its supersymmetric partner. If there are interactions involving some number of particles, there are related interactions involving their supersymmetry partners.

One reason physicists find supersymmetry so exciting is that if it *is* discovered in our world, it will be the first new spacetime symmetry to be found in almost a century. That's why it's "super." I won't give the mathematical explanation, but just knowing that supersymmetry exchanges particles of different spin is enough to deduce a connection. Because their spins are different, bosons and fermions transform differently when they rotate in space. Supersymmetry transformations must involve space and time in order to compensate for this distinction.[23]

But don't think that this means you should be able to picture what a single supersymmetry transformation looks like in physical space. Even physicists understand supersymmetry only in terms of its mathematical description and its experimental consequences. And these, as we'll soon see, could be spectacular.

Superhistory

You can skip this if you like. It's a historical section that won't introduce any concepts that will be essential later on. But the development of supersymmetry is an interesting story, in part because it nicely demonstrates the versatility of good ideas and the way string theory and model building sometimes have a productive, symbiotic relationship. String theory motivated the search for supersymmetry, and the superstring—the best string theory candidate for the real world—was identified only because of insights from supergravity, the supersymmetric theory that includes gravity.

The French-born physicist Pierre Ramond put forward the first supersymmetric theory in 1971. He wasn't working with the four dimensions that we (used to) think we live in, but in two: one of space and one of time. Ramond's goal was to find a way to include fermions in string theory. For technical reasons, the original version of string theory contained only bosons, but fermions are essential to any theory that hopes to describe our world.

Ramond's theory contained two-dimensional supersymmetry and evolved into the fermionic string theory he developed with André Neveu and John Schwarz. Ramond's theory was the first supersymmetric theory to appear in the Western world: Gol'fand and Likhtman in the Soviet Union had simultaneously discovered supersymmetry, but their papers were hidden from the West behind the Iron Curtain.

Since four-dimensional quantum field theory was on much more solid footing than string theory, the obvious question was whether supersymmetry is possible in four dimensions. But because supersymmetry is intricately woven into the fabric of spacetime, it was not a straightforward task to generalize from two to four dimensions. In 1973, the German physicist Julius Wess and the Italian-born physicist Bruno Zumino developed a four-dimensional supersymmetric theory. In the Soviet Union, Dmitri Volkov and Vladimir Akulov independently derived another four-dimensional supersymmetric theory, but once again the Cold War forestalled any exchange of ideas.

Once these pioneers had worked out a four-dimensional supersymmetric theory, more physicists paid attention. However, the Wess-Zumino model of 1973 couldn't accommodate all the Standard Model particles; no one yet knew how to add force-carrying gauge bosons to a four-dimensional supersymmetric theory. The Italian theorists Sergio Ferrara and Bruno Zumino solved this difficult problem in 1974.

On a train trip from Cambridge to London, where we had just attended the Strings 2002 conference, Sergio told me how finding the right theory would have been an impossibly difficult problem had it not been for the formalism of *superspace*, an abstract extension of spacetime that has additional *fermionic dimensions*. Superspace is an extremely complicated concept, and I shall not attempt a description of it. The important point here is that this entirely different type of dimension—which is not a dimension of space—played a crucial role in supersymmetry's development. This purely theoretical device continues to simplify supersymmetry calculations today.

The Ferrara-Zumino theory told physicists how to include electromagnetism and the weak and strong forces in a supersymmetric theory. However, supersymmetric theories did not yet include gravity. So the remaining question for a supersymmetric theory of the world was whether it could incorporate this remaining force. In 1976, three physicists, Sergio Ferrara, Dan Freedman, and Peter van Nieuwenhuizen, solved this problem by constructing *supergravity*, a complicated supersymmetric theory that contains gravity and relativity.

The interesting thing is that while supergravity was being formulated, string theory was marching forward independently. In one of the key theoretical developments in string theory, Ferdinando Gliozzi, Joel Scherk, and David Olive discovered a stable string theory as an outgrowth of the fermionic string theory that Ramond, along with Neveu and Schwarz, had developed. Fermionic string theory turned out to contain a type of particle that no one had previously encountered except in supergravity theories. The new particle had identical properties to the supersymmetric partner of the graviton known as the *gravitino*, and this is indeed what it turned out to be.

Because of the concurrent development of supergravity, physicists seized on and pursued this common element of the two theories, and soon realized that supersymmetry was present in fermionic string theory. With that, the superstring was born.

We'll return to string theory and the theory of the superstring in the following chapter. For now, we'll focus on the other important application of supersymmetry: its consequences for particle physics and the hierarchy problem.

The Supersymmetric Extension of the Standard Model

Supersymmetry would be most economical and compelling if it paired known particles with each other. However, for this to be true the Standard Model would have to contain equal numbers of fermions and bosons—but it doesn't satisfy this criterion. That tells us that if our universe is supersymmetric, it must contain many new particles. In fact, it must contain at least twice the number of particles that experimenters have so far observed. All the fermions of the Standard Model—the three generations of quarks and leptons—must be paired with new, as yet undiscovered bosonic superpartners. And the gauge bosons—the particles that communicate the forces—must have superpartners, too.

In a supersymmetric universe, the partners of quarks and leptons would be new bosons. Physicists, who enjoy whimsical (but systematic) nomenclature, call them *squarks* and *sleptons*. In general, the bosonic supersymmetric partner of a fermion has the same name as the fermion, but with an "s" at the beginning. Electrons are paired with *selectrons*, for example, and top quarks with *stop squarks*. Every fermion has a bosonic superpartner, its allied sfermion.

The properties of these particles and their superpartners are rigidly aligned to one another: the bosonic superpartners have the same masses and charges as their fermionic counterparts, and they also have related interactions. For example, if the electron has charge -1, so does the selectron; and if a neutrino interacts via the weak force, so does the sneutrino.

If the universe is supersymmetric, bosons must also have superpartners. The known bosons in the Standard Model are the force carriers: the photon, the charged Ws, the Z, and the gluons, all of which have spin-1. The nomenclature of supersymmetry dictates that the new fermionic superpartners have the same name as the boson with which they are paired with "-ino" tacked on at the end. So the fermionic partners of gauge particles are called *gaugino* particles, the fermionic partners of gluons are *gluinos*, and the fermionic partner of the Higgs particle is a *Higgsino*. As was true for bosonic superpartners, fermionic superpartners have the same charges, the same interactions, and—if supersymmetry is exact—the same mass as the boson with which they are paired (see Figure 64).

You might find it remarkable that physicists take the possibility of supersymmetry as seriously as they do, given that no superpartner has ever been found. I'm sometimes surprised how confident some of my colleagues are about it. But even though supersymmetry has not yet been found in nature, there are several reasons to suspect its presence.

	particle	superpartner
	lepton	slepton
example	electron	selectron
	quark	squark
example	top	stop
	gauge boson	gaugino
examples	photon	photino
	W boson	wino
	Z boson	zino
	gluon	gluino
	graviton	gravitino

Figure 64. *Particles and their supersymmetric partners.*

Sergio Ferrara, one of the first to work on supersymmetry, expressed the view of many physicists when he told me on our train ride to London that it would be hard to believe that such a surprising and fascinating theoretical construction played no role in the physics of the world.

Other physicists, less taken with the beauty of the symmetry, believe in supersymmetry primarily because of the benefits of the supersymmetric extensions of the Standard Model. Unlike non-supersymmetric theories, they protect the light Higgs particle and the hierarchy of masses.

Supersymmetry and the Hierarchy Problem

The hierarchy problem in the Standard Model was the question of why the Higgs particle is so light. How can there be a light Higgs particle when there are large quantum contributions to its mass from virtual particles? These large contributions tell us that the Standard Model works only if it contains an enormous and unfortunate fudge.

The big advantage of a supersymmetric extension of the Standard Model is that when there are virtual contributions from both particles and superpartners, supersymmetry guarantees the absence of the large quantum contributions to the Higgs particle's mass that made a light Higgs particle seem so unlikely. Supersymmetric theories can have only those interactions in which bosonic and fermionic interactions are correlated. And because of the constraints this imposes, super-symmetric theories don't have problems with large quantum contributions to particle masses.

In a supersymmetric theory, the virtual Standard Model particles aren't the only virtual particles that contribute to the Higgs particle's mass. Virtual superpartners do, too. And because of the remarkable properties of supersymmetry, the two kinds of contribution always add up to zero. The quantum contributions of virtual fermions and bosons to the Higgs particle's mass are related so precisely that the large contributions made by either bosons or fermions individually are guaranteed to cancel each other out. The value of the fermions' contribution is negative and exactly cancels the bosons' contribution.

One such cancellation is illustrated in Figure 65, which shows two diagrams, one with a virtual top quark, and one with a virtual stop squark. Each of the individual diagrams would lead to a large contribution to the Higgs particle's mass. But because of the special relationships between particles and interactions in supersymmetric theories, the huge quantum contributions to the mass from the top quarks and the stop squarks are obliterated because they add up to zero.

Figure 65. *In a supersymmetry theory, the Higgs particle's mass gets contributions from both particles and supersymmetry particles (in this case, a virtual top quark and virtual antitop quark in one diagram, and a virtual stop quark and virtual antitop quark in the other). The two diagrams look different because the interactions of the fermions and bosons are different. Nonetheless, the contributions to the Higgs particle's mass from the two diagrams cancel when added together.*

In a non-supersymmetric theory, huge quantum contributions to the mass of the Higgs particle would destroy low-energy electroweak symmetry breaking unless a huge and unlikely fudge made all the large contributions to the particle's mass add up to a very small number. But a supersymmetric extension of the Standard Model guarantees that any potentially destabilizing influences, such as the ones shown in these diagrams, will add up to zero. A small value for the classical mass of the Higgs particle guarantees that the true mass—which includes the quantum contributions—will also be small.

Supersymmetry is like a flexible, stable foundation for the Standard Model. If you imagine the Standard Model's fine-tuning as the balancing required to make a pencil stand on end then supersymmetry is like a fine wire holding the pencil in place. Alternatively, if you think of the hierarchy problem as the INS officers overstepping their jurisdiction and delaying too many letters, supersymmetric partners are like civil liberty advocates who restrain the immigration officers and let most of the letters pass right through.

Because ordinary virtual particles' contributions together with the supersymmetric partners' contributions add up to zero, supersymmetry guarantees that quantum mechanical contributions from virtual particles do not eliminate low-mass particles from the theory. In a supersymmetric theory, a particle that is supposed to be light, such as the Higgs particle, will remain light, even when we take virtual contributions into account.

Broken Supersymmetry

Although supersymmetry potentially resolves the problem of large virtual contributions to the Higgs particle's mass, there is a serious problem with supersymmetry as I have presented it so far. The world is manifestly not supersymmetric. How could it be? If there existed superpartners with identical masses and charges to those of the known particles, they too would already have been seen. Yet no one has discovered a selectron or a photino.

This doesn't mean that we have to abandon the idea of supersymmetry. But it does mean that supersymmetry, should it exist in nature, cannot be an exact symmetry. Like the local symmetry that accompanies the electroweak force, supersymmetry must be broken.

Theoretical reasoning shows that supersymmetry can be broken by particles and their superpartners not having identical masses; small supersymmetry-breaking effects can distinguish them. The difference between a particle's mass and that of its corresponding superpartner would be controlled by the degree to which supersymmetry is broken. If supersymmetry is broken only a little, the mass difference will be small, whereas if it is badly broken, the difference will be large. In fact, the difference in mass between particles and their superpartners is one way to describe how badly supersymmetry is broken.

In almost all models of supersymmetry breaking, the superpartners' masses are greater than the masses of the known particles. This is fortunate, since superpartners being heavier than their Standard Model counterparts is critical to the consistency of supersymmetry with experimental observations. It would explain why we haven't yet seen them. Heavier particles can be produced only at higher energies, and, if supersymmetry exists, colliders have presumably not yet achieved sufficiently high energy to produce them. Because experiments have explored energies up to a few hundred GeV, the fact that superpartners have not yet been seen tells us that if they exist, they must have masses that are at least that big.

The specific mass that a superpartner must exceed to have eluded detection depends on that particular particle's charge and interactions. Stronger interactions make particles easier to produce. So to avoid being detected, particles with stronger interactions must be heavier than more weakly interacting ones. Current experimental constraints on most models of supersymmetry breaking tell us that, should supersymmetry exist, all superpartners must have a mass of at least a few hundred GeV to have escaped detection. Those superpartners that are subject to the strong force, such as the squarks, must be even heavier— with masses of at least a thousand GeV.

Broken Supersymmetry and the Higgs Particle Mass

As we've seen, the quantum contributions to the Higgs particle's mass are not problematic in supersymmetric theories because supersymmetry guarantees that they add up to zero. However, we have also just seen that supersymmetry must be broken if it is to exist in the real world. Because superpartners don't have the same mass as their Standard Model counterparts in a model with broken supersymmetry, the quantum contributions to the Higgs particle's mass are not so rigidly balanced as they are when supersymmetry is exact. So when supersymmetry is broken, virtual contributions no longer cancel exactly.

Nonetheless, so long as the quantum contributions to the Higgs particle's mass are not too large, the Standard Model can get by without fine-tuning or fudging. Even when supersymmetry is broken—so long as the effect is small—the Standard Model can contain a light Higgs particle. Supersymmetry, even if broken a little bit, is sufficiently powerful to eliminate the huge Planck scale mass contributions from virtual energetic particles. With only a small amount of supersymmetry breaking, no exceptionally unlikely cancellations would be necessary.

We want supersymmetry breaking to be small enough to make the supersymmetry-breaking mass difference between superpartners and Standard Model particles sufficiently small to avoid fudging. It turns out that the quantum contribution to the Higgs particle's mass from a virtual particle and its superpartner, though nonzero, will never have a magnitude much greater than the supersymmetry-breaking mass difference between the particle and its superpartner. That tells us that the mass differences between all particles and their superpartners should be about the weak scale mass. In that case the quantum contributions to the Higgs particle's mass would also be about the weak scale mass, which is about the right size for the mass of the Higgs particle.

Because known particles in the Standard Model are light, the mass difference between a superpartner and a Standard Model particle will be comparable to the superpartner's mass. Therefore, if supersymmetry solves the hierarchy problem, the superpartner masses should not be much greater than the weak scale of about 250 GeV.

If the superpartner masses are about the same as the weak scale mass, the quantum contribution to the Higgs particle's mass will not be very large. Unlike the non-supersymmetric case, in which quantum contributions to the Higgs particle's mass were sixteen orders of magnitude too big, so that intolerable fudging was required to maintain a light Higgs particle, a supersymmetric world with super-symmetry-breaking masses of a few hundred GeV would generate no excessively large quantum contributions to the Higgs particle's mass.

The requirement that the Higgs particle, and therefore the super-partners, not be much heavier than a few hundred GeV (so as not to reintroduce large quantum contributions to the Higgs particle's mass), coupled with the fact that experiments have already looked for super-partners with masses of about a couple of hundred GeV, tells us that if supersymmetry exists in nature and solves the hierarchy problem, then supersymmetric partners must have masses that are about a few hundred GeV. This is quite exciting because it suggests that experimental evidence of supersymmetry could be just around the corner and could appear at particle colliders some time very soon. Only a small increase in energy over the existing collider, the Tevatron, should be sufficient to reach the energies at which superpartners would have to appear.

The Large Hadron Collider will explore this energy range. If super-symmetry is not discovered at the LHC, which will search for particles up to a few thousand GeV in mass, it will mean that superpartners are too heavy to solve the hierarchy problem, and the supersymmetry solution will be ruled out.

But if supersymmetry solves the hierarchy problem, it will be an experimental windfall. A particle accelerator that explores energies of about a TeV (1,000 GeV) will find, in addition to the Higgs particle, a host of supersymmetric partners of Standard Model particles. We should see gluinos and squarks, as well as sleptons, winos (pronounced "weenos," not like Bowery bums), a zino, and a photino. The new particles would have all the same charges as Standard Model particles, but would be heavier. With sufficient energy and collisions, these particles would be hard to miss. If supersymmetry is right, we will soon see it confirmed.

Supersymmetry: Weighing the Evidence

This leaves us with the outstanding question: does supersymmetry exist in nature? Well, the jury is still out. Without more facts, any answer is mere conjecture. At the moment both the defense and the prosecution have compelling arguments in their favor.

We have already mentioned two of the strongest reasons to believe in supersymmetry: the hierarchy problem and the superstring. A third compelling piece of evidence in favor of supersymmetry is the potential unification of forces in supersymmetric extensions of the Standard Model. As discussed in Chapter 11, the interaction strengths of the electromagnetic, weak, and strong forces depend on energy. Although Georgi and Glashow originally found that the Standard Model forces unify, better measurements of these three forces showed that unification in the Standard Model doesn't quite work. A plot of the three interaction strengths as a function of energy is presented in the upper graph in Figure 66.

However, supersymmetry introduces many new particles that interact via these three forces. This changes the distance (or energy) dependence of the forces because supersymmetric partners also appear as virtual particles. These additional quantum contributions enter the renormalization group calculation and influence how the interaction strengths of the electromagnetic, weak, and strong forces depend on energy.

The lower graph in Figure 66 shows how the strengths of the forces depend on energy when the effect of virtual superpartners is included. Remarkably, with supersymmetry the three forces appear to unify more precisely than ever. This is more significant than the earlier unification attempts because we now have much better measurements of the interaction strengths. The intersection of three lines could be coincidence. But it might also be taken as evidence in support of supersymmetry.

Another nice feature of supersymmetric theories is that they contain a natural candidate for dark matter. Dark matter is the nonluminous matter that pervades the universe and has been discovered through its gravitational influence. Even though about one-quarter of the energy in the universe is stored in dark matter, we still don't know what it is.* A supersymmetric particle that does not decay and has the right mass and interaction strength would be a suitable dark matter candidate. And indeed, the lightest supersymmetric particle doesn't decay, and could have the right mass and the right interactions to be the particle of which dark matter is composed. This lightest superpartner could be the photino, the partner of the photon. Or, in the extra-dimensional scenario that we'll consider later on, it could be the wino, the partner of the W gauge boson.

However, the case for supersymmetry is not airtight. The strongest argument against supersymmetry is that neither the Higgs particle nor its supersymmetric partners have yet been found. Although the discovery of supersymmetric partners might be imminent, it is not entirely clear why, if supersymmetry solves the hierarchy problem, they haven't already been observed. Experimenters have reached energies of a few hundred GeV. Although superpartners could certainly be a bit heavier, there really is no reason for them to be. In fact, lighter superpartners are better from the perspective of solving the hierarchy problem. Why, if supersymmetry solves the hierarchy problem, haven't superpartners already been found?

On the theoretical side, supersymmetry is not completely compelling because big questions remain about how it is broken. We know that it must be broken spontaneously, but, as in the case of the Standard Model and weak force symmetry, we don't yet know which particles are responsible. Many fascinating ideas have been suggested, but a completely satisfactory four-dimensional theory has yet to be proposed.

When I first learned about supersymmetry, it almost seemed too easy from a model building perspective. It looked as though supersymmetric theories could contain random unrelated masses, since quantum contributions were absent. Even if we didn't know why very disparate masses should appear, they wouldn't cause any trouble. This was very disappointing from a model building perspective because

---

*The universe contains dark energy (energy that is not carried by any matter), that constitutes 70% of the total energy in the universe. Though it might explain dark matter, neither supersymmetry (nor any other theory) explains dark energy.

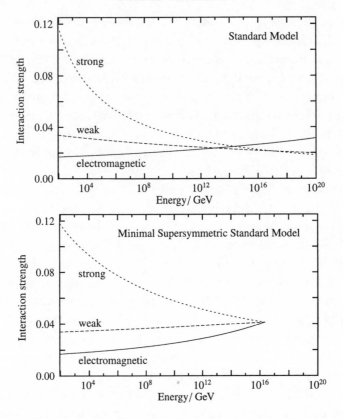

Figure 66. *The upper plot represents the strengths of the electromagnetic, weak, and strong forces as a function of energy in the Standard Model. The curves approach one another, but do not meet in a single point. The lower plot represents the strengths of the same three forces as a function of energy in the supersymmetric extension of the Standard Model. The strength of the three forces is the same at high energy, indicating that the three forces might actually unify into a single force.*

nothing seemed to give any clue about the as yet undetermined underlying theory. And it also was pretty boring, since building models didn't seem to present any challenges.

But then I learned about the supersymmetry *flavor problem*, which tells us that this isn't true; in fact, it's very difficult to make the concrete details of a theory with broken supersymmetry work. The problem is a bit subtle, but it's important nonetheless. The flavor problem is the major obstacle to a simple theory of supersymmetry breaking. All new theories of supersymmetry breaking focus on this problem, and Chapter 17 will show why supersymmetry breaking in extra dimensions is a potential solution.

Recall that the flavors of Standard Model fermions are the three different fermions of the three different generations that have identical charges but different masses: the up, charm, and top quarks, or the electron, muon, and tau, for example. In the Standard Model, the identities of these particles do not change. For example, muons never directly interact with electrons: they interact only indirectly through the exchange of a weak gauge boson. Although muons can decay into electrons, that is only because the decay produces a muon neutrino and an electron antineutrino as well (see Figure 53, p. 175). The muon never converts to an electron directly without the emission of the associated neutrinos.

A physicist's way of expressing this definite identity of a particular lepton type is to say that electron or muon number is conserved. We assign positive electron number to an electron and an electron neutrino, and a negative electron number to a positron and an electron antineutrino. And we assign a positive muon number to a muon and a muon neutrino, and a negative muon number to an antimuon and a muon antineutrino. If muon and electron numbers are preserved, a muon could never decay into an electron and a photon, since we would start with a positive muon number and a zero electron number and end up with a positive electron number and a zero muon number. And in fact, no one has ever seen such a decay. So far as we can tell, electron and muon number are preserved by all particle interactions.

In a supersymmetric theory, electron and muon number conservation would tell us that although an electron and a selectron can interact via the weak force, as can a muon and a smuon, an electron would never interact directly with a smuon. If, for any reason, an electron were paired with a smuon or a muon with a selectron, interactions would be generated that are not seen in nature, such as the muon decaying into an electron and a photon.

The problem is that although such flavor-changing *interactions* don't occur in a truly supersymmetric theory, once supersymmetry is broken, nothing guarantees that muon and electron number remain conserved. Supersymmetry interactions in a theory with broken supersymmetry can change the number of electrons and muons—contradicting what we know from experiments. This is because massive bosonic superpartners do not have the strong sense of identity of their partner fermions. The masses they have in a supersymmetric theory allow the bosonic superpartners to get all mixed up. Not only a smuon, but also a selectron would be paired with a muon, for example. But the pairing of a selectron and a muon would yield all sorts of decays that we know don't occur. In any correct theory of nature, interactions that change muon or electron number would have to be very weak (or nonexistent) since such interactions have never been observed.

The quarks would suffer similar problems. Quark flavor would not be conserved when supersymmetry is broken, and would lead to the dangerous intermingling of generations that Ike feared in the opening story. Some mixing of quarks does happen in nature, but to a far lesser degree than supersymmetry-breaking theories would predict.

Theories of supersymmetry breaking face the formidable challenge of explaining why such flavor-changing interactions do not occur far too often. Unfortunately for supersymmetric theories, most of them cannot explain the absence of flavor-changing effects like these. This is impermissible: such mixing must be forbidden if the theories are to correspond to nature.

If this problem seems obscure to you, you might take some comfort from the fact that many physicists originally felt the same way and also didn't consider the supersymmetry flavor problem to be all that important. To simplify enormously, the split in thinking fell along geographic lines: Europeans didn't care as much as Americans. Those of us who had already spent years thinking about the flavor problem in other contexts knew how difficult it could be to solve. But many others originally ignored the anarchic principle's implications and didn't see why we should worry. In fact, after returning from the International Supersymmetry Conference in Ann Arbor, Michigan, in 1994, David B. Kaplan, a wonderful physicist (and my first collaborator in graduate school) now at the Institute for Nuclear Theory in Seattle, described to me how frustrated he was after he explained his proposed solution to the flavor problem to the audience there, but only afterwards discovered how few people thought there was a problem in the first place!

This all changed rather quickly. Most people now acknowledge the severity of the flavor problem. It is very difficult to find theories of supersymmetry breaking that give all the necessary superpartner masses without compromising particle identities. How to break supersymmetry but prevent flavor changing is a crucial challenge if supersymmetry is to succeed in addressing the hierarchy problem. The loss of muon and electron (and quark) number conservation might sound technical, but it really is the bugaboo of supersymmetry breaking. It is just very difficult to prevent superpartners from turning into each other. Symmetries are in general powerless to prevent it.

So once again we return to our theme: theories with symmetry are elegant, but the broken symmetry that describes the world we see should be equally elegant. How and why is supersymmetry broken? We will have completed the theoretical challenge of understanding supersymmetric theories only when we have a compelling model of supersymmetry breaking.

This is not to say that supersymmetry is necessarily wrong, or even that it has nothing to do with the hierarchy problem. It does, however, mean that an additional ingredient is required for supersymmetric theories of the world to be successful. We'll soon see that the extra ingredient might be extra dimensions.

What to Remember

- *Supersymmetry* essentially doubles the particle spectrum. For every boson in the theory, supersymmetry introduces a partnered fermion, and for every fermion, it introduces a partnered boson.

- Quantum mechanical effects make it difficult (without supersymmetry) to keep the Higgs particle light enough for the Standard Model to work. Until the advent of extra-dimensional theories, supersymmetry was the only known way to address this problem.

- Supersymmetry won't necessarily tell us why the Higgs particle is light, but it does address the hierarchy problem by making a light Higgs particle a plausible assumption.

- The large virtual contributions that Standard Model particles and their superpartners make to the Higgs particle's mass add up to zero. Therefore a light Higgs particle is not problematic in a supersymmetric theory.

- Even though supersymmetry might solve the hierarchy problem, it cannot be exact. If it were, superpartners would have the same masses as Standard Model particles, and we would have already found experimental evidence of supersymmetry.

- *Superpartners*, should they exist, must be more massive than their Standard Model partners. Because high-energy colliders can produce particles only up to a certain mass, these colliders might not yet have had enough energy to produce them. This would explain why we have not yet seen them.

- Once supersymmetry is broken, *flavor-changing interactions* can occur. These are processes that change quarks or leptons into quarks or leptons of another generation (that is, ones that are heavier or lighter) with the same charges. These are very strange processes—they change the identity of known particles, and they occur only rarely in nature. But most theories of broken supersymmetry predict that they should occur very often—more often than we see in experiments.

14

Allegro (Ma Non Troppo)
Passage for Strings

I've got the world on a string.

Frank Sinatra

Fast forward a millennium.

Icarus Rushmore XLII was trying out his new Alicxvr Device, Model 6.3, that he had recently purchased from the Spacernet. (Icarus III's interest in speed and gadgets had apparently been passed down through many generations.) The Alicxvr was designed to let the user view things of any size, from the very small to the very large. Ike was pretty sure that all of his friends who had purchased the Alicxvr Device would first try the large settings, of many megaparsecs, so they could see into outer space beyond the known universe. But Ike thought, "I know just as little about what is happening at extremely tiny distances," and decided to investigate a minuscule size instead.

However, Ike was an impatient sort. He couldn't be bothered to read the extensive instruction manual accompanying his device and instead decided to plunge right in. Blithely ignoring the red indicator overlapping the smallest sizes, he adjusted his dial to the 10^{-33} cm setting and pressed the button labeled "Go."

To his horror, he found himself space-sick in a wildly oscillating, precipitous landscape filled with strings. Space was no longer the smooth, anonymous background he was accustomed to. Instead it was jiggling rapidly in places, heading into pointy sections in others, or wandering off into loops that pinched off or later rejoined the surface. Ike fumbled desperately for the "Stop" button and only just managed to press it in time to return to normal with his senses intact.

After recovering his stability, Ike decided he probably should have read the manual after all. He turned to the "Warning" section and read: "Your new Alicxvr Device Model 6.3 works only for sizes larger than 10^{-33} cm. We have not yet incorporated the latest string theory developments, whose predictions physicists and mathematicians connected to the physical world only last year."

Ike was very disappointed when he realized that only the newer Model 7.0 included the latest results. But Ike then caught up with the most recent string theory developments, souped up his Alicxvr, and never got space-sick again.

Einstein's theory of general relativity was monumental. With it, physicists understood the gravitational field more deeply and could calculate gravity's influence with unprecedented precision. Relativity gave physicists the tools to predict the evolution of all gravitational systems—even that of the entire universe. However, despite all of its successful predictions, general relativity cannot be the final word on gravity. General relativity fails when it is applied to extremely short distances. At very tiny length scales, only a new gravitational paradigm can succeed. Many physicists believe that that paradigm must be string theory.

If string theory is correct, it embraces the successful predictions of general relativity, quantum mechanics, and particle physics. But it also extends physics to distance and energy domains that these other theories are not equipped to handle. String theory is not yet sufficiently developed for us to evaluate its high-energy predictions and validate its efficacy in these elusive distance and energy regimes. But string theory does have several remarkable features that lend credence to this promising picture.

We'll now take a look at string theory and how this dramatic new theory evolved, culminating in the "superstring revolution" of 1984, when physicists demonstrated that pieces of string theory fit together miraculously well. The superstring revolution was only the beginning of an intense research program that actively engages many physicists today. In this and the following chapters, we'll review the history of string theory and some of the recent exciting string theory developments. We'll see that string theory has made remarkable advances and has numerous promising aspects. But we'll also see that string theory faces many crucial challenges that physicists will have to resolve before using it to make predictions about our world.

Incipient Unrest

Quantum mechanics and general relativity peacefully coexist over a wide range of distances, including all those that are accessible to experiments. Although both theories should apply on all length scales, the two theories have a mutual understanding about which of them dominates at measurably long and short distances. Quantum mechanics and general relativity can peacefully share territory because each respects the other's authority in its designated domain. General relativity is important for massive extended objects, such as stars or the galaxy. But gravity's influence on an atom is negligible, so you can safely study an atom ignoring general relativity. Quantum mechanics, on the other hand, is critical at atomic distances because its predictions for an atom are substantial and differ significantly from those of classical physics.

However, quantum mechanics and relativity do not have an entirely harmonious relationship. These two very different theories never adequately negotiated the extremely tiny distance known as the Planck scale length, 10^{-33} cm. From Newton's gravitational force law, we know that the strength of gravity is proportional to masses and inversely proportional to distance squared. Even though on atomic scales, gravity is weak, the gravitational force law tells us that on even tinier scales, the force of gravity is enormous. Gravity is important not only for very massive extended objects, but also for objects that are in extremely close proximity, separated by the minuscule Planck scale length. If we try to make predictions about this unmeasurably small distance, both quantum mechanics and general relativity would contribute significantly, and the two theories' contributions would be incompatible. Neither quantum mechanics nor gravity can be neglected in this contested territory, where quantum mechanical and general relativity calculations fail to cooperate, and predictions are bound to fail.

General relativity works only when there are smooth gravitational fields encoded in a gradually curving spacetime. But quantum mechanics tells us that anything that can probe or influence the Planck scale length has huge momentum uncertainty. A probe with sufficient energy to probe the Planck scale length would induce disruptive dynamical processes, such as energetic eruptions of virtual particles, that would dash any hope of a general relativity description. According to quantum mechanics, at the Planck scale length, instead of a gradually undulating geometry, there should be wild fluctuations and loops and handles of spacetime branching off, the sort of topography that the futuristic Ike encountered. General relativity cannot be used in such untamed territory.

Nor does general relativity step aside to give quantum mechanics free rein, for at the Planck scale length gravity exerts a substantial force. Although gravity is feeble at the particle physics energies we are accustomed to, it is enormously powerful at the high energies required to explore the Planck scale length.* The Planck scale energy—the energy needed to explore the Planck scale length—is exactly the energy at which gravity is no longer dismissible as a feeble force. At the Planck scale length, gravity cannot be ignored.

In fact, at the Planck scale energy, gravity constructs barriers that make conventional quantum mechanical calculations impossible. Anything sufficiently energetic to probe 10^{-33} cm would be snapped up into a black hole that imprisons whatever enters. Only a quantum theory of gravity can tell us what is really going on inside.

At tiny distances, quantum mechanics and gravity cry out for a more fundamental theory. Given the conflict between them, there is no choice but to bring in an external arbiter as an alternative to both. The new regime must allow quantum mechanics and general relativity free rein in their undisputed home territories, but have adequate

---

*Keep in mind that the quantum mechanical relations tell us that while the Planck scale length is minuscule, the Planck scale energy is enormous.

authority to govern the disputed region where neither of the older theories is in control. String theory might be the answer.

The incompatibility of quantum mechanics and gravity also reveals itself through conventional gravity's nonsensical predictions for the high-energy interactions of a particle called the *graviton*—the particle that communicates the gravitational force in a quantum theory of gravity.

According to classical gravitational theory, gravity is communicated between massive objects through a gravitational field in much the same way that, according to Maxwell's classical electromagnetic theory, electromagnetism is communicated from one charged particle to another via a classical electromagnetic field. But quantum electrodynamics (QED), the quantum field theory of electromagnetism, reinterprets this classical electromagnetic force in terms of the exchange of a particle, the photon.* QED, the theory of the photon, is the extension of the classical theory of electromagnetism that incorporates quantum mechanical effects.

Quantum mechanics dictates that, similarly, there must be a particle to transmit the gravitational force. That particle is the graviton. In a quantum theory of gravity, the exchange of a graviton between two objects reproduces Newton's law of gravitational attraction. Although gravitons have not been directly observed, physicists believe that they exist because quantum mechanics tells us they do.

Later on, the distinctive spin of the graviton will be important to us. Because gravitons communicate gravity, a force inherently connected with space and time, they have a different spin from all other known force carriers such as the photon. We will not delve into the reasons here, but the graviton is the only known massless particle whose spin is 2—not 1, as for other gauge bosons, or $\frac{1}{2}$, as for quarks and leptons. The fact that it has spin-2 is important when it comes to finding convincing evidence of extra-dimensional theories. And, as we will soon see, the graviton's spin was also the key to recognizing string theory's potential implications.

However, the quantum field theory description of gravity cannot be complete. No quantum field theory for the graviton can predict its

---

*In actuality it is a virtual photon—not a real physical photon—that is exchanged.

interactions at all energies. When a graviton is as energetic as the Planck scale energy, quantum field theory breaks down. Theoretical reasoning demonstrates that extra graviton interactions that wouldn't make a difference at low energies become important at high energies, but the logic of quantum field theory is not sufficient to tell us what they are or how to include them. If we incorrectly used a quantum field theory of gravity, ignoring the interactions that don't matter at low energies, and attempted to make predictions for extremely energetic gravitons, we would conclude that graviton interactions occur with probability greater than one—something which is clearly impossible. At the Planck scale energy, or equivalently (according to quantum mechanics and special relativity) at the Planck scale length, 10^{-33} cm, the quantum mechanical description of the graviton obviously breaks down.

The Planck scale length, nineteen orders of magnitude smaller than the size of a proton, would be much too small for physicists to care about were it not for the fundamental issues that a more comprehensive theory can potentially address. For example, current theories of cosmology conjecture that the universe began as a tiny ball, a Planck scale length in size. But we have no understanding of the "Bang" of the Big Bang. We understand many aspects of the universe's later evolution, but not how it began. Deducing the physical laws that apply to sizes less than the Planck scale length should shed light on the earliest stage of the evolution of our universe.

Furthermore, there are many mysteries about black holes. Important unresolved questions include what exactly is happening at the black hole's *horizon*, the place of no return beyond which nothing can escape, and at the *singularity*, the place in the center of the black hole where general relativity no longer applies. Another unanswered question is how information about objects that fall into a black hole is stored. Unlike the gravitational force we experience, gravitational effects inside a black hole are strong, as strong as effects from objects with the Planck scale energy in ordinary flat space. We will never solve these black hole mysteries until we resolve the problem of finding a single theory that consistently includes both quantum mechanics and general relativity—a theory of *quantum gravity* on the Planck scale length, 10^{-33} cm. Black holes exemplify some of the questions about strong gravitational effects that will be resolved only by a quantum theory of gravity. String theory is the best known candidate for such a theory.

String Training

String theory's view of the fundamental nature of matter differs significantly from that of traditional particle physics. According to string theory, the most basic indivisible objects underlying all matter are *strings*—vibrating, one-dimensional loops or segments of energy. These strings, unlike violin strings, say, are not made up of atoms which are in turn made up of electrons and nucleons which are in turn made up of quarks. In fact, exactly the opposite is true. These are fundamental strings, which means that everything, including electrons and quarks, consists of their oscillations. According to string theory, the yarn a cat plays with is made of atoms that are ultimately composed of the vibrations of strings.

String theory's radical hypothesis is that particles arise from the resonant oscillation modes of strings. Each and every particle corresponds to the vibrations of an underlying string, and the character of those vibrations determines the particle's properties. Because of the many ways in which strings can vibrate, a single string can give rise to many types of particle. Theorists initially thought there was only a single type of fundamental string that is responsible for all known particles. But that picture has changed in the last few years, and we now believe that string theory can contain different, independent types of string, each of which can oscillate in many possible ways.

Strings extend along a single dimension. At any given time, you need only one number to identify a point along a string, so according to our definition of dimensionality, strings are one-(spatial) dimensional objects. Nonetheless, like real, physical pieces of string, they can curl up and loop around. In fact, there are two types of string: *open strings*, which have two endpoints, and *closed strings*, which are loops with no ends (see Figure 67).

Which particles a string actually produces depends on the string's

Figure 67. *An open string and a closed string.*

energy and on the precise vibrational modes that are excited. Modes of a string are like the resonant modes of a violin string. You can think of the oscillations as elementary units that can be combined to form all known particles. In this language, particles are chords and their interactions are harmonies. The string of string theory doesn't always produce all particles, just as a violin string doesn't produce any sound until someone applies a bow. But just as a bow excites the modes of a violin, energy will excite the modes of a string. And when the string has enough energy, it will produce different particle types.

For both open and closed strings, the resonant modes are those that oscillate an integer number of times along the string's length. A few such modes are depicted in Figure 68. For these modes, the wave oscillates up and down some number of times, with all oscillations completed over the length of the string. For an open string, the wave vibrations hit the end of the string and turn around, going back and forth, whereas waves on closed strings oscillate up and down as they

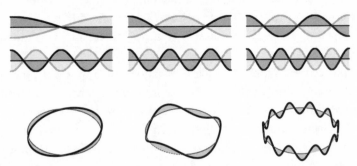

Figure 68. *Some string oscillation modes for (above) an open and (below) a closed string.*

wind around the closed string loop. Any other waves—those that don't complete an integer number of oscillations—won't occur.

Ultimately, the precise way that the string oscillates determines all of a particle's properties, such as its mass, spin, and charge. In general, there will be many copies of particles with the same spin and charge, all with different masses. Because of the infinite number of such modes, a single string can give rise to an infinite number of heavy particles. Known particles, which are relatively light, arise from strings with the fewest oscillations. A mode with no oscillations could be a familiar light particle, such as an ordinary quark or lepton. But an energetic string can oscillate in many ways, so string theory is distinguished by its heavier particles, which arise from higher vibrational modes.

However, more oscillations require more energy. The extra particles from string theory that arise from more oscillations are likely to be extremely heavy—an enormous amount of energy would be required to produce them. So even if string theory is correct, its novel consequences are likely to be extremely difficult to detect. Since we don't expect to produce any of the new heavy particles at accessible energies, we expect string theory and particle physics to give rise to the same observable consequences at the energies we see. This picture might change if some of the recent developments about extra dimensions are correct. But for now, let's review the conventional string theory picture. We'll catch up with extra-dimensional models later on.

String Theory's Origins

By the future Ike XLII's time, string theory could boast quite a long history. But for scientific purposes, we'll confine our story to the twentieth and early twenty-first centuries. We now think of string theory as a theory that might reconcile quantum mechanics and gravity. Originally, however, it had a completely different application. The theory first emerged in 1968 as an attempt to describe the strongly interacting particles known as the hadrons. That theory was not successful; as we saw in Chapter 7, we now know that hadrons are made from quarks held together through the strong force. Nonetheless, string theory survived—not as a theory of hadrons, but as a theory of gravity.

Despite its failure to describe hadrons, we can learn a little about the good features of the string theory of gravity by examining a few of the problems that hadronic string theory faced. Remarkably, the failures of the string theory of hadrons were redeeming features (or at least not obstacles) for the string theory of quantum gravity.

The first problem with the original version of string theory was that it contained a *tachyon*. People initially thought of tachyons as particles traveling faster than the speed of light (the term comes from the Greek *tachos*, meaning "speed"). But we now know that a tachyon indicates an instability in a theory that contains it. Regrettably for science fiction fans, tachyons are not real physical particles that appear in nature. If your theory seems to contain a tachyon, you are analyzing it incorrectly. A system that contains a tachyon can (and will) transform itself into a related system with lower energy in which the tachyon is absent. The system with the tachyon doesn't last long enough for it to have any physical effects; it's only a feature of the incorrect theoretical description. You need to find a theoretical description of the related stable configuration without a tachyon before you can identify true physical particles and forces. Unless it contains such a configuration, your theory is incomplete.

String theory with a tachyon didn't seem to make sense. But no one knew how to formulate the theory in a way that eliminated it. This meant that the predictions from string theory, including those for particles other than the tachyon, were not reliable. You might think that this should have been reason enough to abandon hadronic string theory. But physicists held out hope that the tachyon wasn't real; some thought it might simply be a problem with the mathematical approximations that were made when formulating the theory, but that wasn't very likely.

However, Ramond, Neveu and Schwarz discovered an alternative supersymmetric version of string: the *superstring*. Superstring theory's critically important advantage over the original version of string theory was that it contained spin-$\frac{1}{2}$ particles, giving it the potential to describe the Standard Model fermions such as the electron and the different types of quark. But an added bonus of superstring theory was that it did not contain the tachyon that had plagued the original version of string theory. The superstring, which seemed a more promising theory in any case, didn't have the tachyon instability that would have threatened to hamper its progress.

A second problem for the original string theory of hadrons was that it contained a massless spin-2 particle. Calculations showed that there was no way to eliminate it, but no experimenter had ever discovered this pesky particle. Since experimenters should have been able to observe any massless particle that interacted as strongly as a hadron, hadronic string theory appeared to be in trouble.

Scherk and Schwarz turned string theory on its head when they showed that the "bad" spin-2 particle that confounded hadronic string theory might in fact be the crowning glory of a string theory of gravity; the spin-2 particle could actually be the graviton. They went on to show that the spin-2 particle behaved just as a graviton should. The critical observation that string theory contained a candidate for the graviton made string theory a potential theory of quantum gravity. With a particle description, no one had figured out how to formulate a consistent theory of gravity that worked at all energies. A string theory description, on the other hand, looked like it might do the trick.

There was another indication that although a string theory of hadrons wouldn't work, Scherk and Schwarz might be on the right track with a string theory of gravity. As we saw in Chapter 7, Friedman, Kendall, and Taylor at the Stanford Linear Accelerator Center (SLAC) showed that electrons dramatically scatter from nuclei, implying the existence of hard, pointlike objects—namely quarks—inside. This experiment was similar in spirit to the Rutherford scattering experiment described in Chapter 6. The dramatic scattering results in that case pointed to a hard atomic nucleus, and in this case to pointlike quarks inside the nucleons—not to fluffy, extended strings.

However, the predictions of string theory did not agree with the SLAC experiment's results. Strings would never lead to the dramatic scattering that only a hard, compact object could cause. Because only pieces of the strings would interact at any given time, strings would collide more softly. This quiet, relatively undramatic, scattering was the death knell for the string theory of hadrons. However, from the vantage point of quantum gravity it looked like it could be a very promising property.

In a particle theory of the graviton, the graviton interacts far too strongly at high energies. A better theory would be one in which energetic gravitons don't interact so fiercely. And that is what happens in a string theory of gravity. String theory, which replaces pointlike particles with extended strings, guarantees that the graviton interacts much less dramatically at high energies. Strings—unlike quarks— have no hard scattering processes. They have more "mushy" inter- actions that take place over an extended region.[24] This property means that string theory could potentially solve the problem of the graviton's ridiculously high interaction rate, and correctly predict high-energy graviton interactions. Strings' softer high-energy collisions were another important indicator that a string theory of gravity might be correct.

In summary, superstring theory contains fermions, force-carrying gauge bosons, and the graviton—all the types of particle we know about. It doesn't contain a tachyon. Furthermore, superstring theory includes a graviton whose quantum description potentially makes sense at high energies. String theory looked like it could potentially describe all known forces. It was a promising candidate theory of the world.

The Superstring Revolution

Superstring theory was an extremely bold step, even to solve a problem as deep as quantum gravity. A string theory of gravity predicts an infinitely large number of particles beyond those we know. Moreover, string theory is extremely difficult to analyze with computations. What a steep price to pay for solving the problem of quantum gravity: a theory with infinitely many new particles and a potentially intractable mathematical description. Working on string theory in the 1970s required individuals who were either very determined or somewhat crazy. Scherk and Schwarz were among the very few who negotiated this risky path.

After Scherk's untimely death in 1980, Schwarz persevered with string theory. He collaborated with another (perhaps the only) convert at that time, the British physicist Michael Green, and together they worked out the consequences of the superstring. Schwarz and Green discovered a bizarre feature of the superstring: it makes sense only in ten dimensions, nine of space and one of time. In any other number of dimensions, unacceptable vibrational modes of the string give rise to manifestly nonsensical predictions, such as negative probabilities for processes involving modes of the string that should not exist. In ten dimensions, all the unwanted modes are eliminated. A string theory in any other number of dimensions made no sense.

To clarify, the string itself extends along a single spatial dimension and travels through time. Those were the two dimensions that Ramond had studied when he first discovered supersymmetry. But just as we know that a pointlike object—which has no extent in any spatial dimensions, and therefore has zero spatial dimensions—can move about in three dimensions of space, a string—which has one spatial dimension—can move around in a space with many more dimensions than it itself possesses. Strings could conceivably move around in three, four, or more dimensions. Calculations indicated that the correct number (including time) was ten.

Having too many dimensions was not a novel feature of the super-string. The earlier version of string theory (the one without fermions or supersymmetry) had twenty-six dimensions: one of time and twenty-five of space. But the earlier version of string theory had other problems, like the tachyon. Superstring theory, on the other hand, was sufficiently promising to be worth pursuing.

Even so, string theory was largely ignored until 1984, the year that Green and Schwarz demonstrated a startling feature of the superstring which convinced many other physicists that they were on a promising track. This discovery, together with two other developments that we'll get to soon, are what put string theory into the mainstream of physics.

Green and Schwarz's work addressed a phenomenon known as *anomalies*. As the name suggests, anomalies came as a big surprise when they were first discovered. The first physicists who worked on quantum field theory took for granted that any symmetry of a classical theory would also be preserved by its quantum mechanical extension—the more comprehensive version of the theory that also includes the effects of virtual particles. But that is not always the case. In 1969, Steven Adler, John Bell, and Roman Jackiw showed that even when a classical theory preserves a symmetry, quantum mechanical processes involving virtual particles sometimes violate that symmetry. Such symmetry violations are called *anomalies*, and the theories that contain anomalies are labeled *anomalous*.

Anomalies are extremely relevant to the theories of forces. In Chapter 9 we saw that a successful theory of forces requires the existence of an internal symmetry. These symmetries must be exact, or else there's no way to eliminate the unwanted polarization of the gauge boson, and the theory of forces will then make no sense. The symmetry associated with a force must therefore be *anomaly-free*—the sum of all symmetry-breaking effects must be zero.

This is a powerful constraint on any quantum theory of forces. For example, we now know that it is one of the most compelling explanations for the existence of both quarks and leptons in the Standard Model. Individually, virtual quarks and leptons would lead to anomalous quantum contributions that would break the symmetries of the Standard Model. However, the sum of the quantum contributions from the quarks and the leptons adds up to zero. This miraculous cancellation is what makes the Standard Model hold together; both leptons and quarks are necessary if the forces of the Standard Model are to make sense.

Anomalies were potentially a problem for string theory, which, after all, includes forces. In 1983, when the theorists Luis Alvarez-Gaume and Edward Witten showed that such anomalies occur not only in quantum field theory but also in string theory, it looked as if this discovery would consign string theory to the annals of interesting but overly far-reaching ideas. String theory didn't seem as if it would preserve the requisite symmetries. In the skeptical environment created by string theory's potential for anomalies, Green and Schwarz made quite a splash when they showed that string theory could satisfy the constraints that were needed to avoid anomalies. They computed the quantum contribution to all possible anomalies and showed that for particular forces, anomalies would miraculously add up to zero.

One of the things that made Green and Schwarz's result so surprising is that string theory allows many worrisome quantum mechanical processes, each of which looks as if it could create symmetry-breaking anomalies. But Green and Schwarz showed that the sum of the quantum mechanical contributions to all these possible symmetry-breaking anomalies in ten-dimensional superstring theory is zero. This meant that the many cancellations that were required in string theory calculations actually occur, and, furthermore, that the cancellations happen in ten dimensions, the number of dimensions that was already known to be special for superstring theory. This discovery was sufficiently miraculous for many physicists to decide that such conspiracies could not be coincidental. Anomaly cancellation was a powerful argument in favor of the ten-dimensional superstring.

Furthermore, Green and Schwarz completed their work at a felicitous moment. Physicists had been searching unsuccessfully for theories that could extend the Standard Model to incorporate supersymmetry and gravity, and they were ready to consider something new. They could not ignore Green and Schwarz's discovery of a supersymmetric theory that could potentially reproduce all the particles and forces of the Standard Model. Even though the additional structure of string theory was a nuisance, the superstring had succeeded where other potentially more economical theories had failed.

Two further significant developments soon ensured string theory's inclusion in the physics canon. One was from the Princeton collaboration of David Gross, Jeff Harvey, Emil Martinec, and Ryan Rohm, who in 1985 derived a theory that they named the *heterotic string*. This word is derived from the word "heterosis," which in botany means "hybrid vigor," a term used to refer to hybrid organisms with properties superior to those of their progenitors. In string theory, a vibrational mode can move either clockwise or counterclockwise along the string. The name "heterotic" was used because waves moving to the left were treated differently from those waves moving to the right, and consequently the theory included more interesting forces than did the versions of string theory that were already known.

The discovery of the heterotic string was further confirmation that the forces that Green and Schwarz had discovered to be anomaly-free and acceptable in ten dimensions were truly special. They had found several sets of forces, including all of those that had already been shown to be possible in string theory, as well as another set of forces that had never before been discovered (theoretically) to be part of string theory. The forces of the heterotic string were precisely the new ones that Green and Schwarz had shown were free from anomalies. With the heterotic string, this additional set of forces, which could include those of the Standard Model, was shown not only to be a true string theory possibility, but one that could be realized explicitly. Physicists considered the heterotic string a real breakthrough in the attempt to relate string theory to the Standard Model.

There was one final development that cemented string theory's prominence. This discovery dealt with the extra dimensions essential to the superstring. It is all well and good to show that superstring theory is internally consistent and embodies the forces of the Standard Model, but this is not very interesting if you are stuck with the wrong number of dimensions of space. Superstring theory stipulates ten dimensions. The world around us appears to contain only four (including time). Something needs to be done about the superfluous six.

Physicists now think that one answer might be compactification—rolled-up dimensions of an imperceptibly small size, as described in Chapter 2. At first, however, this curling up of extra dimensions didn't seem the right way to treat the extra dimensions of string theory. The problem was that theories with rolled-up dimensions could not reproduce the important (and surprising) feature of the weak force discussed in Chapter 7: the weak force treats left- and right-handed particles differently. This is not a mere technical detail. The entire structure of the Standard Model relies on left-handed particles being the only ones that experience the weak force. Otherwise, few predictions of the Standard Model would work.

Although ten-dimensional string theory could treat left- and right-handed particles differently, it appeared that this would no longer be true once the six extra dimensions were rolled up. The resulting four-dimensional effective theory always contained neatly matched pairs of left- and right-handed particles. All of the forces that acted on left-handed fermions also acted on right-handed ones, and vice versa. If string theory could not find a way out of this impasse, it would have to be scrapped.

In 1985, Philip Candelas, Gary Horowitz, Andy Strominger, and Edward Witten recognized the significance of a more subtle and complicated way to curl up the extra dimensions, namely a compactification known as *Calabi-Yau manifolds.* The details are complicated, but basically Calabi-Yau manifolds leave a four-dimensional theory that can distinguish left from right and potentially produce the particles and forces of the Standard Model, including the parity-violating weak force. Furthermore, rolling up the extra dimensions into a Calabi-Yau manifold preserves supersymmetry.* With the Calabi-Yau breakthrough, superstring theory was in business.

In many physics departments, superstring theory superseded particle physics, and the superstring revolution was more like a coup. Because superstring theory incorporates quantum gravity and could contain the known particles and forces, many physicists went so far as to think of it as the ultimate theory that underlies everything. Indeed, in the 1980s string theory was dubbed the "Theory of Everything" (or "TOE"). String theory was more ambitious even than Grand Unified Theories: with string theory, physicists hoped to unify all forces (including gravity) at an energy higher even than the energy associated with GUTs. Even without any observations that supported string theory, many physicists decided that string theory's potential for reconciling quantum mechanics and gravity was reason enough to support its claim to prominence.

The Endurance of the Old Regime

If string theorists are right, and the world is ultimately composed of fundamental oscillating strings, must all of particle physics then be abandoned? The answer is a resounding "No." The goal of string

---

*In fact, compactification on a Calabi-Yau manifold preserved just the right amount of supersymmetry for the theory to reproduce features of the Standard Model. Too much supersymmetry, and you couldn't have left-handed particles that had different interactions from right-handed ones.

theory is to reconcile quantum mechanics and gravity at distances smaller than the Planck scale length, where we believe that a new theory takes over. Therefore, in conventional string theory (as opposed to the variants suggested by extra-dimensional models), a string should be about the Planck scale length in size. That tells us that in conventional string theory, the differences between particle physics and string theory should appear only at this tiny Planck scale length or, equivalently, at the ultra-high Planck scale energy, where gravity is expected to be strong. This size is so tiny, and this energy so high, that strings would in no way obviate the particle description at experimentally accessible energies.

For energies below the Planck scale energy a particle physics description is in fact quite adequate. If a string is so small that its length is undetectable, the string might as well be a particle; no experiment could tell the difference. Particles and Planck-length strings are indistinguishable. The string's one-dimensional extent is just as invisible to us as the tiny curled-up extra dimensions we considered earlier. Unless we have instruments that can handle sizes of order 10^{-33} cm, such a string is much too small to see.

It makes sense that string theory and particle physics look the same at achievable energies. The uncertainty principle tells us that the only way to study small distances is with high-momentum particles, which are very energetic. Therefore, without sufficient energy, you have no way of seeing that the string is long and skinny, rather than point-like.

In principle, we could find evidence to support string theory by searching for the many new particles it predicts—the particles that correspond to the many possible oscillations of the string. The problem with this strategy is that most new string-induced particles would be extremely heavy, with a mass as big as the Planck scale mass, 10^{19} GeV. This mass is huge compared with the mass of particles that have been detected experimentally, the heaviest of which is about 200 GeV.

The extra particles that would arise from the oscillations of the string would be so heavy because the string's *tension*—its resistance to stretching that determines how readily a string will oscillate and produce heavy particles—would be large. The Planck scale energy determines the strings' tension; this tension is required for string theory to reproduce the correct interaction strength for the graviton, and hence for gravity itself.[25] The higher the string tension, the more energy is required to generate oscillations (just as it's harder to pluck, or displace, a tight bowstring than a loose one). And this large energy translates into a large mass for the extra string-derived particles. These Planck-mass particles are too massive to be produced at any particle experiment operating today (or, most probably, in the future).

So, even if string theory is correct, we are unlikely to find the many additional heavy particles it predicts. The energy of current experiments is sixteen orders of magnitude too low. Because the extra particles are so extraordinarily heavy, the prospects for discovering evidence of strings experimentally is very poor, with the possible exception of the extra-dimensional models I'll discuss later on.

In most string theory scenarios however, because the string length is so tiny and the string tension is so high, we won't see any evidence to support string theory at the energies achievable in accelerators, even if the string description is correct. Particle physicists who are interested in predicting experimental results can safely apply conventional four-dimensional quantum field theory, ignore string theory, and still get the correct results. As long as you look only at sizes greater than 10^{-33} cm, (or, equivalently, energies below 10^{19} GeV), nothing we have considered earlier about the low-energy consequences of particle physics would change. Given that the size of a proton is about 10^{-13} cm and that the maximum energy reach of current accelerators is about a thousand GeV, it's a pretty safe bet that particle-theory predictions will suffice.

Even so, particle physicists who concentrate on low-energy phenomena have good reasons to pay attention to string theory. String theory introduces new ideas, both mathematical and physical, that no one would otherwise have considered, such as branes and other extra-dimensional notions. Even in four dimensions, string theory has paved the way to an improved understanding of supersymmetry, quantum field theory, and the forces a quantum field theory model might contain. And of course, if string theory does give a fully consistent quantum mechanical description of gravity, that would be a formidable achievement. These benefits make string theory very worthwhile, even to those exclusively concerned with experimentally accessible phenomena. Although it will be very difficult (if not impossible) to detect strings, the theoretical ideas illuminated by string theory might be pertinent to our world. We'll soon see what some of these might be.

Aftermath of the Revolution

In 1984, at the height of the "superstring revolution," I was a graduate student at Harvard. It rapidly became apparent that in research, a beginning physicist had two choices. She could adopt string theory, following in the footsteps of Ed Witten and David Gross, who were then both at Princeton. Or she could remain a particle physicist with more immediate contact with experimental results, in the school of Howard Georgi and Sheldon Glashow, both then at Harvard. It might seem incredible that physicists interested in the same problems could have been so divided, but the notions of how to make progress were very different in the two camps.

The excitement at Harvard remained with particle physics, and many physicists there largely dismissed string theory. A number of problems in particle physics and cosmology remained unsolved— why not answer these questions before delving into the mathematical minefield that string theory was threatening to become? Was it acceptable for physics to extend into unmeasurable domains? With the many brilliant people and many exciting ideas about how to go beyond the Standard Model of particle physics using more traditional methods, there was not much motivation to jump ship.

Elsewhere, however, physicists were convinced that all of the questions about superstring theory would soon be solved, and that string theory was the physics of the future (and of the present). Superstring theory was in its early stages. Some believed that with enough man-hours devoted to it (and they were primarily man-hours), string theorists would ultimately derive known physics. In the 1985 paper about the heterotic string, Gross and his colleagues wrote, "Although much work remains to be done there seem to be no insuperable obstacles to deriving all of known physics from the . . . heterotic string."* String theory promised to be the Theory of Everything. Princeton was in the vanguard of this effort. Physicists there were so certain that string theory was the road to the future that the department no longer contained any particle theorists who didn't work on string theory—a mistake that Princeton has yet to correct.

Today, we can't say whether or not the obstacles facing the theory are "insuperable," but they are certainly challenging. Many major unanswered questions remain. Addressing the unresolved problems of string theory appears to require a mathematical apparatus or a fundamental new approach that goes well beyond the tools that physicists and mathematicians have so far developed.

Joe Polchinski, in his widely used string theory textbook, writes that "string theory may resemble the real world in its broad outline,"† and so it does in some respects. String theory can include the particles and forces of the Standard Model, and can be reduced to four dimensions when other dimensions are curled up. However, although there is tantalizing evidence that string theory could incorporate the Standard Model, the program for finding the ideal Standard Model candidate is nowhere near completion after twenty years of searching.

Physicists initially hoped that string theory would make a unique prediction for what the world should be like, one that would be borne out by the world that we see. But there are now many possible models that can arise in string theory, each containing different forces, different dimensions, and different combinations of particles. We want to find the set that corresponds to the visible universe and the reason that this set is special. Right now, no one knows how to choose among the possibilities. And in any case, none of them look quite right.

For example, Calabi-Yau compactification can determine the number of generations of elementary particles. One possibility is indeed the three generations of the Standard Model. But there is not a unique Calabi-Yau candidate. Although string theorists originally

---

*D. Gross, J. Harvey, E. Martinec, and R. Rohm, "Heterotic string theory (I): The free heterotic string," *Nuclear Physics* B, vol. 256, pp. 253–84 (1985).
†Joseph Polchinski, *String Theory, Vol. 1: An Introduction to the Bosonic String* (Cambridge: Cambridge University Press, 1998).

hoped that Calabi-Yau compactification would single out a preferred shape and unique physical laws, they were quickly disappointed. Andy Strominger described to me how within a week of discovering a Calabi-Yau compactification and thinking it was unique, his collaborator Gary Horowitz found several other candidates. Andy later learned from Yau that there were tens of thousands of Calabi-Yau candidates. We now know that string theories based on Calabi-Yau compactification can contain hundreds of generations. Which Calabi-Yau compactification, if any, is correct? And why? Even though we know that some of string theory's dimensions must curl up or otherwise disappear, string theorists have yet to determine the principles that tell us the size and shape of the curled-up dimensions.

Moreover, in addition to the new heavy string particles arising from waves that oscillate many times along the string, string theory contains new low-mass particles. And we would expect that if they existed and were as light as string theory naively predicts, those particles would be visible to experiments in our world. Most models based on string theory contain many more light particles and forces than we observe at low energies, and it is not clear what singles out the right ones.

Getting string theory to match the real world is an enormously complicated problem. We have yet to learn why the gravity, particles, and forces derived from string theory should agree with what we already know to be true in our world. But these problems with particles, forces, and dimensions pale in comparison with the real elephant in the room—the gross overestimate for the energy density of the universe.

Even in the absence of particles, the universe can carry energy known as vacuum energy. According to general relativity, this energy has a physical consequence: it stretches or shrinks space. Positive vacuum energy accelerates the expansion of the universe, while negative energy makes it collapse. Einstein first proposed such an energy in 1917 in order to find a static solution to his equations of general relativity in which the gravitational effect of the vacuum energy would cancel that of matter. Although he had to abandon this idea for many reasons, including Edwin Hubble's observed expansion of the universe in 1929, there is no theoretical reason that such vacuum energy should not exist in our universe.

Indeed, astronomers have recently measured the vacuum energy in our cosmos (it also known as *dark energy* or the *cosmological constant*) and found a small positive value. They have observed that distant supernovas are dimmer than you'd expect unless they were accelerating away. The supernova measurements and the detailed observations of relic photons created during the Big Bang tell us that the expansion of the universe is accelerating, which is evidence that the vacuum energy has a small positive value.

This measurement is very exciting. But it also introduces a significant puzzle. The acceleration is very slow, which tells us that the value of the vacuum energy, though nonzero, is extremely tiny. The theoretical problem with the observed vacuum energy is that it is far smaller than anyone would estimate. According to string theory estimates, the energy should be much bigger. But if it were, this energy wouldn't just lead to the hard-to-measure supernova acceleration. If the vacuum energy were big, the universe would long ago have collapsed (if negative) or quickly expanded away to nothing (if positive).

String theory has yet to explain why the universe's vacuum energy is as small as we know it must be. Particle physics has no answer to this problem either. However, unlike string theory, particle physics does not purport to be a theory of quantum gravity—it's less ambitious. A particle physics model that cannot explain the energy is unsatisfying, but a string theory that gets the energy wrong is ruled out.

The question of why the energy density is so extraordinarily tiny is an entirely unsolved problem. Some physicists believe that there is no true explanation. Although string theory is a single theory with a single parameter—the tension of the plucked string—string theorists cannot yet use it to predict most features of the universe. Most physical theories contain physical principles which allow you to decide which of the many possible physical configurations a theory would actually predict. For example, most systems will settle down into the configuration that has the lowest energy. But that criterion doesn't seem to work for string theory, which looks as if it might give rise to an infinite number of different configurations that don't have the same vacuum energy—and we don't know which of them, if any, is preferred.

Some string theorists no longer try to find a unique theory. They look at the possible sizes and shapes of rolled-up dimensions and the different options for the energy a universe could contain, and conclude that string theory can only delineate a landscape that describes the huge number of possible universes in which we might live. These string theorists don't think that string uniquely predicts the vacuum energy. They believe that the cosmos houses many different disconnected regions with different values of the vacuum energy, and we live in the portion of the cosmos that contains the right one. Of the many possible universes, only the one that can give rise to structure could (and does) contain us. Those physicists think that we live in a universe with such an incredibly unlikely value for the vacuum energy because any larger value would have prevented the formation of galaxies and structure in the universe—and hence prevented us.

This reasoning has a name: the anthropic principle. The anthropic principle diverges substantially from the original string theory goal of predicting all the features of the universe. It says that we don't have to explain the small energy. Disconnected universes with many possible values of the vacuum energy exist, but we live in one of the few where structure can form. The value of the energy in this universe is ridiculously small, and only exceptional versions of string theory would predict this minuscule value, but we could exist only in a universe with minuscule energy. This principle might be discredited by future advances, or it could be vindicated by more thorough investigations. Unfortunately, however, it will be difficult (if not impossible) to test. A world in which the anthropic principle is the answer would certainly be a disappointing and not very satisfying scenario.

In any case, string theory in its current state of development certainly does not predict the features of the world, even though it is a single theory in its underlying formulation. Once again, we are faced with the question of how to connect a beautiful symmetric theory to the physical realities of our universe. The simplest formulation of the theory is too symmetric: many dimensions and many particles and forces that we know must be different appear to be on the same footing. And to make the connection to the Standard Model, and the world we see, this huge order must be disturbed. After symmetry breaking, the single string theory can manifest itself in many different guises, according to which of the symmetries get broken, which particles become heavy, and which dimensions distinguish themselves.

It is as if string theory is a beautifully designed suit that doesn't quite fit. In its current state, you can hang it on a rack and admire its fine stitching and intricately woven pattern—it really is beautiful—but you can't wear it until you make the necessary adjustments. We'd like string theory to accommodate everything we know about the world. But "one size fits all" rarely looks good on anybody. Right now, we don't even know whether we have the right tools to tailor string theory correctly.

Since we don't really know all of the theory's implications, and it is not clear that we ever will, some physicists simply define string theory as whatever resolves the paradox of quantum mechanics and general relativity at small distances. Certainly most string theorists believe that string theory and the correct theory are the same, or at least very closely connected.

There's clearly a lot left to learn. It is still too early to decide the ultimate merits of a string theory description of the world. Perhaps more elaborate mathematical machinery will permit physicists to truly understand string theory, or perhaps physical insights garnered from applying string theory's implications to the surrounding universe will provide the critical clues. Addressing the unresolved problems of string theory appears to require a fundamentally new approach that goes well beyond the tools that mathematicians and physicists have so far developed.

Nonetheless, string theory is a remarkable theory. It has already led to important insights into gravity, dimensions, and quantum field theory and it's the best candidate we know of for a consistent theory of quantum gravity. Furthermore, string theory has led to incredibly beautiful mathematical advances. But string theorists have yet to make good on the promises they made in the 1980s to connect string theory to the world. We still don't know most of string theory's implications.

In fairness, questions in particle physics were not immediately answered either. Many of the particle physics problems that were known in the 1980s have still not been solved. These questions include an explanation of the origin of the disparate masses for the elementary particles and determining the correct solution to the hierarchy problem. Moreover, model builders are still waiting for the experimental clues that tell us which of the myriad possibilities correctly describe physics beyond the Standard Model. Until we explore energies higher than a TeV, we are unlikely to know with certainty the answers to the questions we care most about.

Today, both the string theory and the particle physics communities have a more sober view of their level of understanding than they did in the 1980s. We are trying to address difficult questions, and they will take time to answer. But this is an exciting time, and despite (or perhaps because of) the many unsolved problems, there is good reason to be optimistic. Physicists now have a better grasp of many consequences of both particle physics and string theory, and open-minded physicists today stand to profit from the achievements of both schools. That is the middle ground that some physics colleagues and I prefer— and it has led to many of the exciting results that we will shortly encounter.

What to Remember

- The *graviton* is the particle that communicates the gravitational force, much as the photon communicates the electromagnetic force.

- According to string theory, the fundamental objects of the world are *strings*, not pointlike particles.

- Later models of extra dimensions won't explicitly use string theory; at distances greater than the minuscule Planck scale length (10^{-33} cm), particle physics suffices.

- Nonetheless, string theory is important to particle physics, even at low energies, because of the new concepts and analytic tools it introduces.

15

Supporting Passages:
Brane Development

Insane in the membrane
Insane in the brain.   Cypress Hill

Ike Rushmore XLII decided to dive down once again to the minuscule Planck scale. Happily, his souped-up Alicxvr worked perfectly and he smoothly arrived in a ten-dimensional universe filled with strings. Eager to explore his new environment, Ike cranked up the hyperdrive attachment he had just purchased from Gbay. He watched with fascination as strings collided and tangled in mesmerizing ways.

Although Ike worried that the Alicxvr might break down, he was curious to learn more about this novel world. So he increased the pressure on the hyperdrive lever. At first strings collided together even more frequently. But when he cranked up the lever still more, he entered a new, completely unrecognizable environment. Ike couldn't even tell whether spacetime was intact. But he kept cranking up the hyperdrive, and, strangely enough, emerged unscathed.[26]

However, his surroundings were now quite different. Ike was no longer in the ten-dimensional universe he had started off in. He was instead in an eleven-dimensional universe filled with particles and branes. And, odd as it seemed, nothing in this new universe interacted very much. When Ike looked back at his controls, he discovered that the hyperdrive lever had mysteriously been reset to low. Confused and rather exasperated, Ike cranked up the lever once again, only to find himself back where he started. When Ike checked the controls, he discovered the hyperdrive lever was once again back at low.

Ike thought his Alicxvr was probably malfunctioning. But when he checked his up-to-date manual he discovered that his device was operating perfectly—high hyperdrive in ten-dimensional string theory was the same as low hyperdrive in an alternate eleven-dimensional world. And vice versa.

The manual didn't say what should happen when the hyperdrive wasn't very low or very high, so Ike entered the spacernet and put himself on the wait-list for an improved version that would solve the problem. But the Alicxvr designers promised only that the release date would be some time within the millennium.

In today's physics world, you might say that "string theory" is a misnomer. In fact, the theorist Michael Duff facetiously refers to "string theory" as "the theory formerly known as strings." String theory is no longer just the theory of strings extending in one spatial direction, but also the theory of branes that can extend in two, three, or more dimensions.[27] We now know that branes, which can extend in any number of dimensions up to the number that superstring theory contains, are just as much a part of superstring theory as are strings themselves. Theorists ignored them earlier on because they studied strings when the string interaction strength "lever" was low and brane interactions were less important. Branes turned out to be the missing piece that miraculously completed several jigsaw puzzles.

In this chapter, I'll describe the evolution of branes from an amusing, neglected curiosity into a central player in the string theory story. We will see several ways in which branes helped to resolve some bewildering aspects of string theory since the mid-1990s. Branes helped physicists to understand the origin of mysterious particles in string theory that couldn't possibly arise from strings. And when physicists included branes, they discovered *dual theories*—pairs of theories that seem very different from each other but have the same physical consequences. The opening story refers to one remarkable example of duality that this chapter will explore: an equivalence between ten-dimensional superstring theory and eleven-dimensional supergravity, which is a theory that contains branes but no strings.

This chapter will also introduce *M-theory*, an eleven-dimensional theory that embraces both superstring theory and eleven-dimensional supergravity, and whose existence was inferred using the insights from branes. No one really knows what the "M" stands for—the term's originator, Edward Witten, deliberately left it ambiguous—but suggestions have included "membrane," "magic," and "mystery." At this point, I'll just say that M-theory is still a "Missing theory" which is postulated but not fully understood. However, even though M-theory still leaves many questions unanswered, the advances made with branes revealed theoretical connections that called for M-theory's more complex, more enveloping structure. That is why string theorists study it today.

This chapter updates the string theory picture that began in the 1980s, presenting some aspects of the more modern viewpoint that physicists developed in the 1990s. Much of this material will not be central to branes' applications to particle physics, and later brane-world conjectures won't explicitly rely on any of the phenomena described below. You should therefore feel free to skip ahead if you choose. But if you like, take this opportunity to get acquainted with some of the remarkable developments in string theory that were in large part responsible for placing branes squarely on string theory's theoretical map.

Nascent Branes

In Chapter 3 we saw that branes extend over some, but not necessarily all, of space's dimensions. For example, a brane might extend only over three dimensions of space, even if the bulk space contains many more. Extra dimensions might terminate on branes; in other words, branes can bound extra-dimensional space. We also know that a brane can house particles that move only along its dimensions. Even if there were many additional spatial dimensions, particles confined to a brane would move only along the more limited region occupied by that brane; they wouldn't be free to explore the full extra-dimensional bulk.

We will now see that branes are more than just a location; they are objects in their own right. Branes are like membranes, and, like membranes, they are real things. Branes can be slack, in which case they can wiggle and move, or they can be taut, in which case they will probably sit still. And branes can carry charges and interact via forces. Furthermore, branes influence how strings and other objects behave. All these properties tell us that branes are essential to string theory; any consistent string theory formulation must include branes.

In 1989, Jin Dai, Rob Leigh, and Joe Polchinski, all then at the University of Texas, and independently the Czech physicist Petr Hořava, mathematically discovered a particular type of brane called a D-brane in the equations of string theory. Whereas closed strings loop around, open strings have two free ends. These ends have to be somewhere, and in string theory the allowed locations for open string ends are D-branes (the "D" refers to Peter Dirichlet, a nineteenth-century German mathematician). The bulk can contain more than one brane, so not all strings necessarily end on the same brane. But Polchinski, Dai, Leigh, and Hořava discovered that all open strings have to end on branes, and string theory tells us what dimensions and properties these branes will have.

Some branes extend in three dimensions, but others extend in four or five or more dimensions. In fact, string theory contains branes that extend in any number of dimensions up to nine. The string theory convention for labeling branes is to use the number of dimensions of space—not of spacetime—in which they extend. For example, a 3-brane is a brane that extends through three dimensions of space (but four dimensions of spacetime). When we come to look at the consequences of branes for the visible world, 3-branes will be very important. However, for the applications of branes discussed in this chapter, branes with other numbers of dimensions will also play a role.

Different types of brane arise in string theory. They are distinguished not only by their dimensionality—the number of dimensions in which they extend—but also by their charges, their shape, and an important characteristic called *tension* (which we'll get to soon). We don't know whether branes exist in the real world, but we do know the types of brane that string theory says are possible.

Branes were just a curiosity at the time they were discovered. Back then, no one saw any reason to include branes that interacted or moved. If strings interacted only weakly, as string theorists initially assumed, D-branes would be so taut that they would just sit there and not contribute to string motion or interactions. And if branes don't respond to strings in the bulk, they would just be an unnecessary complication. They would be a place or location, but they would be no more relevant to the motions and interactions of strings than the Great Wall of China is to your daily existence. Moreover, physicists didn't want to include branes in a physical realization of string theory because branes violated their intuition that all dimensions are created equal. Branes distinguish certain dimensions—those along the brane are different from those that extend off it—whereas the known laws of physics treat all directions the same. Why should string theory be different?

We also expect physics at any one point in space to be the same as it is at any other. But branes don't respect this symmetry either. Although branes extend infinitely far along some dimensions, they are situated at a fixed position in the other directions. That is why they don't span all of space. But in those directions in which the brane's position is fixed, an inch from the brane is not the same as a yard or a half-mile from the brane. Imagine a brane that was drenched in perfume. You would definitely be able to tell whether you were near it or far from it.

For these reasons, string theorists initially ignored branes. But about five years after branes were discovered, their status in the theoretical community dramatically improved. In 1995 Joe Polchinski irreversibly changed the course of string theory when he showed that branes were dynamical objects that were integral to string theory and were likely to play a critical role in its ultimate formulation. Polchinski explained what types of D-brane are present in superstring theory, and demonstrated that these branes carry charge[28] and therefore interact.

Moreover, the branes in string theory have finite tension. Brane tension is akin to the tension of the surface of a drum that returns to its taut position after you pinch it or punch it. If a brane's tension were zero, any small touch would have an enormous effect since the brane would have no resistance. On the other hand, if a brane's tension were infinite, you couldn't have any effect on it in the first place, for it would be a stationary object, not a dynamical one. Because the tension of branes is finite, branes can move and fluctuate and respond to forces, just like any other charged object.

Branes' finite tension and nonzero charge tell us that they are not merely places, they are also things: their charges tell us that they interact, and finite tension tells us that they move. Like a trampoline—a surface that interacts with its environment when it is depressed and when it springs back—a brane can move and interact. For example, both trampolines and branes can be distorted. And both trampolines and branes can influence their surroundings, trampolines by pushing on people and air, and branes by pushing on charged objects and the gravitational field.

If branes exist in the cosmos, their violation of spacetime symmetries should be no more disturbing than the violation of spatial symmetries caused by the Sun or the Earth. The Sun and the Earth are also located in particular locations; when measured with respect to the Sun or the Earth, not all positions in three-dimensional space are the same. Nonetheless, physical laws preserve the spacetime symmetries of three-dimensional space, even if the state of the universe does not. Branes would be no worse than the Sun or the Earth in this respect. Branes, like all other objects at definite places in space, break some symmetries of spacetime.

A moment's reflection reveals that this is not such a bad thing. After all, if string theory is the true description of nature, then not all dimensions are created equal. The three familiar spatial dimensions look alike, but the extra dimensions must be different; if they weren't, they wouldn't be "extra." From the vantage point of the physical universe, the violation of spacetime symmetries could help explain why extra dimensions are different: branes might correctly distinguish string theory's extra dimensions from the three spatial dimensions we experience and know.

In later chapters, I will consider branes with three spatial dimensions and describe some of their potentially radical implications for the real world. But for the rest of this chapter we'll concentrate on why branes are so significant in string theory—so important, in fact, that they catalyzed the "second superstring revolution" of 1995. The next section gives a few reasons why branes have remained at the forefront of string theory for the past decade, and why we now think they're here to stay.

Mature Branes and the Missing Particles

While Joe Polchinski was hard at work investigating D-branes, Andy Strominger, then his colleague at Santa Barbara, was pondering *p-branes*—fascinating solutions to Einstein's equations. They expand infinitely far in some spatial directions, but in the remaining dimensions they act as black holes, trapping objects that come too close. D-branes, on the other hand, are surfaces on which open strings end.

Andy told me how he and Joe would discuss their research progress every day over lunch. Andy would talk about *p*-branes, and Joe would discuss D-branes. Although they were both studying branes, like all other physicists they initially thought that their two types of brane were two different things. Joe eventually realized that they were not.

Andy's work demonstrated that the *p*-branes he was studying are critically important in string theory because in some spacetime geometries they give rise to new types of particle. Even if string theory's non-intuitive and remarkable premise is true, and particles arise as oscillation modes of strings, string oscillations don't necessarily account for all particles. Andy showed that there could still be additional particles that arise independently of strings.

Branes come in different shapes, forms, and sizes. Although we've focused on branes as the places where strings end, branes themselves are independent objects that can interact with their environment. Andy considered *p*-branes that wrap around a very tiny curled-up region of space, and he found that these tightly wrapped branes can act like particles. A wrapped *p*-brane that acts like a particle can be compared to a tightly cinched lasso. Just as a loop of rope becomes tiny once you pull it tightly around a pole or a bull's horn, a brane can wrap around a compact region of space. And if that region of space is tiny, then the brane that is wrapped around it will be tiny as well.

These small branes, like more familiar macroscopic objects, have a mass that grows with their size. More of something (like lead pipe or dirt or cherries) is heavier, and less of it is lighter. Because a brane wrapped around a tiny region of space is so small, it will also be extremely light. And Andy's calculations showed that in the extreme case when the brane is as minuscule as you can imagine, this tiny brane looks like a new massless particle. Andy's result was crucial in that it showed that even the most basic hypothesis of string theory—that everything arises from strings—is not always correct. Branes, too, contribute to the particle spectrum.

Joe's remarkable observation of 1995 was that these new particles that arise from tiny p-branes could also be explained with D-branes. In fact, in his paper establishing the relevance of D-branes, Joe showed that D-branes and p-branes were actually the same thing. At energies where string theory makes the same predictions as general relativity, D-branes morph into p-branes. Joe and Andy, though they didn't realize it at first, had actually been studying the same objects. This result meant that the significance of D-branes could no longer be questioned: they were no less important than the p-branes that had preceded them, and those p-branes were essential to the string theory spectrum of particles. Furthermore, there is a beautiful way to understand why p-branes are equivalent to D-branes. It is based on the subtle and important notion of *duality*.

Mature Branes and Duality

Duality is one of the most exciting concepts of the last ten years in particle physics and string theory. It has played a major role in recent advances in both quantum field theory and string theory, and, as we will soon see, it has especially important implications for theories with branes.

Two theories are dual when they are the same theory with different descriptions. In 1992 the Indian physicist Ashoke Sen was one of the first to recognize duality in string theory. In his work, which followed up the idea of duality that the physicists Claus Montonen and David Olive had originally introduced in 1977, he showed that a particular theory remained exactly the same if the particles and strings of the theory were interchanged. In the 1990s, the Israeli-born physicist Nati Seiberg, who was then at Rutgers University, also demonstrated remarkable dualities between different supersymmetric field theories with superficially different forces.

To understand duality's significance, it helps to know a little about how string theorists generally do calculations. String theory's predictions depend on the string's tension. But they also depend on the value of a number called the *string coupling*, which determines the strength with which strings interact. Do they brush past each other, corresponding to a weak coupling, or do they collude about their mutual fates, corresponding to a strong coupling? If we knew the value of the string coupling, we could study string theory for only that particular value. But because we don't yet know the value of the string coupling, we can hope to understand the theory only when we can make predictions for any string interaction strength. Then we can find out which one works.

The problem was that since the inception of string theory, the strongly coupled theory had appeared to be intractable. In the 1980s only string theory with weakly interacting strings was understood. (I'm using the adjective "weak" to describe the strength of string interactions, but don't be misled by the word—this has nothing to do with the weak force.) When strings interact very strongly, it's enormously difficult to calculate anything. Just as it's simpler to untie a loose knot than a tight one, a theory with only weak interactions is much more manageable than a theory with strong ones. When strings interact with each other very strongly, they can get into a tangled mess that is too difficult to unravel. Physicists have tried various ingenious approaches for calculations involving strongly interacting strings, but have found no methods that they could usefully apply to the real world.

In fact, not only string theory, but all fields of physics are easier to understand when interactions are weak. That's because if the weak interaction is only a small *perturbation*, or alteration, to a solvable theory—usually a theory with no interactions—then you can use a technique called *perturbation theory*. Perturbation theory lets you creep up on the answer to a question in the weakly interacting theory by starting from the theory with no interactions and calculating small improvements in incremental stages. Perturbation theory is a systematic procedure that tells you how to refine a calculation in successive steps until you reach any desired level of precision (or until you get tired, whichever comes first).

Using perturbation theory to approximate a quantity in an unsolvable theory might be compared to mixing paint to approximate a desired color. Suppose you're striving for a subtle blue with hints of green that resembles the Mediterranean at its most beautiful. You might start with blue, and then mix in smaller and smaller amounts of green, alternating at times with a bit more blue, until you've achieved (almost) the precise color you're after. Perturbing your paint mixture in this fashion is a way of proceeding in stages to obtain as close an approximation as you want to the desired color. Similarly, perturbation theory is a method for closely approximating the correct answer to whatever problem you're studying by making incremental progress, starting from a problem you already know how to solve.

Trying to find the answer to a problem about a theory with strong coupling, on the other hand, is more like trying to reproduce a Jackson Pollack painting by randomly pouring paint. Each time you poured some paint, the picture would change completely. Your painting would be no closer to the desired goal after twelve iterations than it would be after eight. In fact, each time you poured the paint you would as likely as not cover up much of your previous attempt, changing the picture so much that you would essentially be starting afresh each time.

Perturbation theory is similarly useless when a solvable theory is perturbed by a strong interaction. As with futile attempts to reproduce a modern splattered masterpiece, systematic attempts to approximate a quantity of interest in a strongly interacting theory will not succeed. Perturbation theory is useful and calculations are under control only when interactions are weak.

Sometimes, in certain exceptional situations, even when perturbation theory is useless, you can still understand the qualitative features of a strongly interacting theory. For example, the physical description of your system might resemble the weakly interacting theory in gross outline, even though the details are likely to be rather different. More often, however, it is impossible to say anything at all about a theory with strong interactions. Even the qualitative features of a strongly interacting system are often completely different from those of a superficially similar, weakly interacting system.

So, there are two things you might expect for strongly interacting ten-dimensional string theory. You might believe that no one can solve it and you can't say anything about it at all, or you might expect strongly interacting ten-dimensional string theory to look, at least in gross outline, like the weakly coupled string theory. Paradoxically, in some cases neither of these options turns out to be correct. In the case of a particular type of ten-dimensional string theory called IIA, the strongly interacting string looks nothing like the weakly interacting string. But we can nonetheless study its consequences because it is a tractable system in which calculations are possible.

At Strings '95, a conference held at the University of Southern California in March of that year, Edward Witten flabbergasted the audience by demonstrating that at low energies, a version of ten-dimensional superstring theory with strong coupling was completely equivalent to a theory that most people would have thought was entirely different: eleven-dimensional supergravity, the eleven-dimensional supersymmetric theory that contains gravity. And the objects in this equivalent supergravity theory interacted weakly, so perturbation theory could be usefully applied.

This meant, paradoxically, that you could use perturbation theory to study the original strongly interacting, ten-dimensional superstring theory. You would not use perturbation theory in the strongly interacting string theory itself, but in a superficially entirely different theory: weakly interacting, eleven-dimensional supergravity. This remarkable result, which Paul Townsend of Cambridge University had previously also observed, meant that despite their different packaging, at low energies, ten-dimensional superstring theory and eleven-dimensional supergravity were in fact the same theory. Or, as physicists would say, they were dual.

We can illustrate the idea of duality with our paint analogy. Suppose that we started off with blue paint, but then "perturbed" it by adding green. A good description of our paint mixture would then be blue paint with a hint of green. But suppose instead that the green paint we added wasn't a small perturbation: suppose that we added an enormous amount of green paint. If that amount far exceeded the amount of the original blue paint, a better, "dual" description of the mixture would be green paint with a hint of blue. The preferred description entirely depends on the quantities of each color involved.

Similarly, a theory might have one description when a coupling of an interaction is small. But when that coupling is sufficiently large, perturbation theory is no longer useful in the original description. Nonetheless, in certain remarkable situations, the original theory can be completely repackaged in such a way that perturbation theory applies. That would be the dual description.

It's as if someone presented you with the ingredients for a five-course meal. Even with all the ingredients, you might not know where to start. To make the meal work, you'd have to figure out which ingredients are intended for which course, how the spices interact with the food and with one another, and what to cook and when. But if caterers delivered the same ingredients pre-organized and prepared into salad, soup, appetizer, main course, and dessert, I expect that anyone could manage to turn that into a meal. With the same ingredients organized the right way, making a dinner goes from a complicated to a trivial problem.

Duality in string theory works in this way. Although strongly interacting, ten-dimensional superstring theory looked completely intractable, the dual description automatically organizes everything into a theory in which perturbation theory can be applied. Calculations which are difficult in one theory become manageable in the other. Even when the coupling in one theory is too big to use perturbation theory, the coupling of the other is sufficiently small to allow you to carry out perturbation calculations. However, we have yet to understand duality fully. For example, no one knows how to compute anything when the string coupling is neither very small nor very large. But when one of the couplings is either very small or very large (and the other is, respectively, very large or very small), then we can do calculations.

The duality of strongly coupled superstring theory and weakly coupled, eleven-dimensional supergravity theory tells you that you can calculate everything you would want to know in a strongly interacting, ten-dimensional superstring theory by performing calculations in a theory that is superficially entirely different. Everything predicted by the strongly interacting, ten-dimensional superstring theory can be extracted from weakly interacting, eleven-dimensional supergravity theory. And vice versa.

The feature of this duality that makes it so incredible is that both descriptions involve only *local interactions*—interactions with nearby objects. Even if corresponding objects exist in both descriptions, duality is only a truly surprising and interesting phenomenon if both descriptions have local interactions. After all, a dimension is more than a collection of points: it is a way of organizing things according to whether they are nearby or far apart. A computer dump might contain everything I want to know and be equivalent to an organized set of files and documents, but it wouldn't be a simple description unless the information were coherently organized with the relevant information contiguous. The local interactions in both the ten-dimensional superstring theory and the eleven-dimensional supergravity theory are what makes the dimensions in both theories—and therefore the theories themselves—meaningful and useful.

The equivalence between ten-dimensional superstring theory and eleven-dimensional supergravity vindicated Paul Townsend at Cambridge and Michael Duff, then at Texas A&M. For a long time, string theorists had largely rebuffed and maligned their work on eleven-dimensional supergravity—they couldn't understand why Duff and Townsend were wasting their time with this theory when string theory was so obviously the physics of the future. After Witten's talk, string theorists had to concede that eleven-dimensional supergravity was not only interesting, it was equivalent to string theory!

I realized how much attention this surprising result about duality was receiving when I was on a plane returning from London. A fellow passenger, who turned out to be a rock musician, saw that I was reading some physics papers. He came over and asked me whether the universe had ten or eleven dimensions. I was a little surprised. But I did answer and explained that in some sense, it is both. Since the ten- and eleven-dimensional theories are equivalent, either one can be considered correct. The convention is to give the number of dimensions in whichever of the theories has weakly interacting strings and thus a lower physical value of the string coupling.

But unlike the couplings associated with Standard Model forces, whose strength we can measure, we don't yet know the size of the string coupling. It might be weak, in which case perturbation theory can be applied directly; or it might be strong, in which case you would be better off using perturbation theory in the dual description. Without knowing the value of the string coupling, we have no way of knowing which, if either, of the two descriptions is the simpler way to describe string theory as it applies to the world.*

And there were more duality surprises at Strings '95. Until then, most string theorists had thought there were five versions of superstring theory, each of which included different forces and interactions. At Strings '95, Witten (and before that, Townsend and another British physicist, Chris Hull) demonstrated dualities between pairs of versions of superstring theory. And over the course of 1995 and 1996, string theorists showed that all these versions of ten-dimensional theories were dual to one another, and, furthermore, were dual to eleven-dimensional supergravity. Witten's talk had triggered a veritable duality revolution. With the extra input from the nature of branes, the five apparently distinct superstring theories were shown to be the same theory in different guises.

Because the various versions of string theory are actually the same, Witten concluded that there must exist a single theory that encompasses eleven-dimensional supergravity and the different manifestations of string theory, whether or not they contain only weak interactions. He named the new eleven-dimensional theory M-theory—the theory I mentioned at the beginning of this chapter. You can get any known version of superstring theory from M-theory. But M-theory also extends beyond the known versions to domains we have yet to understand. M-theory has the potential to give a more unified, coherent picture of the superstring and to fully realize string theory's potential as a theory of quantum gravity. However, more pieces or patterns are needed before string theorists understand M-theory sufficiently well to pursue these goals. If the known versions of superstring theory are shards taken from an archeological dig, M-theory is the sought-after enigmatic artifact that would piece them

---

*Although we can do perturbation theory when the coupling is very weak, or when there is a weakly coupled dual description of a theory with strong interactions, we have no way of using perturbation theory when the interaction strength is in the middle—that is, about 1. That means that even when there is a dual description, we don't have a complete solution to the theory.

all together. No one yet knows the best way to formulate M-theory. But string theorists now think of it as their primary goal.

More On Duality

This section gives a little more detail about the particular duality I mentioned above, the one between ten-dimensional superstring theory and eleven-dimensional supergravity. I won't use these explanations later, so feel free to skip ahead to the next chapter if you like. But since this is a book about dimensions, a digression about the duality between two theories with different dimensions doesn't seem entirely out of place.

One feature that makes duality a little more reasonable is that one of the two theories always contains strongly interacting objects. If the interactions are strong, only rarely can you directly deduce the physical implications of the theory. Although it's strange to think of a theory that looks ten-dimensional being best described by another, totally different, eleven-dimensional theory, it seems less strange when you remind yourself that your ten-dimensional theory contained such strongly interacting objects that you couldn't predict what was happening there in the first place. All bets were off anyway.

There are nonetheless many baffling features about a duality between theories with different numbers of dimensions. And in the particular case of the duality between ten-dimensional superstring theory and eleven-dimensional supergravity, at first glance there appears to be an extremely basic problem. Ten-dimensional superstring theory contains strings, whereas eleven-dimensional supergravity does not.

Physicists used branes to solve this puzzle. Even though eleven-dimensional supergravity does not contain strings, it does contain 2-branes. But unlike strings, which have only one spatial dimension, 2-branes have two (as you might have guessed). Now, suppose that one of the eleven dimensions is rolled up into an extremely tiny circle. In that case, a 2-brane that encircles the rolled-up circular dimension looks just like a string. The rolled-up brane appears to have only one spatial dimension, as illustrated in Figure 69. This means that eleven-dimensional supergravity theory with a rolled-up dimension appears to contain strings, even though the original eleven-dimensional theory does not.

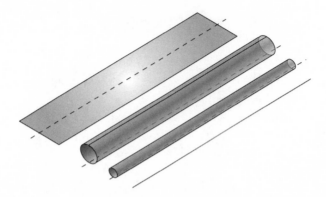

Figure 69. *A brane with two spatial dimensions that is rolled up on a very small circle looks like a string.*

This might sound like a cheat, since we have already argued that a theory with a curled-up dimension always appears to have fewer dimensions at long distances and low energies, so you wouldn't be surprised to find that an eleven-dimensional theory with a rolled-up dimension acts like a ten-dimensional theory. If you want to show that these ten- and eleven-dimensional theories are equivalent, why should it be sufficient to study the eleven-dimensional theory when one of the dimensions is rolled up?

The key to the answer is that in Chapter 2 we only showed that a rolled-up dimension is invisible at long distances or low energies. Edward Witten went further at Strings '95. He demonstrated the equivalence of the ten- and eleven-dimensional theories by showing that the eleven-dimensional supergravity theory with one of the dimensions curled up is completely equivalent to ten-dimensional superstring theory, even at short distances. When a dimension is curled up, you can still distinguish points at different locations along this dimension if you look closely enough. Witten showed that everything about the dual theories is equivalent, even those particles which have enough energy to probe distances smaller than the curled-up dimension's size.

Everything about the eleven-dimensional supergravity theory with a curled-up dimension—even short distances and high-energy processes and objects—has a counterpart in the ten-dimensional superstring theory. Furthermore, the duality holds true for a dimension curled up in a circle of any size, no matter how large. Earlier, when we looked at a rolled-up dimension, we argued only that a small curled-up dimension would not be noticed.

But how can theories with different numbers of dimensions possibly be the same? After all, the number of dimensions of space is the number of coordinates that we need to specify the position of a point. The duality could be true only if superstring theory always uses an additional number to describe pointlike objects.

The key to the duality is that in superstring theory, there are special new particles that you can uniquely identify only by specifying the value of the momentum in nine spatial dimensions and also the value of a charge, whereas in eleven-dimensional supergravity you need to know momentum in ten spatial dimensions. Notice that even though you have nine dimensions in one case and ten in the other, in both cases you would need to specify ten numbers: nine values of momentum and a charge in one case, and ten values of momentum in the other.

Ordinary uncharged strings don't pair with objects in the eleven-dimensional theory. Because you need to know eleven numbers to locate an object in the spacetime of the eleven-dimensional theory, only particles that carry charge have eleven-dimensional mates. And the partners of the objects in the eleven-dimensional theory particles turn out to be branes—namely, charged, pointlike branes called Do-branes. String theory and eleven-dimensional supergravity are dual because for every Do-brane of a given charge in ten-dimensional superstring theory,* there is a corresponding particle with a particular eleven-dimensional momentum. And vice versa. The objects of the ten- and eleven-dimensional theories (and their interactions as well) are exactly matched.

Although charge might seem very different from momentum in a particular direction, if every object with a particular momentum in the eleven-dimensional theory matches onto an object with a

---

*Really it is a bound state of Do branes.

particular charge in the ten-dimensional one (and vice versa), it is up to you whether you want to call that number momentum or charge. The number of dimensions is the number of independent directions of momentum—that is, the number of different directions in which an object can travel. But if momentum along one of the dimensions can be replaced by a charge, the number of dimensions isn't really well defined. The best choice is determined by the value of the string coupling.

This astonishing duality was one of the first analyses in which branes proved instrumental. Branes were the additional ingredients that were needed for different string theories to match onto each other. But the critical feature of string theory branes that is important for their application in physical theories is that they can house particles and forces. The next chapter explains why.

What to Remember

- String theory is a misnomer: string theory also contains higher-dimensional branes. *D-branes* are a type of brane in string theory on which open strings (strings that don't loop back on themselves) must end.

- Branes played a role in many of the important string theory developments of the last decade.

- Branes were critical in demonstrating *duality*, which showed that superficially different versions of string theory are in fact equivalent.

- At low energies, ten-dimensional superstring theory is dual to eleven-dimensional *supergravity*—an eleven-dimensional theory that contains supersymmetry and gravity. Particles in one theory match onto branes in the other.

- The results about branes from this chapter will not be relevant later on. These results do, however, explain some of the excitement about branes in the string theory community.

16

Bustling Passages: Braneworlds

Welcome to where time stands still.
No one leaves and no one will.

Metallica

Icarus III was becoming increasingly disillusioned with Heaven. He had expected it to be a liberal, forgiving environment. But instead, gambling was prohibited, metal silverware was forbidden, and smoking was no longer allowed. The most restrictive constraint of all was that Heaven was stuck on a Heavenbrane; its residents were forbidden to travel into the fifth dimension.

Everyone on the Heavenbrane knew about the fifth dimension and the existence of other branes. In fact, the righteous Heavenbraners often whispered about the unsavory characters sequestered on a Jailbrane not too far away. However, the Jailbraners couldn't hear any of the slander that Heavenbraners spread about them, so all remained peaceful in the bulk and on the branes.

From the perspective of the "duality revolution," you might think that branes were a great boon to people trying to connect string theory to the visible universe. If all of the different formulations of string theory are in fact one and the same, physicists would no longer be faced with the daunting task of finding the rules by which nature chooses among them. There is no need to play favorites if all of the string theories appearing in different guises are actually the same.

But nice as it would be to think that we are closer to finding the connection between string theory and the Standard Model, the task is not so simple. Although branes were critical to the dualities that reduced the number of distinct manifestations of string theory, they actually increased the number of ways in which the Standard Model might emerge. That is because branes can house particles and forces that string theorists didn't take into account when they originally developed string theory. Because of the many possibilities as to what types of brane exist and where they are situated in the higher-dimensional space of string theory, there are conceivably many new ways to realize the Standard Model in string theory that no one had thought about before. The forces of the Standard Model do not necessarily arise from a single fundamental string: they could instead be new forces that emerge from strings extended among different branes. Although dualities tell us that the original five versions of superstring theory are equivalent, the number of braneworlds that are conceivable in string theory is stupendous.

Finding a unique Standard Model candidate looked like it would be as hard as ever. String theorists' euphoria about duality was tempered by this realization. However, those of us who were looking for new insights into observable physics were in heaven. With the new possibilities for forces and particles confined to branes, it was time for us to rethink the starting point for particle physics.

The feature of branes that is essential to their potentially observable applications is that they can trap particles and forces. The purpose of this chapter is to give you a flavor of how this works. We'll begin by explaining why the branes of string theory confine particles and forces. We'll consider the braneworld idea and the first known braneworld, one which was derived from duality and string theory. In the chapters that follow, we'll proceed to those aspects of braneworlds and their potential physical applications that I find most exciting.

Particles and Strings and Branes

As Ruth Gregory, a general relativist from Durham University, puts it, branes in string theory come "fully loaded" with particles and forces. That is, certain branes always have particles and forces that are trapped on them. Like housebound cats that never venture beyond the walls of their domicile, those particles that are confined to branes never venture off them. They can't. Their existence is predicated on the presence of the branes. When they move, they move only along the spatial dimensions of the brane; and when they interact, they interact only on the spatial dimensions spanned by the brane. From the perspective of brane-bound particles, if it weren't for gravity or bulk particles with which they might interact, the world might as well have only the dimensions of the brane.

Let's now see how string theory can confine particles and forces on branes. Imagine that there is only one D-brane, suspended somewhere in a higher-dimensional universe. Because, by definition, both ends of an open string must be on a single D-brane, this D-brane would be where all open strings begin and end. The ends of each open string wouldn't be stuck in any particular location, but they would have to lie somewhere on the brane. Like train tracks that confine wheels but allow them to roll, the branes act as fixed surfaces in which the ends of the string are confined but can nonetheless move.

Because the vibrational modes of open strings are particles, the modes of an open string with both ends confined to a brane are particles that are confined to this brane. Those particles would travel in and interact only along the dimensions spanned by the brane.

It turns out that one of these particles arising from a brane-bound string is a gauge boson that can communicate a force. We know this because it has the spin of a gauge boson (which is 1), and because it interacts just as a gauge boson should. Such a brane-bound gauge boson would communicate a force that would act on other brane-bound particles, and calculations show that the particles on the receiving end are always charged under this force. In fact, the endpoint of any string ending on the brane would act like a charged particle. The presence of the brane-bound force and these charged particles is what tells us that a D-brane of string theory comes "loaded" with charged particles and a force that acts upon them.

In setups with more than one brane, there will be more forces and more charged particles. Suppose, for example, that there were two branes. In that case, in addition to the particles confined to each of the branes, there would be a new type of particle arising from strings whose two ends were on the two different branes (see Figure 70).

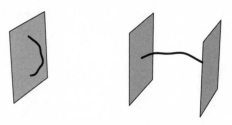

Figure 70. *A string that begins and ends on a single brane can give rise to a gauge boson. A string with each end on a different brane gives rise to a new type of gauge boson. When the branes are separated, the gauge boson has nonzero mass.*

It turns out that if the two branes are separated from each other in space, the particles associated with the string that extends between them will be heavy. The mass of the particles arising from the vibrational modes of this string grows with the distance between the branes. This mass is like the energy that gets stored when you stretch a spring—the more it is stretched, the more energy it contains. Similarly, the lightest particle that arises from a string stretched between two branes will have a mass that increases in proportion to the brane separation.

However, when a spring is relaxed in its rest position, it doesn't store any energy. Similarly, if the two branes are not separated—that is, if they are in the same place—the lightest string particle arising from the string with an end on each brane is massless.

Let's now assume that the two branes coincide, so that they produce some massless particles. One of these massless particles would be a gauge boson—not one of the gauge bosons that arises from strings with both ends on a single brane, but a distinct, new one. This new massless gauge boson, which arises only when there are coincident branes, communicates a force that acts on particles on either one or both of the two branes. Furthermore, as with all other forces, the forces on the brane are associated with a symmetry. In this case the symmetry transformation would be the one that exchanges the two branes (which a punning Igor might enjoy).[29]

Of course, if two branes really were in the same place, you might think it a little odd to refer to them as two distinct objects. And you would be right: if two branes are in the same place, you can just as well imagine them as a single brane. This new brane exists in string theory. It is secretly two coincident branes, and has the properties those branes would have. It houses all the different types of particles discussed above: the particles that arise from open strings ending on each brane in the original two-brane description, as well as the strings whose ends are both on a single brane.

Now imagine that many branes are superimposed. There would then be many new types of open string because the two string ends can be confined to any of the branes (see Figure 71). Open strings that

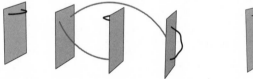 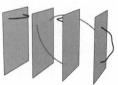

Figure 71. *Each string that begins and ends on the same brane or that extends between branes gives rise to gauge bosons. When the branes coincide, there are new massless gauge bosons, corresponding to each of the ways in which a string can begin and end on each of the coincident branes.*

extend between different branes, or the strings that begin or end on any single one of the branes, imply new particles, composed of the vibrational modes of these strings. Once again, these new particles include new types of gauge boson and new types of charged particle. And once again the new forces are associated with new symmetries that interchange the various superimposed branes.

So indeed, branes come "loaded" with forces and particles; many branes mean rich possibilities. Furthermore, even more intricate situations can arise involving separated batches of branes. Branes situated in different places would carry entirely independent particles and forces. The particles and forces that are confined to one group of branes would be entirely different from the particles and forces confined to the others.

For example, if the particles of which we are composed, together with electromagnetism, are all confined to one brane, we would experience the electromagnetic force. However, particles that are confined to distant branes would not; those foreign particles would be insensitive to electromagnetism. On the other hand, particles confined to distant branes could experience novel forces to which we are completely insensitive.

An important property of such a setup, which will be relevant later on, is that particles on separated branes don't interact with each other directly. Interactions are local: they can take place only among particles in the same place; particles on separated branes would be too far apart to interact with each other directly.

You might compare the bulk, the full higher-dimensional space, to a huge tennis stadium with separate matches going on throughout. The ball on any one court would go back and forth across the net and could move anywhere on that court. However, each match would proceed independently of the others, and each ball would stay on its own isolated court. Just as the ball in a given court should stay there and only the two tennis players on that court would have access to it, brane-confined gauge bosons or other brane-confined particles interact only with objects on their own brane.

However, particles on separate branes can communicate with one another if there are particles and forces that are free to travel throughout the bulk. Such bulk particles would be free to enter and leave a brane. They might occasionally interact with particles on a brane, but they can also travel freely in the full higher-dimensional space.

A setup with separated branes and bulk particles that communicate between them would be like a stadium with separate simultaneous matches in which the players in the separate games have the same coach. The coach, who might well want to keep an eye on several games going on at the same time, would travel from one court to another. If one player wanted to communicate something to a player on another court, he could tell the coach who could carry the message over. The players wouldn't communicate directly during their matches, but they could nonetheless communicate via a person who travels between their respective courts. Similarly, bulk particles could interact with particles on one brane and subsequently interact with particles on a distant brane, thereby permitting the particles that are confined to separated branes to communicate indirectly.

In the next section we will see that the graviton, the particle that communicates the gravitational force, is one such bulk particle. In a higher-dimensional setup, it would travel throughout the higher-dimensional space and interact with all particles everywhere, whether they are on a brane or not.

Gravity: Different Again

Gravity, unlike all other forces, is never confined to a brane. Brane-bound gauge bosons and fermions are the result of open strings, but in string theory, the graviton—the particle that communicates gravity—is a mode of a closed string. Closed strings have no ends, and therefore there are no ends to pin down on a brane.

Particles that are the vibrational modes of closed strings have unrestricted license to travel in the full higher-dimensional bulk. Gravity, the force we know to be communicated by a closed-string particle, is thus once again singled out from the other forces. The graviton, unlike gauge bosons or fermions, *must* travel though the entire higher-dimensional spacetime. There is no way to confine gravity to lower dimensions. In later chapters we will see that, amazingly, gravity can be localized near a brane. But one can never truly confine gravity on a brane.

This means that although braneworlds could trap most particles and forces on branes, they will never confine gravity. This is a nice property. It tells us that braneworlds will always involve higher-dimensional physics, even if the entire Standard Model is stuck on a four-dimensional brane. If there is a braneworld, everything on it will still interact with gravity, and gravity will be experienced everywhere in the full higher-dimensional space. We'll soon see why this important distinction between gravity and other forces might help to explain why gravity is so much weaker than the other known forces.

Model Braneworlds

Very soon after physicists recognized the importance of branes to string theory, branes became the focus of intense study. In particular, physicists were eager to learn about their potential relevance to particle physics and our conception of the universe. As of now, string theory doesn't tell us whether branes exist in the universe and, if they do, how many there are. We know only that branes are an essential theoretical piece of string theory, without which it wouldn't fit together. But now that we know that branes are part of string theory, we have also begun to ask whether they could be present in the real world. And if they are, what are the consequences?

The potential existence of branes opens up many new possibilities for the composition of the universe, some of which might even be relevant to the physical properties of matter that we observe. The string theorist Amanda Peet, upon hearing Ruth Gregory's expression "fully loaded" branes, interjected that branes "blasted open the field of string-based model building." After 1995, branes became a new model building tool.

Towards the end of the 1990s, many physicists, myself included, expanded their horizons to include the possibility of branes. We asked ourselves, "What if there were a higher-dimensional universe in which the particles and forces we know about don't travel in all dimensions, but are confined to fewer dimensions on a lower-dimensional brane?"

Brane scenarios introduced many new possibilities for the global nature of spacetime. If Standard Model particles are confined to a brane, then we are as well, since we and the cosmos that surrounds us are composed of these particles. Furthermore, not all particles have to be on the same brane. There might therefore be entirely new and unfamiliar particles that experience different forces and interactions from the ones we know. The particles and forces we observe might be only a small part of a much larger universe. Two physicists from

Cornell, Henry Tye and Zurab Kakushadze, coined the term "brane-worlds" to label such scenarios. Henry told me that he used the term so that he could, in one fell swoop, describe all of the many ways in which the universe could include branes without being wedded to any particular possibility.

Although the proliferation of potential braneworlds might be frustrating to string theorists trying to derive a single theory of the world, it is also thrilling. These are all real possibilities for the world in which we live, and one of them might truly describe it. And because the rules of particle physics would be somewhat different in a higher-dimensional universe than particle physicists have assumed, extra dimensions introduce new ways of trying to address some of the puzzling features of the Standard Model. Although these ideas are speculative, braneworlds that address problems in particle physics should soon be testable in collider experiments. This means that experiments, rather than our prejudice, could ultimately decide whether these ideas apply to our world.

We are about to investigate some of these new braneworlds. We'll ask what they might look like and what their consequences could be. We will not restrict ourselves to braneworlds derived explicitly from string theory, but will consider model braneworlds that have already introduced new ideas into particle physics. Physicists are so far from understanding the implications of string theory that it would be premature to exclude models just because no one has yet found a string theory example with a particular set of particles or forces or a particular distribution of energy. These braneworlds should be thought of as targets for string theory explorations. In fact, the warped hierarchy model I'll talk about in Chapter 20 was derived from string theory only after Raman Sundrum and I introduced it as a braneworld possibility.

The following chapters will present several different braneworlds. Each of them will illustrate a completely new physical phenomenon. The first will show how braneworlds can evade the anarchic principle; the second will show that dimensions can be much larger than we previously thought; the third will show that spacetime can be so curved that we would expect objects to have very different sizes and masses; and the last two will show that even infinite extra dimensions can be invisible when spacetime is curved, and that spacetime might even appear to have different dimensions in different places.

I'm presenting several models because they are all real possibilities. But just as important, each of them contains some new feature that physicists recently thought impossible. I'll summarize the significance of each model and how it violated conventional wisdom at the end of each of the chapters. Feel free to read these bullet summaries first to get the big picture, a quick résumé of the significance of the particular model that chapter explains.

Before proceeding to these braneworlds, I'll now briefly present the first known braneworld, one which was derived directly from string theory. Petr Hořava and Edward Witten hit upon this braneworld—called "HW" after their initials—in the course of exploring string theory duality. I'm presenting this model because it is interesting in its own right, but also because it has several properties that foreshadow features of the other braneworlds we will soon encounter.

Hořava-Witten Theory

The HW braneworld is pictured in Figure 72. It's an eleven-dimensional world bounding two parallel branes, each of which has nine spatial dimensions bounding a bulk space that has ten spatial dimensions (eleven of spacetime). The HW universe was the original braneworld theory; in HW, each of the two branes contains a different set of particles and forces.

The forces on the two branes are the same as those of the heterotic string that was introduced in Chapter 14; that was the theory that David Gross, Jeff Harvey, Emil Martinec, and Ryan Rohm discovered, in which oscillations moving to the left or the right along the string interact differently. Half of those forces are confined to one of the two boundary branes, and the other half are confined to the other. There are enough forces and particles confined to each of the two branes that either one of them could conceivably contain all the particles of the Standard Model (and therefore us). Hořava and Witten assumed that the particles and forces of the Standard Model reside on

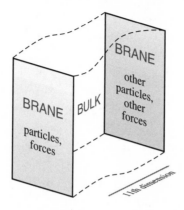

Figure 72. *Schematic drawing of the Hořava-Witten braneworld. Two branes with nine spatial dimensions (represented schematically by two-dimensional branes) are separated along the eleventh spacetime dimension (the tenth spatial dimension). The bulk includes all spatial dimensions: those nine that extend in the spatial directions along the two branes, and the additional one that extends between them.*

one of the two branes, whereas gravity and other particles that are part of the theory, but which we haven't observed in our world, are free to travel on the other brane or off the branes in the full eleven-dimensional bulk.

In fact, the HW braneworld didn't just have the same forces as the heterotic string—it *was* the heterotic string, albeit with strong string coupling. This is another example of duality. In this case, an eleven-dimensional theory with two branes bounding the eleventh dimension (the tenth dimension of space) is dual to the ten-dimensional heterotic string. That is to say, when the interactions of the heterotic string are very strong, the theory is best described as an eleven-dimensional theory with two boundary branes and nine spatial dimensions. This is not unlike the duality between ten-dimensional superstring theory and eleven-dimensional supergravity that was discussed in the previous chapter. But in our present example, the eleventh dimension is not rolled up, but is instead bounded between two branes. Once again, an eleven-dimensional theory can be equivalent to a ten-dimensional one, albeit when one theory has strong interactions and the other has feeble ones.

Of course, even if Standard Model particles are confined to a brane, the theory would still have more dimensions than we see around us. If the Hořava-Witten braneworld is to correspond to reality, six of its dimensions must be unseen. Hořava and Witten assumed that six dimensions were curled up into a tiny Calabi-Yau shape.

Once six dimensions are curled up, you can think of the HW universe as a five-dimensional effective theory with four-dimensional boundary branes. This picture of a five-dimensional universe with two boundary branes is an interesting one that many physicists have investigated. Raman and I applied some of the techniques that two physicists, Burt Ovrut and Dan Waldram, used to study the HW effective theory to the different five-dimensional theories that I'll discuss in Chapters 20 and 22.

One fascinating element of the Hořava-Witten braneworld is that it can accommodate not only the Standard Model particle and forces, but also a full Grand Unified Theory. And because gravity originates in higher dimensions, it's possible for gravity and other forces to have the same strength at high energy in this model.

The HW braneworld illustrates three reasons why braneworlds can matter for real-world physics. First, it involves more than a single brane. This means that it can contain forces and particles that interact with each other only weakly because of the distance between the two branes on which they are bound. The only way that particles confined to different branes can communicate is through common interactions with bulk particles. This first feature will be significant in the sequestering models that we'll look at in the next chapter.

The second important braneworld feature is that any braneworld introduces new length scales into physics. These new scales, like the size of the additional dimensions, might be relevant to unification or the hierarchy problem. Problems in both of these theories center around why there should be very different energy and mass scales in a single theory, and why quantum effects don't tend to equate the two.

Finally, branes and the bulk can carry energy. This energy can be stored by the branes and by the higher-dimensional bulk; it doesn't depend on the particles that are present. Like all forms of energy, it curves the bulk spacetime. We will soon see that such curvature of spacetime caused by energy spreading throughout the space can be very important to braneworlds.

The HW braneworld certainly has many tantalizing features. But it also suffers from the problems all realizations of string theory seem to have in reproducing known physics. Hořava-Witten theory is very difficult to test experimentally because its dimensions are so small. The many unseen particles have to be heavy enough to have avoided detection, and six of its dimensions have to be curled up, even though neither the size nor the shape of the curled-up dimensions has been determined.

Proceeding along these lines, one might accidentally hit on the version of string theory that correctly describes nature; this possibility is not definitively ruled out. For this to happen, we would have to be very lucky indeed. But problems in particle physics also beckon, and it is worth investigating how these problems might be resolved in a world with extra dimensions of space and branes that extend along only a restricted subset of these dimensions. That is what the rest of this book is about.

What to Remember

- Braneworlds are possible within the framework of string theory. Particles and forces in string theory can be trapped on branes.

- Gravity is different from other forces. It is never confined to a brane and always spreads through all dimensions.

- If string theory describes the universe, it could contain many branes. Braneworlds are very natural in this context.

17

Sparsely Populated Passages: Multiverses and Sequestering

Just turn around now
('cause) you're not welcome anymore.

Gloria Gaynor

Despite its explicit prohibition on the Heavenbrane, Icarus III ulti-mately returned to gambling. After ignoring repeated reprimands, he was sentenced to confinement on the Jailbrane, a distant brane separated from the Heavenbrane along a fifth dimension. Even after he was sequestered on the Jailbrane, Ike doggedly tried to contact his former buddies. But the distance between their two branes made communication difficult. He was reduced to flagging down passing bulk mail carriers, many of whom ignored his entreaties altogether. The few who did stop always conveyed his messages to the Heavenbrane, but at a frustratingly leisurely pace.

Meanwhile, back on the Heavenbrane, disaster loomed. The guardian angels, who had so bravely rescued the hierarchy, had no respect for the other residents' family values and were on the verge of creating intergenerational instability. Heaven's fallen angels considered all pairings acceptable and encouraged everyone to mix with a trophy partner from another generation.

When Ike learned of the threat, he was aghast and he resolved to redeem the situation. Ike realized that by using the slow and deliberate manner with which he was constrained to communicate with the Heavenbrane, he could judiciously feed the massive egos of the unruly angels living there. Thanks to Ike's helpful intervention, the angels stopped threatening the social order. Although Icarus III still had to serve his sentence, the relieved residents on the Heavenbrane honored him forevermore in urban myth.

This chapter is about *sequestering*, one of several reasons that extra dimensions could prove to be important for particle physics. Sequestered particles are physically separated on different branes. By confining different particles to different environments, sequestering might explain the distinctive properties that distinguish one particle from another. Sequestering might also be the reason that the anarchic principle, which says that everything should interact, doesn't always hold true. If particles are separated in extra dimensions, they are less likely to interact with one another.

In principle, particles could have been sequestered in three spatial dimensions. But as far as we can tell, all directions and all places in three-dimensional space are the same. The known laws of physics tell us that any particle can be anywhere in the three dimensions we see, so sequestering in three dimensions isn't an option. However, in higher-dimensional space, photons and charged objects cannot necessarily be just anywhere. Extra dimensions introduce a way to separate particles. Distinct particle types might be restricted to separate regions of space occupied by different branes. Because not all points in extra dimensions look the same, extra dimensions introduce a way to separate particles by confining different particle types to separated branes.

Theories that sequester particles have the potential to solve many problems. The story about Ike refers to my first foray into extra dimensions—the application of sequestering to supersymmetry breaking. Whereas four-dimensional theories face serious problems because supersymmetry-breaking models generally introduce unwanted interactions, sequestered supersymmetry-breaking models appear to be far more promising. Sequestering might also explain why particles have different masses from one another, and why proton decay does not occur in extra-dimensional models. In this chapter, we'll explore sequestering and a few of its particle physics applications. We'll see how even ideas, such as supersymmetry, that we thought applied to four-dimensional spacetime might be more successful in an extra-dimensional context.

My Passage to Extra Dimensions

We physicists are fortunate to have many conference opportunities to meet and share stimulating research ideas with colleagues. But such an overwhelming number of conferences and workshops in particle physics are held each year that choosing which invitations to accept can be difficult. Some are major gatherings that provide an opportunity to hear about others' recent work and to share your latest results. Some are relatively short conferences, lasting two or three days, in which physicists report major new results in a highly specialized field. Other meetings are longer workshops where physicists begin or complete collaborations with colleagues. Sometimes conferences are held in such spectacular locations they are just too good to miss.

Although Oxford is a very nice place, the supersymmetry conference that I attended there in early July of 1998 fits best into the first category. Supersymmetry, which for many years was considered the only possible way out of the hierarchy problem, has evolved over time into a major research area, and every year physicists gather to discuss recent progress in the field.

The Oxford conference held a surprise, however. The most interesting topic was not supersymmetry, but the newly emergent idea of extra dimensions. One of the most stimulating talks was about large extra dimensions, the subject of Chapter 19. Other talks were on the fate of string theory's extra dimensions, and still others discussed potential experimental implications of extra dimensions. The novelty and speculative nature of such ideas was clear from the title of Chicago theorist Jeff Harvey's talk: he and several later speakers jokingly named their talks after *Fantasy Island*. Joe Lykken, a theorist from Fermilab, even had a slide with a little man pointing to "Da brane. Da brane." (Needless to say, the joke about Tattoo, famous for welcoming "da plane" to Fantasy Island, was lost on those who had not experienced the joys of American seventies TV.)

Despite the jokes, I returned from the Oxford supersymmetry conference thinking about extra dimensions and why problems in particle physics might be solved in an extra-dimensional world. Although I was skeptical about the large extra dimensions that was one of the hot topics, and did not plan to work on them myself, I was fairly convinced that branes and extra dimensions could be important model building tools with the potential to explain some of the mysterious particle physics phenomena that have defied simple four-dimensional explanations.

That year I was planning to spend the rest of the summer in Boston. This was not the norm for me at the time; most of the Boston theoretical physics community, myself included, travel for a large part of each summer, attending assorted conferences and workshops. But I had decided to stay at home to relax and think about new ideas.

Raman Sundrum, who was then a postdoctoral fellow at Boston University, had also decided to remain in Boston that summer. I had often met Raman at conferences or when we visited each other's institutions, and we had even briefly overlapped as postdoctoral fellows at Harvard. Since Raman had already thought about extra dimensions, I decided that it could be useful to discuss my ideas and questions with him.

Raman is an interesting character. Whereas most physicists in the early stages of their career work on relatively safe problems—questions of general interest in which they are likely to make progress—Raman insisted on focusing on whatever he considered most important, even when it was an extremely difficult problem or diverged considerably from other people's interests. Despite his obvious talent, his idiosyncratic approach had kept him from a faculty job and brought him to his third postdoctoral position. But at that time, Raman was thinking about extra dimensions and branes; his interests and those of the rest of the physics community had begun to converge.

Our collaboration began at MIT's branch of Toscanini's (now sadly closed), an ice cream shop in the MIT student center that served great ice cream and very good coffee. Toscanini's was the ideal venue for discussing ideas without constraints or interruptions, as well as indulging in the delicious research stimulants that were available there.

From those early days, chatting over coffee, our research evolved and jelled as the summer progressed. By August it had reached the point where we needed bigger and bigger blackboards to hold all the details we were discussing. Since the blackboard in my office at MIT, where I was then a professor, was rather small, we would wander the "infinite corridor" (the very long hallway that runs the length of MIT's main building) searching for empty classrooms.

The particular research problem we focused on was the application of sequestering to supersymmetry breaking. The idea was to sequester particles responsible for supersymmetry breaking from the Standard Model particles and thereby prevent unwanted interactions between them (see Figure 73). We chose the word "sequester" to distinguish models in which particles are separated on different branes from the so-called "hidden sector" models of supersymmetry breaking that were fashionable at the time. In hidden sector models, supersymmetry-breaking particles interacted feebly with Standard Model particles,

Figure 73. *In this model for supersymmetry breaking, there are two branes. Standard Model particles are on one brane, and particles that break supersymmetry are sequestered on the other. The two branes each have three spatial dimensions and are separated in a fifth spacetime dimension, which is the fourth dimension of space.*

but weren't actually hidden (despite the name), and therefore could interact in ways that are not acceptable in the real world.

In the beginning I was very enthusiastic about our ideas and Raman was skeptical, although our roles alternated over time. But with one enthusiast and one skeptic, we quickly covered a lot of ground and got to the heart of the physics we were thinking about. Sometimes we even dismissed ideas too quickly, but usually one or the other of us maintained a point of view long enough to make progress.

Francis Bacon, who along with Galileo is considered one of the founders of the modern scientific method, described the difficulty of making progress while nonetheless retaining the skepticism necessary to ensure the correctness of your results.* How can you take an idea seriously enough to delve into its consequences, while simultaneously suspecting that it might be incorrect? Given enough time, a single person can fluctuate between these two attitudes and arrive at the correct answer. But with two of us taking opposing attitudes, it was often a matter of hours or even minutes before we discarded an intriguing but faulty idea.

Nonetheless, the idea we started with, sequestering to prevent unwanted interactions in supersymmetric theories, seemed to me as if it had to be right. Nothing in four dimensions worked in a sufficiently compelling way, and extra dimensions seemed to have all the necessary ingredients for a successful model. However, it was not until the end of the summer that Raman and I understood sequestering and its consequences for supersymmetry breaking well enough to finally see eye to eye and converge on its merits.

Naturalness and Sequestering

The reason that sequestering could be important is that it is a way to prevent the problems caused by the anarchic principle, the unofficial rule that says that in four-dimensional quantum field theory, anything that can happen will happen. The problem with the anarchic principle

*Francis Bacon, *On Scientific Inquiry*.

is that theories end up predicting interactions and relationships among masses that are not seen in nature. Even interactions that don't occur in a classical theory (the one without quantum mechanics taken into account) will occur once virtual particles are included; virtual particle interactions induce all possible interactions.

Here's an analogy that illustrates why. Suppose you told Athena that it would snow tomorrow, and Athena then told Ike. Even though you had no direct communication with Ike, your communication would nonetheless influence what Ike would wear the next day—he would put on a parka because of your virtual advice.

Similarly, if a particle interacts with a virtual particle, and that virtual particle interacts in turn with a third particle, the net effect is that the first and the third particles interact. The anarchic principle tells us that processes involving virtual particles are bound to occur, even if they don't happen classically. And those processes often induce unwanted interactions.

Many of the problems in particle physics theories stem from the anarchic principle. For example, the quantum contributions to the Higgs particle's mass that result from virtual particles are the root of the hierarchy problem. Any path that the Higgs particle takes can be temporarily interrupted by heavy particles, and these interventions increase the Higgs particle's mass.

We saw another example involving the anarchic principle in Chapter 11. In most theories with broken supersymmetry, virtual particles induce unwanted interactions—interactions that we know from experiments do not take place. Those interactions would change the identity of the known quarks and leptons. Such *flavor-changing interactions* either don't occur in nature or occur very rarely. If we want a theory to work, we must somehow eliminate these interactions— which the anarchic principle tells us will arise.

Virtual particles don't necessarily lead to these unwanted predictions. The theory won't predict these unwanted interactions in the unlikely event that there are enormous cancellations between the classical and quantum mechanical contributions to a physical quantity. Even though the classical and quantum contributions would individually be much too big, the two together could conceivably add up to an acceptable prediction. But this way of getting around the problem is almost certainly a stopgap measure substituting for a true solution. None of us really believe that such precise, accidental cancellations are the fundamental explanation for the absence of certain interactions. We grudgingly employ the fortuitous cancellations as a crutch so that we can ignore these problems and proceed to investigate other aspects of our theories.

Physicists believe that interactions are absent from a theory only if the interactions were eliminated in a way that fits the physicists' notion of what's natural. In the everyday world, the word "natural" refers to things that happen spontaneously, without human intervention. But for particle physicists, "natural" means more than something that happens—it means something that, if it should happen, would not present a puzzle. For physicists, it is only "natural" to expect the expected.

The anarchic principle and the many undesirable interactions that quantum mechanics will induce tell us that some new concepts must enter into any model of physics that underlies the Standard Model if this model is to have a chance of being correct. One reason that symmetries are so important is that they are the only natural way in a four-dimensional world to guarantee that unwanted interactions do not occur. Symmetries essentially provide an extra rule about which interactions can conceivably happen. You can readily understand this phenomenon with the help of an analogy.

Suppose that you prepare six table settings, but you have to prepare them so that all six settings are the same. That is, your settings permit a symmetry transformation that interchanges every pair of settings. Without such a symmetry, you could in principle have given one person two forks, another three, and someone else a pair of chopsticks. But with the symmetry constraints, you can only make settings in which all six people have the same number of forks, knives, spoons, and chopsticks—you could never give one person two knives and another person three.

Similarly, symmetries tell you that not all interactions can occur. Even if many of the particles interact, quantum contributions generally won't produce interactions that violate a symmetry if the classical interactions preserve that symmetry. If you don't start with symmetry-violating interactions, you won't ever generate any (aside from the rare known anomalies mentioned in Chapter 14), even when you include all possible interactions involving virtual particles. By imposing symmetry on your table settings, you will always end up with identical settings, no matter how many changes you make, such as adding grapefruit spoons or steak knives. Similarly, interactions that are inconsistent with a symmetry will not be induced, even when quantum mechanical effects are taken into account. If a symmetry weren't already violated in the classical theory, there would be no path that a particle could take that could induce a symmetry-violating interaction.

Until recently, physicists thought that symmetries were the only way to avoid the anarchic principle. But as Raman and I saw, once we'd eaten enough ice cream, separated branes are another way. A crucial reason why extra dimensions initially appeared so promising to me is that they suggested a reason, apart from symmetry, that restricted or unusual types of interactions could be natural. Sequestering unwanted particles can prevent unwanted interactions because they won't generally occur between particles that are separated on different branes.

Particles on different branes don't interact strongly because interactions are always local—only particles in the same place interact directly. Sequestered particles can make contact with particles on other branes, but only if there are interacting particles that can travel from one brane to the other. Like Ike on the Jailbrane, particles on different branes have only restricted means of communication with each other because they have no options apart from invoking an intermediary. Even if such indirect interactions do occur, they are often extremely small, since intermediary particles in the bulk, particularly ones with mass, only rarely travel long distances.

This suppression of interactions between particles sequestered in different places would be similar to the suppression of international information in a country, which I'll call Xenophobia, where the government carefully controls the borders and the media. In Xenophobia, information not provided locally could be acquired only from foreign visitors who manage to enter, or from newspapers or books that get smuggled in.

Similarly, separated branes provide a platform from which to evade the anarchic principle, thereby doubling the set of tools at nature's disposal for guaranteeing the absence of unwanted interactions. A further merit of sequestering is that it can even protect particles from the effects of symmetry breaking. So long as symmetry breaking happens sufficiently far away from those particles, it will have very little effect on them. When symmetry breaking is sequestered, it is quarantined, much as a contagious disease is contained when everyone with the disease is kept within a restricted region. Or, to use our other analogy, dramatic events that occur outside Xenophobia would have no effect in Xenophobia without an intervening communicator. Without porous borders, Xenophobia could function independently of the rest of the world.

Sequestering and Supersymmetry

The particular problem that Raman and I investigated in the summer of 1998 dealt with how sequestering might operate in nature to yield a universe with broken supersymmetry that has the properties of the universe we see. We have seen that supersymmetry can elegantly protect the hierarchy and guarantee that all the large quantum mechanical contributions to the Higgs particle's mass add up to zero. But, as we saw in Chapter 13, even if supersymmetry exists in nature, it must be broken in order to explain why we've observed particles but not their superpartners.

Unfortunately, most models with broken symmetry predict interactions that don't occur in nature, and such models cannot possibly be right. Raman and I wanted to find a physical principle that nature might use to protect itself from these unwanted interactions so that we could incorporate it into a more successful theory.

We focused on supersymmetry breaking in a braneworld context. Braneworlds can preserve supersymmetry. But just as in four dimensions, supersymmetry can be spontaneously broken when some part of the theory contains particles that don't preserve supersymmetry. Raman and I realized that if all the particles responsible for supersymmetry breaking were separated from the Standard Model particles, the model with broken supersymmetry would be less problematic.

We therefore assumed that Standard Model particles were confined to one brane, and that the particles responsible for supersymmetry breaking were sequestered on another. We observed that with such a setup, the dangerous interactions that quantum mechanics could induce don't necessarily occur. Apart from the supersymmetry-breaking effects that might be communicated via intermediary particles in the bulk, the interactions of Standard Model particles would be the same as in a theory with unbroken supersymmetry. So just as in a theory with exact supersymmetry, unwanted flavor-changing interactions that are inconsistent with experiments should not happen. Bulk particles that interact with particles on both the supersymmetry-breaking brane and the Standard Model brane would determine precisely which interactions are possible—and they wouldn't necessarily include the forbidden ones.

Of course, some supersymmetry breaking has to be communicated to the Standard Model particles. Unless supersymmetry breaking is communicated to them, nothing will raise the superpartners' masses. Although we don't know the exact values for the superpartners' masses, experimental constraints, combined with supersymmetry's role in protecting the hierarchy, tell us approximately what the superpartners' masses should be.

The constraints tell us the qualitative relationships among the masses of the superpartners. Roughly speaking, all the superpartners have about the same mass, and those masses are all approximately the weak scale mass, 250 GeV. We needed to ensure that the masses of the superpartners fell in this range, while still preventing unwanted interactions from occurring. All the pieces had to fit for the theory of sequestered supersymmetry breaking to have a chance of being right.

The key to our model's success was finding the intermediary particle that could carry the news of supersymmetry breaking to the Standard Model particles and give the superpartners the masses they needed to have. But we also wanted to be sure that our intermediary would not incite impossible interactions.

The graviton, a bulk particle that interacts with energetic particles no matter where they are, looked like the perfect candidate. The graviton interacts with particles on the supersymmetry-breaking and on the Standard Model branes. Furthermore, the interactions of the graviton are known—they follow from the theory of gravity. We could show that the graviton's interactions, while generating the necessary superpartner masses, do not generate the interactions that would cause quarks or leptons to confuse their identities—the interactions that are known not to occur in nature. The graviton therefore looked like a promising choice.

When Raman and I worked out the superpartner masses that would follow from a mediating messenger graviton, we found that, despite the simple elements, the calculation was surprisingly subtle. Classical contributions to supersymmetry-breaking masses turn out to be zero, and only quantum mechanical effects communicated supersymmetry breaking. When we realized this, we called the graviton-induced communication of supersymmetry-breaking *anomaly mediation*. We chose the name because, like the anomalies I discussed in Chapter 14, the specific quantum mechanical effects broke a symmetry that would otherwise be present. The great thing was that since the masses of the superpartners depended on known quantum effects in the Standard Model, rather than unknown higher-dimensional interactions, we could predict the relative sizes of the superpartners' masses.

It took a few days to get it all straight, which meant that I could go from disappointment to relief in the same day. I remember startling my dinner companion one evening when I became completely distracted because I recognized an error and solved a problem that had worried me earlier in the day. In the end, Raman and I discovered that if gravity communicates supersymmetry breaking, sequestered supersymmetry breaking works surprisingly well. All the superpartners had the right masses, and the relationship between the gaugino and squark masses was in the range where we wanted it to be. Although not everything worked quite as simply as we had initially hoped, important relations among the superpartners' masses fell into place without inducing the impossible interactions that are problematic for other supersymmetry-breaking theories. And with only slight modifications, everything worked.

And, best of all, thanks to the distinctive predictions for the superpartners' masses, our idea can be tested. A very significant feature of sequestered supersymmetry breaking is that, even though the extra dimension could be extraordinarily tiny, something like 10^{-31} cm in size, only about a factor of a hundred bigger than the minuscule Planck scale length, there would still be visible consequences. This goes against standard wisdom, which says that only much larger dimensions could have visible consequences, through either a modified gravitational force law or new heavy particles.

Although it is indeed true that we won't see either of the above experimental consequences when the extra dimension is small, the graviton communicates supersymmetry breaking to the gauginos in a very particular way that we could calculate from the known gravitational interactions and the known interactions that occur in a theory with supersymmetry. The sequestered supersymmetry-breaking model predicts distinctive mass ratios for the gauginos, the partners of the gauge bosons, and those masses can be measured.[30]

This is very exciting. If physicists discover superpartners, they can then determine whether the relationships among their masses agree with what we predict. An experiment to search for these gauge superpartners is under way right now at the Tevatron—the proton-antiproton collider at Fermilab in Illinois. If we are very lucky, we will see results in the next few years.

In the end, Raman and I were both reasonably confident that we had discovered something interesting. But we both had some residual concerns. I was a little afraid that such an interesting idea, if true, couldn't have been overlooked, and that we still needed to ensure that we hadn't missed some hidden flaw in our model. Raman, too, thought the idea too good to have been overlooked. But he was confident it was right, and was afraid only that we might have missed a similar idea in the physics literature.

He wasn't far from the truth. Anomaly mediation of supersymmetry breaking was independently discovered around the same time by Gian Giudice at CERN, Markus Luty at Maryland, Hitoshi Murayama at Berkeley, and Riccardo Rattazzi at Pisa, who had been working together that same summer. They released a paper the day after ours came out. Their research was amazing to me. I couldn't see how two groups of physicists could have traced the same tortuous journey through ideas in a single summer, but Raman had correctly guessed that others might have had similar interests. In fact, we were both right in a way. Although the other group had similar ideas, they developed them independently of the extra-dimensional motivation—without which anomaly-mediated masses were just a curiosity. As Riccardo generously said to the physicist Massimo Porrati, a mutual friend, Raman and I had done it better, not because our version of anomaly mediation was more correct, but because we had a reason anyone would care in the first place! That reason was extra dimensions. Without extra dimensions, supersymmetry breaking wouldn't be sequestered and anomaly-mediated masses would be swamped by larger effects.

Other physicists have since gone on to investigate sequestered models of supersymmetry breaking. They have found ways to join this with other, older ideas to make even more successful models, ones that might represent the real world. People have even found ways to extend the lesson of sequestering back to four dimensions.

There are too many models to enumerate, but let me just mention two ideas I found particularly interesting. The first idea arose from a collaboration between Raman and Markus Luty. They used the insights from the warped geometry (described in Chapter 20) to reinterpret the consequences of sequestering in four dimensions. With these ideas, they developed a new class of four-dimensional symmetry-breaking models.

Another interesting idea was called *gaugino mediation*. The idea was to communicate supersymmetry breaking not through the graviton, but instead through gauginos, the supersymmetric partners of the gauge bosons. For this to work, gauge bosons and their partners couldn't be stuck on a brane; they would have to be free to travel in the bulk. Raman reminded me that gaugino mediation was actually one of the many ideas we had dismissed early on. But the excellent model builders David E. Kaplan, Graham Kribs, and Martin Schmaltz, and, separately, Zacharia Chacko, Markus Luty, Ann Nelson, and Eduardo Ponton, demonstrated that we had been too hasty, and that gaugino mediation might work beautifully in communicating supersymmetry-breaking masses while preserving all the advantages of sequestered supersymmetry breaking.*

Sequestering and Shining Masses

Sequestered symmetry breaking is a powerful tool for model building. The real world could contain separated branes, and by constructing models with this assumption, physicists can explore the range of possibilities.

The previous section explained how problems with flavor-changing interactions might be solved in theories with supersymmetry. But another question challenging the model builder is why there should be different flavors of quarks and leptons with different masses in the first place. The Higgs mechanism gives particles their masses, but the precise values are different for each flavor. This can be true only if each of the flavors interacts differently with whatever plays the role of the Higgs particle. Given that the three flavors of each particle type, such as the up, charm, and top quarks, have exactly the same gauge interactions, it's mysterious that they should all have different masses. Something has to distinguish them, but the particle physics of the Standard Model doesn't tell us what.

We can try to make models that explain different masses. But almost invariably, any model would also contain unwanted interactions that would change flavor identities. What we need is something that can safely distinguish flavors without producing these problematic interactions.

Nima Arkani-Hamed and the German-born physicist Martin Schmaltz assumed that different Standard Model particles were housed on separate branes and that they could explain some masses. Nima and Savas Dimopoulos found another, even simpler possibility. They assumed that there was a brane on which particles of the Standard Model were confined, and that the interactions among particles on this brane treated all flavors identically. But with only

*John Ellis, Costas Kounnas, and Dmitri Nanopoulos had also considered related ideas in string theory earlier on.

flavor-symmetric interactions, which treat all the flavors the same, all particles would have exactly the same mass. Clearly, we can explain the different masses only if something treats the particles differently.

Nima and Savas assumed that other particles responsible for flavor-symmetry breaking were sequestered on other branes. As with sequestered supersymmetry breaking, flavor-symmetry breaking could then be communicated to Standard Model particles only via interactions with particles in the bulk. If there were many bulk particles interacting with the Standard Model, each of which communicated flavor-symmetry breaking from a different brane at a different distance, their model could explain the different masses of the Standard Model flavors. Symmetry breaking communicated from distant branes would induce smaller masses than symmetry breaking communicated from nearby branes. Nima and Savas named their idea *shining* to emphasize this fact. Just as light looks dimmer when its source is further away, the effect of symmetry breaking is smaller when it originates on a more distant brane. In their scenario, different flavors of quarks and leptons would be different because they each interact with a different brane at a different distance.

Extra dimensions and sequestering are novel and exciting ways to address problems in particle physics. And it doesn't necessarily stop there. Recently we have shown that sequestering could even play an important role in cosmology, the science of the evolution of our universe. It's clear that we have yet to discover all the merits of a universe (or multiverse) that contains sequestered particles, and new ideas are still to come.

What's New

- Particles can be sequestered on different branes.

- Even tiny extra dimensions can have consequences for the properties of observable particles.

- Sequestered particles are not necessarily subject to the anarchic principle. Not all interactions necessarily occur, since distant particles cannot directly interact.

- In a model in which particles that play a role in supersymmetry breaking are sequestered from Standard Model particles, supersymmetry can be broken without introducing interactions that would change particles into other flavors.

- Sequestered supersymmetry breaking is testable. If gauginos are produced at high-energy colliders, we can compare gaugino masses and see whether they agree with predictions.

- Sequestered flavor-symmetry breaking might help to explain disparate particle masses.

18

Leaky Passages:
Fingerprints of Extra Dimensions

I was peeking
But it hasn't happened yet
I haven't been given
My best souvenir
I miss you
But I haven't met you yet.   Bjork

Athena had to admit that she missed Ike. Even though she had often found him annoying, she was pretty lonely without him. She was looking forward to spending time with K. Square, an exchange student who was planning to visit. But she was appalled by the closed-mindedness of her neighbors, who were all apprehensive about K. Square's impending arrival. It didn't matter that he spoke the same language and behaved the same way as everyone else. In the current climate, K. Square's foreign origin alone was enough to make them wary.

When Athena asked her neighbors why they were so anxious, they replied, "What if he sends for his heavier relatives? What if they're not so well behaved as he is and stick to their foreign laws? And when they all arrive together, what will happen then?"

Unfortunately, Athena heightened their suspicions by telling them that K. Square and his relatives couldn't possibly stay long in any case, since they were all very unstable and the K. Square family could visit only during the commotion of energetic gatherings. Recognizing her unfortunate choice of words, Athena reassuringly added that the foreigners would stick to local laws during their brief and exciting visits. Convinced, her neighbors then joined Athena in welcoming the K. Square clan.

Earlier in this book I explained how extra dimensions might be hidden. They could be rolled up or hemmed in by branes so as to be imperceptibly small. But can an extra-dimensional universe really hide its nature so completely that none of its features distinguishes it from a four-dimensional world? That would be hard to believe. Even if compactified dimensions are so small that we could be lulled into believing that the world is four-dimensional, a higher-dimensional world must contain some new elements that distinguish it from a truly four-dimensional one.

If there are extra dimensions, such fingerprints of extra dimensions are sure to exist. Such fingerprints are particles called *Kaluza-Klein (KK) particles.** KK particles are the additional ingredients of an extra-dimensional universe. They are the four-dimensional imprint of the higher-dimensional world.

Should KK particles exist and be sufficiently light, high-energy colliders will produce them and they will leave their mark in experimental data. The extra-dimensional detectives—the experimenters—will piece together these clues, transforming data into forensic evidence of a higher-dimensional world. This chapter is about these Kaluza-Klein particles, and why, in a higher-dimensional world, you can be confident of their existence.

Kaluza-Klein Particles

Even if a bulk particle travels in higher-dimensional space, we still should be able to describe its properties and interactions in four-dimensional terms. After all, we don't see extra dimensions directly, so everything should appear to us as if it is four-dimensional. Just as Flatlanders, who see only two spatial dimensions, could observe only

---

*K. Square in the story. KK particles are also known as Kaluza-Klein modes, where "modes" refers to their quantized momenta.

two-dimensional disks when a three-dimensional sphere passed through their world, we can see only particles that look like they travel in three spatial dimensions, even if those particles originated in higher-dimensional space. These new particles that originate in extra dimensions, but appear to us as extra particles in our four-dimensional spacetime,* are Kaluza-Klein (KK) particles. If we could measure and study all their properties, they would tell us everything there is to know about the higher-dimensional space.

Kaluza-Klein particles are the manifestation of a higher-dimensional particle in four dimensions. Just as you can reproduce any sound a violin string could make by the superposition of many resonant modes, you can reproduce a higher-dimensional particle's behavior by replacing it with appropriate KK particles. The KK particles fully characterize higher-dimensional particles and the higher-dimensional geometry in which they travel.

In order to mimic the behavior of higher-dimensional particles, KK particles would have to carry extra-dimensional momentum. Every bulk particle that travels through the higher-dimensional space gets replaced in our effective four-dimensional description by KK particles that have the correct momenta and interactions to mimic that particular higher-dimensional particle.[31] A higher-dimensional universe hosts both familiar particles and their KK relatives that carry extra-dimensional momenta that are determined by the detailed properties of the curled-up space.

However, a four-dimensional description doesn't include information about extra-dimensional position or momentum. Therefore, the extra-dimensional momentum of the KK particles must be called something else when viewed from our four-dimensional perspective. The relationship between mass and momentum imposed by special relativity tells us that extra-dimensional momentum would be seen in the four-dimensional world as mass. KK particles are therefore particles like the ones we know, but with masses that reflect their extra-dimensional momenta.

The KK particles' masses are determined by the higher-dimensional

---

*This is our usual counting of spacetime dimensions. Our previous discussion of Flatland in Chapter 1 preceded relativity, so we only counted spacial dimensions there.

geometry. However, their charges are the same as those of known four-dimensional particles. That is because if known particles originate from higher-dimensional spacetime, higher-dimensional particles have to carry the same charges as known particles. That's also true for the KK particles that mimic the higher-dimensional particles' behavior. So for each particle we know about, there should be many KK particles with the same charge, each with different mass. For example, if an electron travels in higher dimensions, it would have KK partners that have the same negative charge. And if a quark travels in higher dimensions, it would have KK relatives that, like the quark, experience the strong force. KK partners have identical charges to the particles we know, but masses that are determined by extra dimensions.

Determining Kaluza-Klein Masses

Understanding the origin and masses of KK particles requires taking a step beyond the intuitive picture of invisible curled-up dimensions that we looked at earlier on. For simplicity, we'll first consider a universe without branes, in which every particle is fundamentally higher-dimensional and is free to move in all directions—including any additional ones. To be concrete, we'll imagine a space with only one additional dimension which has been rolled up into a circle and elementary particles that travel inside that space.

Had we lived in a world where classical Newtonian physics was the final word, Kaluza-Klein particles could have had any amount of extra-dimensional momentum and therefore any mass. But because we live in a quantum mechanical universe, this is not the case. Quantum mechanics tells us that, just as only the resonant violin modes contribute to the sounds the violin strings can make, only quantized extra-dimensional momenta contribute when KK particles reproduce the motion and interactions of a higher-dimensional particle. And just as the notes of a violin string depend on its length, the quantized extra-dimensional momenta of the KK particles depend on the extra dimensions' sizes and shapes.

The extra-dimensional momenta that the KK particles carry would appear to us in our apparently four-dimensional world as a distinctive pattern of KK particle masses. If physicists discover KK particles, these masses will tell us about the geometry of the extra dimensions. For example, if there is a single extra dimension that is curled into a circle, these masses would tell us the extra dimension's size.

The procedure for finding the allowed momenta (and hence masses) for KK particles in a universe with a curled-up dimension is very similar to the method you use to mathematically determine resonant violin modes, and also to the method that Bohr used to determine quantized electron orbits in an atom. Quantum mechanics associates all particles with waves, and only those waves that can oscillate an integer number of times over the extra-dimensional circle are allowed. We determine the allowed waves, and then use quantum mechanics to relate wavelength to momentum. And the extra-dimensional momenta tell us the allowed KK particles' masses, which is what we want to know.

The constant wave—the one that doesn't oscillate at all—is always allowed. This "wave" is like the surface of a perfectly still pond, without any visible ripples, or a violin string that has not yet been plucked. This probability wave has the same value everywhere in the extra dimension. Because of the constant value of this flat probability wave, the KK particle associated with this wave doesn't favor any particular extra-dimensional location over any other. According to quantum mechanics, this particle carries no extra-dimensional momentum and, according to special relativity, has no additional mass.

The lightest KK particle is therefore the one associated with this constant probability value in the extra dimension. At low energies this is the only KK particle that can be produced. Since it has neither momentum nor structure in the extra dimension, it is indistinguishable from an ordinary four-dimensional particle with the same mass and charge. With only a low energy, the higher-dimensional particle is not able to wiggle around at all in the compact rolled-up dimension. In other words, low energy won't produce any of the additional KK particles that distinguish our universe from one with more dimensions. Low-energy processes and the lightest KK particles will therefore tell us nothing about the existence of an extra dimension, never mind its size or shape.

However, if the universe contains additional dimensions, and particle accelerators achieve sufficiently high energies, they will create heavier KK particles. These heavier KK particles, which carry nonzero extra-dimensional momenta, will be the first real evidence of extra dimensions. In our example, those heavier KK particles are associated with waves that have structure along the circular additional dimension; the waves vary as they wind around the rolled-up dimension, oscillating up and down an integer number of times along its length.

The lightest such KK particle would be the one whose probability function has the largest wavelength. And the largest wavelength for which the oscillation fits in a circle is the one that oscillates up and down exactly once as the wave winds around the rolled-up dimension. That wavelength is determined by the size of the extra dimension's circumference (it's approximately the same size). A larger wavelength would not fit; the wave would be mismatched when it returned to a single point along the circle. The particle with this probability wave is the lightest KK particle that "remembers" its extra-dimensional origin.

It makes sense that the wavelength of the wave associated with this lightest particle with nonzero extra-dimensional momentum would be about the same as the extra dimension's size. After all, intuition tells us that only something sufficiently small to probe features or interactions on a tiny scale would be sensitive to a curled-up dimension's existence. Trying to investigate an extra dimension with a bigger wavelength would be like trying to measure the location of an atom with a ruler. For example, if you were trying to detect an extra dimension with light or some other probe of a particular wavelength, the light would have to have a wavelength smaller than the size of the extra dimension. Because quantum mechanics associates probability waves with particles, the above statements about the wavelengths of probes translate into statements about particle properties. Only particles with sufficiently small wavelength and therefore (from the uncertainty principle) sufficiently high extra-dimensional momentum and mass could be sensitive to an extra dimension's existence.

Another attractive feature of the lightest of the KK particles with nonzero extra-dimensional momentum is that its momentum (and hence its mass) is smaller when the extra dimension is bigger. A larger extra dimension should be more accessible and give more readily observable consequences because lighter particles are easier to produce and discover.

If extra dimensions do exist, this lightest KK particle would not be the sole evidence for them. Other, higher-momentum particles would leave even sharper fingerprints of extra dimensions at particle colliders. These particles would have probability waves that oscillate more than once when traversing the curled-up dimension. Because the nth such particle would correspond to a wave that oscillated n times as it wound around the rolled-up dimension, the masses of these KK particles would all be integer multiples of the lightest one. And the higher the momenta, the sharper the fingerprints of extra dimensions that the KK particles will leave at particle colliders. Figure 74 schematically shows the values of KK particle masses, which are proportional to the inverse size of the extra dimension, and a couple of waves that correspond to these massive particles.

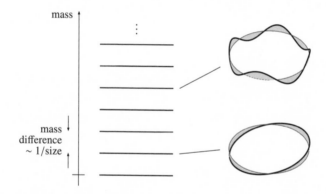

Figure 74. *Kaluza-Klein particles correspond to the waves that oscillate an integer number of times around the curled-up dimension. Waves with more oscillations correspond to heavier particles.*

The many successively heavier KK particles resemble the multiple generations of an immigrant family. The members of the youngest generation who were born in the USA fully assimilate American culture, speak English perfectly, and don't betray their foreign roots at all. That isn't as true for the previous generation, the parents of this youngest generation: perhaps they speak with a trace of an accent, and occasionally tell a few proverbs from the old country. The generation that is older still would sound even more foreign, and wear clothing and tell stories that originated in their homeland. These earlier generations might be said to add cultural dimensions to what would otherwise be a less colorful, uniform society.

Similarly, the lightest KK particles are indistinguishable from particles in a fundamentally four-dimensional world; only the more massive "older relations" would reveal evidence of extra dimensions. Although the lightest of the KK particles would appear to be four-dimensional, their provenance would become apparent once sufficient energy to produce the more massive "elders" was achieved.

If experimenters discover new heavy particles with the same charges as familiar ones and masses that are similar to one another, those particles will be strong evidence of extra dimensions. If such particles share the same charges and occur at regular intervals of mass, it would very likely mean that a simple curled-up dimension has been discovered.

But more complicated extra-dimensional geometries will yield more complicated patterns of masses. If enough such particles are discovered, the KK particles would then reveal not only the existence of extra dimensions, but also the extra dimensions' sizes and shapes. No matter what the geometry of the hidden dimensions, the KK particles' masses would depend on it. In all cases, the KK particles and their masses could tell us quite a lot about extra-dimensional properties.

Experimental Constraints

Until recently, most string theorists assumed that extra dimensions are no bigger than the minuscule Planck scale length. This is because gravity becomes strong at the Planck scale energy, and a theory of quantum gravity, which could be string theory, should take over at that point. But the Planck scale length is far smaller than any length we can study experimentally. The tiny Planck scale length corresponds (according to quantum mechanics and special relativity) to the enormous Planck scale mass (or energy)—ten thousand trillion times the reach of current particle accelerators. Planck-mass KK particles would be so heavy that they would be well out of range of any conceivable experiment.

However, perhaps extra dimensions are bigger and KK particles are lighter. Why not ask instead what experimental tests tell us about an extra dimension's size? What do we really know, theoretical prejudice aside?

If the world is higher-dimensional and there are no branes, then all familiar particles—the electron, for example—would have KK partners.[32] These would be particles that have exactly the same charge as familiar particles, but carry momenta in the additional dimensions. The electron's KK partners would be negatively charged like the electron, but heavier. If an extra dimension is rolled up into a circle, the mass of the lightest such particle would differ from the electron's mass by an amount inversely proportional to the extra dimension's size. That means that, the larger the extra dimension, the smaller the particle's mass. Because a bigger dimension would give rise to lighter KK particles, none of which have been seen in experiments, the bounds on KK particles' masses constrain the allowed size of an extra curled-up dimension.

So far there has not been any sign of such charged particles at colliders operating at energies up to about 1,000 GeV. Since the KK particles would be signatures of extra dimensions, our not yet having seen them tells us that extra dimensions cannot be too large. Current experimental constraints tell us that extra dimensions cannot be any larger than 10^{-17} cm (one hundred-thousandth of a trillionth of a centimeter).* This is extremely small—far smaller than anything we can see directly.

This limit on an extra dimension's size is about ten times smaller than the weak scale length. But even though 10^{-17} cm is small, it is huge compared with the Planck scale length, which is 10^{-33} cm, sixteen orders of magnitude smaller. This means that extra dimensions could be much bigger than the Planck scale length and still have evaded detection. The (modern) Greek physicist Ignatius Antoniadis was one

---

*Remember, we have assumed no branes; this limit will change in the following chapters.

of the first to imagine that extra dimensions were not the Planck length, but were instead comparable in size to the length scale associated with the weak force. He was thinking about what new physics might appear when colliders increase their energy even a little. After all, the hierarchy problem tells us that something must be seen at those energies, at which particles will be produced with weak scale energies and masses.

But even the above limit on the size of extra dimensions doesn't necessarily always apply. KK particles are fingerprints of extra dimensions, but they can be wily and surprisingly hard to find. We've recently learned a lot more about KK particles and what they might look like. The following chapters will explain the latest results about why, once branes come into the picture, extra dimensions can be bigger than 10^{-17} cm and still escape detection—even though you'd expect bigger dimensions to give rise to lighter KK particles. Some models with surprisingly large dimensions—dimensions that you might have thought would have very visible consequences—can be invisible yet nonetheless help to explain the mysterious properties of Standard Model particles. And Chapter 22 will present an even more surprising result: an infinitely large extra dimension could give rise to infinitely many light KK particles, yet nonetheless leave no observable trace.

What's New

- Kaluza-Klein (KK) modes are particles that carry extra-dimensional momentum; they are higher-dimensional interlopers in our four-dimensional world.

- KK particles would look like heavy particles with the same charges as known particles.

- KK particles' masses and interactions are determined by the higher-dimensional theory; they therefore reflect the properties of the higher-dimensional spacetime.

- If we could find and measure the properties of all KK particles, we would know the size and shape of the higher dimensions.

- Current experimental constraints tell us that if all particles travel through higher-dimensional space, extra dimensions can be no bigger than about 10^{-17} cm.

19

Voluminous Passages:
Large Extra Dimensions

I couldn't even see the millimeter when it fell.

Eminem

Now that K. Square's brief visit was over, Athena spent a lot of time at the local Internet cafe. She was exhilarated by her recent discovery of some mysterious new websites, the most intriguing of which was xxx.socloseandyetsofar.al. Athena suspected that these suggestive sites were a consequence of the recent AOB (America on Brane)/ Spacetime Warner multimedia merger, but she had to go home before she had time to investigate.

When Athena arrived at her house she rushed to her computer, where she once again sought the exotic hyperlinks that had been so readily accessible at the Internet cafe. To her frustration, however, CyberNanny prevented her from reaching the forbidden dimensionally enhanced sites.* But by cloaking her identity with her secure alias, Mentor, Athena vanquished her cybercensor and succeeded in finally returning to the mysterious hyperlinks.

Athena secretly hoped that K. Square had sent her a message that was hidden in a webpage. But the sites were not easy to understand, and she managed to pick up only a few potentially meaningful signals. She resolved to study their content some more and hoped the merger— unlike the other merger with a similar name—would last long enough for her to figure them out.

---

*Physicists post their papers on a website that begins with "xxx": check out xxx.lanl.gov. Internet filters have occasionally forbidden access to this site as well.

At the Oxford supersymmetry conference in 1998, the Stanford physicist Savas Dimopoulos gave one of the most interesting talks. He reported on work he had done in collaboration with two other physicists, Nima Arkani-Hamed and Gia Dvali. The colorful names of these three match their colorful characters and ideas. Savas gets very excited about his projects; his collaborators tell me that his enthusiasm is always contagious. He was so taken with extra dimensions that he told a colleague that all the new unexplored physics ideas made him feel like a kid in a candy store—he wanted to eat it all before anyone else got any. Gia, a physicist from former Soviet Georgia, takes great risks, both in his approach to physics and in his audacious feats of mountaineering. He was once stuck without any food on a stormy mountaintop in the Caucusus for two nights. Nima, a physicist from an Iranian family, is very energetic, stimulating, and vividly articulate. Now my Harvard colleague, he often roams the hallways enthusiastically explaining his latest research and convincing others to join in.

Ironically, Savas's talk at the supersymmetry conference, which was not about supersymmetry at all but was instead about extra dimensions, stole some of supersymmetry's thunder. He explained that extra dimensions, rather than supersymmetry, could be the physical theory underlying the Standard Model. And if his suggestion was correct, experimenters could expect to find evidence of extra dimensions, rather than supersymmetry, when they explore the weak scale in the near future.

This chapter presents Arkani-Hamed, Dimopoulos, and Dvali's* idea about how very large dimensions might explain the weakness of gravity. In essence, large extra dimensions could dilute the gravitational force so much that gravity's strength would be much weaker than estimates without extra dimensions would have you believe. Their models don't actually solve the hierarchy problem because you still have to explain why the dimensions are so large. But ADD hoped this new and different question would be more tractable.

We'll also consider the related question ADD asked: how big can rolled-up extra dimensions be if Standard Model particles are confined

*For brevity, I'll refer to them collectively as "ADD."

to a brane and are not free to travel in the bulk without contradicting experimental results? The answer they found was extraordinary. At the time they wrote their paper, it looked like extra dimensions could be as big as a millimeter.

Dimensions (Almost) as Large as a Millimeter

In the ADD model, as in the sequestering model I described in Chapter 17, the Standard Model particles are confined to a brane. However, the two models had very different objectives, so their remaining features are completely different. Whereas the sequestering model had one additional dimension that was bounded between two branes, the ADD models all have more than one dimension and those dimensions are curled up. Depending on the details of the implementation, space in their models contains two, three, or more additional curled-up dimensions. Moreoever, the ADD model contains a single brane on which the Standard Model particles are confined, but that brane does not bound space. It simply sits inside the extra curled-up dimensions, as is illustrated in Figure 75.[33]

Figure 75. *Schematic drawing of the ADD braneworld. The universe's extra dimensions are rolled up (and large). We live on a brane (the dotted line along the cylinder), so only gravity experiences the extra dimensions.*

One question ADD wanted to address with their setup was how large extra dimensions could still be hidden if all particles of the Standard Model were trapped on a brane and the only force in the higher-dimensional bulk was gravity. The answer they found surprised most physicists. As opposed to the size of one-hundredth of a thousandth of a trillionth of a centimeter that we considered in the previous chapter, these extra curled-up dimensions could be as large as a millimeter. (Actually, it's a little tricky to give the precise number now because, as we'll discuss further later in the chapter, physicists at the University of Washington have since looked for millimeter-size extra dimensions experimentally and they didn't find them. Based on their results, we now know that extra dimensions must be smaller than about a tenth of a millimeter, or else they would be ruled out. Nonetheless, dimensions that are even one-tenth of a millimeter in size would still be rather shocking.)

You might have thought that if dimensions were as big as a millimeter (or even ten times smaller), we would surely know about them already. After all, anyone who can't see a millimeter-size object needs new glasses. On the scales of particle physics, a millimeter is enormous.

To get an idea of how extraordinary extra dimensions a millimeter—or even one a tenth of a millimeter—in size would be, let us recap the sorts of length scale we have discussed so far. The Planck scale length, well out of any experimental reach, is 10^{-33} cm. The TeV scale, which experiments currently explore, is about 10^{-17} cm; physicists have tested electromagnetism down to distances as small as 10^{-17} cm. The sizes ADD were talking about are huge in comparison. In the absence of branes, millimeter-size extra dimensions would be an absurdity that would have been ruled out.

However, branes make far larger extra dimensions conceivable. Branes can trap quarks, leptons, and gauge bosons so that *only* gravity experiences the full higher dimensionality of space. In the ADD scenario, which assumes that everything other than gravity is confined to a brane, everything that doesn't involve gravity would look exactly the same as it would without the extra dimensions, even if the extra dimensions were extremely large.

For example, everything you see would look four-dimensional. Your eye detects photons, and photons in the ADD model are trapped on a brane. Therefore all objects you see would look as if there were only three spatial dimensions. If photons are trapped on a brane, then no matter how strong your glasses you could never see any evidence of extra dimensions directly.

In fact, you could hope to see evidence of millimeter-size dimensions in the ADD scenario only with an extremely sensitive gravity probe. All of the usual particle physics processes, such as interactions mediated by the electromagnetic force, electron-positron pair creation, and the binding of the nucleus through the strong force, occur only on the four-dimensional brane and would be exactly the same as in a purely four-dimensional universe.

Charged KK particles would not be a problem either. The previous chapter explained that extra dimensions cannot be very big when all particles are in the bulk, because if they were, we would already have seen the KK partners of Standard Model particles. But this is not true in the ADD scenario because all Standard Model particles—the electron, for example—are bound to a brane. So the Standard Model particles, which don't travel in the higher-dimensional bulk, wouldn't carry extra-dimensional momenta. Standard Model particles, which are confined to a brane, therefore wouldn't have KK partners. And since there would be no KK partners, the constraints based on KK particles such as the ones considered in the last chapter wouldn't apply.

In fact, in the ADD model, the only particle that must have KK partners is the graviton, which we know must travel in the higher-dimensional bulk. However, the graviton's KK partners interact far more weakly than the Standard Model KK partners. Whereas Standard Model KK partners interact via electromagnetism, the weak force, and the strong force, the KK partners of the graviton interact only with gravitational strength—as weakly as the graviton itself. The graviton's KK partners would therefore be much harder to produce and detect than the KK partners of Standard Model particles. After all, no one has ever directly seen the graviton. Its KK partners, which interact as weakly as the graviton itself, should be no easier to find.

ADD realized that if the only constraints on extra dimensions came from gravity, the size of the extra dimension in their scenario where

Standard Model particles are stuck to a brane could be much larger than the previous chapter suggests. The reason is that gravity is very feeble, and is therefore extremely difficult to investigate experimentally. For light objects at close distances, gravity is so weak that its effects are readily overwhelmed by other forces.

For example, the gravitational force between two electrons is 10^{43} times weaker than the electromagnetic force. The gravitational force of the Earth dominates only because its net charge is zero. On small scales, not only the net charge matters, but also the way charges are distributed. To test the gravitational force law between small objects, the pull of gravity must be shielded from even the tiniest consequences of the other forces. Although the planets orbiting the Sun, the Moon orbiting the Earth, and the evolution of the universe itself tell us about the form of gravity at very large distances, gravity is hard to test at short distances. We know a lot less about it than we do about the other forces. So if gravity is the only force in the bulk, the existence of surprisingly large extra dimensions would not contradict any experimental results. Dimensions with brane-bound particles are hard to observe.

In 1996, when ADD wrote their paper, Newton' inverse square law had been tested down to distances of only about a millimeter. That meant that extra dimensions could be as large as a millimeter and no one would have seen any evidence of them. As ADD said in their paper, "Our interpretation of M_{Pl} [the Planck energy] as a fundamental energy scale [where gravitational interactions become strong] is then based on the assumption that gravity is unmodified over the 33 orders of magnitude between where it is measured . . . down to the Planck length 10^{-33} cm."* In other words, in 1998 nothing was known about gravity from experiments at distances smaller than about a millimeter. At separations less than that, the gravitational force law could behave differently, with gravitational attraction increasing much more rapidly as objects approached each other, for example—yet no one would have known.

---

*Nima Arkani-Hamed, Savas Dimopoulos, Gia Dvali, "The hierarchy problem and new dimensions at a millimeter," *Physics Letters* B, vol. 429, pp. 263–72 (1998).

Large Dimensions and the Hierarchy Problem

The possibility of large extra dimensions was an important observation. But ADD didn't study large extra dimensions simply to explore abstract possibilities. Their true interest was particle physics, and the hierarchy problem in particular.

As was explained in Chapter 12, the hierarchy problem concerns the large ratio of the weak scale mass and the Planck scale mass, the masses that we associate with particle physics and gravity. Until recently, the main question that particle physicists asked was why the weak scale mass is so small, despite the large (Planck-scale-mass-size)* virtual contributions to the Higgs particle's mass that tend to make it larger. Until physicists started thinking about extra dimensions, all attempts to address the hierarchy problem involved enhancing the Standard Model in the hope of finding a more comprehensive underlying particle physics theory that would explain why the weak scale mass is so much smaller than the Planck scale mass.

But the hierarchy problem involves a large disparity between two numbers. The conundrum is why the Planck scale and the weak scale are so different. So the hierarchy problem can be phrased another way: why is the Planck scale mass so large when the weak scale mass is so small—or, equivalently, why is the strength of gravity acting on elementary particles so weak? Put this way, the hierarchy problem raises the question of whether gravity, and not particle physics, is different from what physicists have assumed.

ADD pursued this train of logic and suggested that attempts at solving the hierarchy problem through extensions of the Standard Model were on the wrong track. They observed that sufficiently large extra dimensions could equally well solve the problem. They proposed that the fundamental mass scale that determines gravity's strength is not the Planck scale mass, but a much smaller mass scale, close to a TeV.

However, ADD were then left with the question of why gravity

*Remember the Planck scale *length* is tiny, but the Planck scale *mass* (or *energy*) is enormous.

should be so weak. After all, the reason that the Planck scale mass is so big is that gravity is weak—gravity's strength is inversely proportional to this scale. A much smaller fundamental mass scale for gravity would make gravitational interactions far too strong.

But this problem wasn't insurmountable. ADD pointed out that it was only higher-dimensional gravity that was necessarily strong. They reasoned that large extra dimensions could dilute the strength of gravity so much that although the gravitational force would be very strong in higher dimensions, gravity in the lower-dimensional effective theory would be very feeble. In their picture, gravity appears feeble to us because it gets diluted in a very large extra-dimensional space. The electromagnetic, strong, and weak forces, on the other hand, would not be feeble because those forces would be confined to a brane and would not be diluted at all. Large dimensions and a brane could therefore conceivably explain why gravity is so much feebler than the other forces.

Nima told me that the turning point in their research was when he and his collaborators understood the precise relationship between the strengths of higher- and lower-dimensional gravity. This relationship was not new. String theorists, for example, always used it to relate the four-dimensional gravitational scale to the ten-dimensional one. And, as I briefly explained in Chapter 16, Hořava and Witten used the relationship between the strengths of ten- and eleven-dimensional gravity when they observed that gravity can be unified with other forces: a large eleventh dimension permits the higher-dimensional gravitational scale, and hence the string scale, to be as low as the GUT scale. But no one before had recognized that higher-dimensional gravity could be sufficiently strong to address the hierarchy problem so long as extra dimensions are large enough to adequately dilute it. After Nima, Savas, and Gia had thought about extra dimensions for a while and learned how to relate higher- and lower-dimensional gravity, they understood this extraordinary implication.

Relating Higher- and Lower-Dimensional Gravity

In Chapter 2 we saw that when you explore only those distances that are larger than the size of curled-up extra dimensions, the extra dimensions are imperceptible. However, that doesn't necessarily mean that additional dimensions don't have physical consequences; even though we don't see them, they can still influence the values of quantities we do see. Chapter 17 gave an example of this phenomenon. In the sequestering model of supersymmetry breaking, in which supersymmetry breaking occurred on a distant brane and the graviton communicated the breaking to the supersymmetric partners of Standard Model particles, the values of superpartner masses reflected the extra-dimensional origin of supersymmetry breaking and its communication via gravity.

We'll now consider another example in which extra dimensions influence the values of measurable quantities. The sizes of the compactified dimensions determine the relationship between the strength of four-dimensional gravity (that is, the one we observe) and the strength of the higher-dimensional gravity from which it derives. Gravity is diluted in extra dimensions and is weaker when curled-up extra dimensions enclose a larger volume.

To see how this works, let's return to the example of Chapter 2, where we considered the three-dimensional garden-hose universe as an analogy for a bulk three-dimensional space bounded by branes. If water were to enter the hose through a pinhole (see Figure 23, p. 47), it would initially spurt out from the hole and spread in all three dimensions. However, once the water reached the width of the hose, it would spread only along the hose's length—which is why the hose appears to be one-dimensional when we measure the gravitational force law at distances greater than the extra dimensions' size.

But even though the water travels only along the single dimension of the hose, its pressure depends on the size of the cross-section. One way to understand this is by imagining what would happen if the width of the hose increased. The water that entered through the pinhole would then spread out over a larger region, and the pressure of the water exiting the hose would be weaker.

If the pressure of water represents gravitational force lines, and the water entering the hose through the pinhole represents the field lines emerging from a massive object, then the force lines from this massive object would initially spread in all three directions, just like the water in the previous example. And when the force lines reach the walls of the universe (the branes), they would bend and run solely along the single large dimension. With the hose, we found that the wider the nozzle, the weaker the water pressure. Similarly, the area of extra dimensions in our toy garden-hose universe would determine how dilute the field lines will be in the lower-dimensional world. The larger the area of the extra dimensions, the weaker the gravitational field strength in the effective lower-dimensional universe would be.

The same argument applies to rolled-up dimensions in a universe with any number of curled-up dimensions. The larger the volume of the extra dimensions, the more dilute the gravitational force and the weaker the strength of gravity. We can see this with a higher-dimensional hose analogous to the one we just considered. Gravitational force lines in a higher-dimensional hose would first spread out in all dimensions, including the extra curled-up dimensions. The force lines would reach the boundary of the curled-up dimensions, after which they would spread out only along the infinite dimensions of the lower-dimensional space. The initial spreading out in the extra dimensions would reduce the density of force lines in the lower-dimensional space, so the strength of gravity experienced there would be weaker.[34]

Back to the Hierarchy Problem

Because of the dilution of gravity in extra dimensions, lower-dimensional gravity is weaker when the volume of the extra-dimensional compactified space is bigger. ADD observed that this dilution of gravity into extra dimensions could conceivably be so large that it could account for the observed weakness of four-dimensional gravity in our world.

They reasoned as follows. Suppose that gravity in a higher-dimensional theory does not depend on the enormous Planck scale mass of 10^{19} GeV, but instead on a much smaller energy, about a TeV, sixteen orders of magnitude smaller. They chose a TeV to eliminate the hierarchy problem: if a TeV or some nearby energy were the energy at which gravity became strong, there would be no hierarchy of masses in particle physics. Everything, both particle physics and gravity, would be characterized by the TeV scale. So maintaining a reasonably light Higgs particle with mass of about a TeV would not be a problem in their model.

According to their assumption, at energies of about a TeV higher-dimensional gravity would be a reasonably strong force, comparable in strength to the other known forces. To have a sensible theory that agrees with what we see, ADD therefore needed to explain why four-dimensional gravity looks so weak. The added ingredient in their model was the assumption that the extra dimensions are extremely large. Ultimately we would want to explain this large size. But according to their proposal, the curled-up dimensions enclose such a large volume. And, in keeping with the logic of the previous section, four-dimensional gravity would be extremely feeble. Gravity in our world would be weak because extra dimensions are large, not because there is fundamentally a big mass responsible for the tiny gravitational force. The Planck scale mass that we measure in four dimensions is large (making gravity appear weak) only because gravity has been diluted in large extra dimensions.

How large would these extra dimensions have to be? The answer depends on the number of extra dimensions. ADD considered different possible numbers of dimensions for their model, since experiments haven't yet decided how many dimensions there are. Notice that we are interested only in the large dimensions at this point. So if you think that you and your local string theorist know that the number of spatial dimensions is nine or ten, you can still consider different possibilities for the number of large dimensions and assume that all the other dimensions are small enough to ignore.

The size of the dimensions in the ADD proposal depends on how many there are, because volume depends on the number of dimensions. If all dimensions were the same size, a higher-dimensional region would enclose more volume than a lower-dimensional one and would therefore dilute gravity more. You can see this easily enough from the fact that lower-dimensional objects fit inside higher-dimensional ones. Or, returning to our sprinkler analogy in Chapter 2, you can see that a plant receives more water from a sprinkler that spreads water only over a line segment of a particular size (one dimension) than one that spreads water over the surface area enclosed by a circle (two dimensions) whose diameter is the same size. When water spreads over a higher-dimensional region, it becomes more diluted.

If there were only one large extra dimension, it would have to be enormous to satisfy the ADD proposal. It would have to be as large as the distance from the Earth to the Sun in order to dilute gravity enough. That's not allowed. If the extra dimension were that big, the universe would behave as if it were five-dimensional at measurable distances. We already know that Newton's gravitational force law applies at these distances; a large extra dimension that would modify gravity at such large distances is clearly ruled out.

However, with as few as two additional dimensions, the size of the dimensions is almost acceptably small. If there were just two additional dimensions, they could be as small as a millimeter and still adequately dilute gravity. That is the reason ADD paid so much attention to the millimeter scale. Not only was it on the verge of experimental probes, but two additional dimensions of this size could be relevant to the hierarchy problem. Gravity would spread throughout these two millimeter-size dimensions and yield the weak gravitational force we know. Of course, a millimeter is still pretty big, but as we said earlier, gravity tests are not nearly as restrictive as you might think. Spurred on by the ADD scenario, people thought harder about looking for rolled-up dimensions of this size.

With more than two additional dimensions, gravity is modified only at a very small distance. With more additional dimensions, it can be sufficiently diluted even if those extra dimensions are relatively small. For example, with six extra dimensions the size need only be about 10^{-13} cm, one ten thousandth of a billionth of a centimeter.

Even with such small dimensions we could, if we're lucky, find evidence of one of these examples some time very soon—not in the direct gravitational tests we'll discuss in the next section, but in the experiments at high-energy particle colliders that we'll consider afterwards.

Looking for Large Dimensions

How would one go about finding differences in gravity at small distances? What should one look for? We know that if there are curled-up dimensions, the strength of gravity at distances less than the size of the extra dimensions would decrease more rapidly with distance than Newton had predicted, because gravity would spread out in more than three spatial dimensions. Whenever objects were separated by less than the extra dimensions' sizes, higher-dimensional gravity would apply. A bug sufficiently small to circle a curled-up dimension would experience the extra dimension, both because it could travel in it and because the gravitational force would spread around it in all dimensions. So if anyone, such as this unusually perceptive bug, could detect the gravitational force at short distances, extra dimensions would have visible consequences.

This tells us that by exploring gravity at distances as small as (or smaller than) a proposed curled-up dimension's size, and studying how gravity's strength depends on the separation of masses at those distances, an experiment could study the behavior of gravity and look for evidence of extra dimensions. However, experiments that are sensitive to gravity at very short distances are formidably difficult to build. Gravity is so weak that it is readily overwhelmed by the other forces, such as electromagnetism. As mentioned earlier, at the time of the ADD proposal, experiments had searched for deviations from Newton's gravitational force law and shown that the law applies at least down to distances of about a millimeter. If anyone could do better and study even shorter distances, they had a chance of discovering the large dimensions of the ADD proposal, which were just on the verge of experimental accessibility.

Experimenters rose to the new challenge. Motivated by the ADD idea, Eric Adelberger and Blayne Heckel, two professors at the University of Washington, designed a beautiful experiment whose purpose was to look for deviations from Newton's law at very short distances. Others have also studied short-distance gravity, but this experiment was the most stringent test of the ADD proposal.

Their apparatus, located in the basement of the University of

Washington physics department, is called the Eöt-Wash experiment. The name refers to a famous physicist who studied gravity, the Hungarian Baron Roland von Eötvös. The Eöt-Wash group's experiment is illustrated in Figure 76. It consists of a ring suspended above two attractor disks, one slightly above the other. Holes are bored into the ring, and the upper and lower disks, and they are aligned in just such a way that if Newton's law is correct, the ring won't twist. However, if there were extra dimensions, the difference in gravitational attraction from the two disks would not agree with Newton's law and the ring would twist.

Figure 76. *The apparatus of the Eöt-Wash experiment. A ring is suspended over two disks. The holes in the ring and the disks guarantee that the ring will not twist if Newton's inverse square law is obeyed. The three spheres near the top of the apparatus are used for calibration purposes.*

Adelberger and Heckel found no twisting and concluded that no extra-dimensional (or other) effects modified the gravitational force at the distances they could study. Their experiment measured the gravitational force at distances smaller than ever before, establishing that Newton's law applies all the way down to about a tenth of a millimeter. This meant that extra dimensions, even those for which Standard Model particles are confined on a brane, cannot be quite as big as the millimeter that ADD had suggested. They have to be at least ten times smaller.

Remarkably, millimeter-size dimensions are also prohibited by observations of outer space. The quantum mechanical uncertainty principle associates a millimeter with an energy of only about 10^{-3} eV, and a tenth of a millimeter with an energy of about 10^{-2} eV—either way, an extremely small energy, orders of magnitude less than that needed to produce an electron, for example.

Particles with such a low mass could be found in the surrounding universe and in celestial objects, such as supernovae or the Sun. These particles would be so light that if they existed, hot supernovae could produce them. Because we know how quickly supernovae cool and we understand the cooling mechanism (via neutrino emission), we know that there can't be too many other low-mass objects emitted. The cooling rate would be too fast if energy leaked out in some other way. In particular, gravitons shouldn't carry away too much energy. Using this reasoning, physicists showed (independently of terrestrial experiments) that extra dimensions should be smaller than about a hundredth of a millimeter.

However, you should bear in mind that, impressive as it is to rule out deviations of gravity at millimeter distances, this doesn't test most of the currently proposed extra-dimensional models. Remember, only the model with two large extra dimensions produces effects that would be visible on the millimeter scale. If a theory with more than two large extra dimensions solves the hierarchy problem (or if one of the models we'll consider in the next chapter applies to the world), deviations from Newton's law would occur only at much shorter distances.

We don't know for sure what the gravitational attraction between two objects less than a tenth of a millimeter apart will look like. No one has ever tested it. So we don't know whether extra dimensions open up at a tenth of a millimeter, which, if you think about it, is not all that small. Relatively large extra dimensions—though not quite as big as a millimeter—remain a viable possibility. To test such models we'll have to wait for collider tests, the subject of the next section.

Collider Searches for Large Extra Dimensions

High-energy particle colliders are well-suited to discover KK particles from large extra dimensions, even if there are more than two of them. In the ADD large extra-dimensions models, the KK partners of the graviton are always incredibly light. If the large-dimension proposal applies to the real world, the graviton KK partners would be light enough to be produced at accelerators, no matter how many extra dimensions there were. That tells us that even if dimensions are smaller than a millimeter, current and future accelerator searches should be able to discover them. Current colliders create more than enough energy to make such low-mass particles. In fact, if the only relevant quantity were energy, KK particles would already have been produced in abundance.

However, there is a catch. The graviton's KK partners interact only incredibly feebly—as feebly, in fact, as the graviton itself. Since a graviton's interactions are so negligible that gravitons are never produced or detected at colliders at a measurable rate, an individual graviton KK partner wouldn't be either.

But the potential for detecting KK particles from higher dimensions is actually much more promising than this dismal assessment might lead you to believe. This is because, if the ADD proposal is correct, there would be so many light KK partners of the graviton that together they could leave detectable evidence of their existence. If the large-dimensions scenario is true, then even though any individual KK particle could be produced only rarely, the probability of producing one of the large number of light KK particles would be measurably large. For example, if there were two extra dimensions, about one hundred billion trillion KK modes would be light enough to be produced at a collider operating at an energy of about a TeV. The rate of producing at least one of these particles would be fairly high, even if the rate of producing any single one of them were extremely low.

It would be as if someone hinted something to you in such a subtle manner that you didn't take it to heart the first time you heard it. But afterwards, fifty people repeated the same thing. Even though you wouldn't take much notice the first time you heard the message, by the fiftieth time the message would register. Similarly, although the light KK particles are light enough to be produced at current accelerators, they interact so weakly that we can't detect any individual one. However, when an accelerator achieves sufficiently high energy to produce a lot of them, KK particles will leave observable signals.

The Large Hadron Collider, which will study TeV-scale energies, could produce KK particles at a measurable rate if the ADD idea is correct. That might sound like a fortunate coincidence—why should an energy of about a TeV be relevant to KK production rates when neither the KK masses nor the mass that determines the interaction strength of the KK particles (that is, M_{Pl}) are a TeV? The answer is that an energy of about a TeV determines the strength of higher-dimensional gravity, and higher-dimensional gravity ultimately determines what a collider will produce. Because the interactions of the many KK partners of the graviton are equivalent to the interaction of a single higher-dimensional graviton, and the higher-dimensional graviton interacts strongly at energies of about a TeV, the sum of the contributions of all KK particles must also be significant at this scale.

Experimenters are already looking for KK particles at the Tevatron at Fermilab. Although the Tevatron doesn't reach energies as high as the LHC will achieve, it does attain energies where it makes sense to start looking. But the LHC will do better, and has a much greater chance of finding ADD KK particles, should they exist.

What would these KK particles look like? The answer is that the collisions that produce graviton KK partners will look like ordinary collider events, except that it will appear that energy is missing. At the LHC, when two protons collide they could produce a Standard Model particle and a graviton KK partner. The Standard Model particle could be a gluon, for example—the protons would collide to produce a virtual gluon, and this virtual gluon could turn into a true physical gluon and a graviton KK partner.

However, any individual KK particle would interact too feebly for it to be detected—remember, graviton KK partners interact very weakly and might be detected only because there are so many of them. But because the detector would register the gluon—or, more accurately, the jet (see Chapter 7) that surrounds the gluon—the event that produced the graviton KK partner would be recorded, even if the graviton KK partner is not. The key to identifying the event's extra-dimensional origin would be that the unseen KK partner carrying away energy into the extra dimensions so that energy would seem to be missing. By studying single jet events where the energy of the emitted gluon is less than the energy that entered the collision, experimenters could deduce that they had produced a graviton KK partner (see Figure 77). This would be similar to the way in which Pauli surmised the existence of the neutrino (as we saw in Chapter 7).

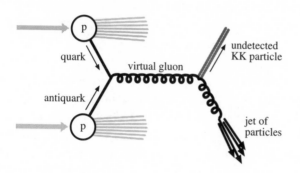

Figure 77. *KK particle production in the ADD model. Protons collide, and a quark and an antiquark annihilate into a virtual gluon. The virtual gluon turns into an undetected KK particle and an observable jet. The gray lines are sprays of additional particles that protons always emit when they collide.*

Because all we would know about the new particle is that it carries away energy, in actuality we wouldn't know for sure that the accelerator had produced a KK particle and not some other particle that interacts too weakly to detect. However, by doing detailed studies of the missing-energy events—how the production rate depends on energy, for example—experimenters could hope to determine whether the KK particle interpretation is correct.

KK particles would be the most accessible extra-dimensional interlopers in our four-dimensional world because they are likely to be the lightest of the objects that could signal extra dimensions. But, if we're lucky, other signatures of the ADD model could appear along with them, including even more exotic objects. If ADD are correct, higher-dimensional gravity would become strong at about a TeV, which is to say at far lower energy than would be true in a conventional four-dimensional world. If that is the case, black holes might be produced at close to a TeV energy, and such higher-dimensional black holes would be a gateway to a better understanding of classical gravity, quantum gravity, and the shape of the universe. If the relevant energies of the ADD proposal are sufficiently low, black hole production could be imminent; they could be formed at the LHC.

The higher-dimensional black holes that would form at colliders would be far smaller than the ones in the universe around us. They would be comparable in size to the very tiny extra dimensions. In case you are worried, rest assured that these small, very short-lived black holes won't pose a danger to us or to our planet: they'll be gone well before they could do any damage. Black holes don't last for ever: they evaporate by emitting radiation through the phenomenon known as *Hawking radiation*. But just as a small drop of coffee evaporates more quickly than a full cup, a small black hole evaporates more quickly than a big one, so the small black holes that could conceivably be produced at colliders will evaporate almost immediately. Nevertheless, if they are produced, these higher-dimensional black holes would last long enough to leave visible signs of their existence at a detector. They would have a very distinctive appearance since they would produce many more particles than you would find in ordinary particle decays, and these particles would go off in all directions.

Furthermore, if the ADD model is correct, black holes and KK partners of the graviton might not be the only exotic new discoveries. If ADD and string theory are both right, colliders could produce strings at very low energies, almost as low as a TeV. Once again, this is because the fundamental gravity scale is so low in the ADD models. Higher-dimensional gravity would become strong at about a TeV, and quantum gravity could contribute measurable effects.

The strings of the ADD theory would not be nearly as massive as the inaccessible Planck scale mass. If you think of strings as notes, the strings of the ADD proposal are far less high-pitched. The low-pitched strings of the ADD models would have mass not much bigger than a TeV. If we're lucky, they'll be light enough for the LHC to produce. Collisions with high enough energy would then produce the light strings of this model in abundance, along with new objects called *string balls*, containing many long strings.

However, despite the appeal of such potential discoveries, you should bear in mind that in all likelihood the energy at the LHC will be close to, but not as high as, the energy needed to make strings and black holes. Whether or not the ADD strings and black holes will be visible depends on the precise energy of higher-dimensional gravity (and, of course, on whether the proposals are correct).

The Fallout

The ADD proposal was fascinating. Who would have thought that extra dimensions could be so large, or that they could have so much bearing on problems of immediate interest (to particle physicists at least), such as the hierarchy problem? However, this proposal did not actually *solve* the hierarchy problem. It turned the hierarchy problem into another question: can additional dimensions be this large? This remains an outstanding question for the ADD scenario. Without some new and as yet undetermined physical principles, dimensions are not expected to be so extraordinarily large. At the very least, according to known theories, you would still need supersymmetry to maintain the large flat space that is needed for the ADD proposal. In essence, supersymmetry would stabilize and reinforce large dimensions that would otherwise collapse. Since one nice feature of ADD seemed to be that it could eliminate the need for supersymmetry, this is a bit disappointing.

The other weakness of the theory is its cosmological implications. For the theory to agree with known facts about the evolution of the universe, some of its numbers have to be very carefully chosen. And the bulk has to contain very little energy, or else cosmological evolution won't agree with observations. Again, this might be possible, but the whole point of a solution to the hierarchy problem is to eliminate the necessity for large fudges.

Nonetheless, many physicists were open to the idea of taking extra-dimensional theories seriously and trying to devise ways to search for them. Experimenters, especially, were excited. As Joe Lykken, a particle physicist working at Fermilab, said to me when describing experimenters' reaction to large extra dimensions, "To them, all this 'beyond the Standard Model' research is kooky. Supersymmetry or extra large dimensions? Who cares? Extra dimensions is no kookier." Experimenters were hungry to search for something new, and extra dimensions provided a very interesting alternative to supersymmetry.

Theorists had a more mixed reaction. On the one hand, large extra dimensions seemed outlandish; no one had considered them before, since no one knew of any reason why extra dimensions should be so large. On the other hand, no one could identify a way to rule them out. In fact, before the first paper about large extra dimensions was written, Gia Dvali, one its the authors, spoke about them at Stanford. The authors, who were aware of the radical nature of their proposal, awaited the talk with trepidation, and were relieved when there were no serious objections. But they were also dismayed—how could people accept this pretty radical idea with such equanimity? Nima told me that when they first posted their paper on the Internet, they had a similar experience. Although they had expected a flood of responses, they received only two. Apparently the Italian physicist Riccardo Rattazzi and I were the only ones to comment on some potential problems. And even these two messages were not really independent: Riccardo and I had just discussed the paper at CERN, where we were both visiting.

Subsequently, as physicists absorbed the implications of the ADD model, they investigated the real-world consequences in more detail, considering tests of gravity, accelerator searches, astrophysical conse-quences, and cosmological implications. Reactions varied according to research interest or style.

Physicists whose research explored details of the Standard Model were happy to accept the possibility of a new idea, one which was in any case interesting. Surprisingly, there was more hostility from some model builders, who were unwilling to forfeit ideas about supersym-metry that had become entrenched over the years. Admittedly, altering the Standard Model so dramatically poses formidable challenges. Any new model would have to reproduce those features of the Standard Model that have already been experimentally verified, and theories that alter the Standard Model too dramatically will have a tough time meeting this challenge. Furthermore, the shining light of supersym-metry—the unification of couplings, the fact that at high energy all forces would have equal strength—would have to be abandoned. However, younger theorists not so wedded to supersymmetry were more excited. The topic of extra dimensions was a new, not-yet-cornered idea, and posed new challenges and open questions.

The reaction from string theorists was mixed as well. When Savas Dimopoulos began his project, he foresaw that work on extra dimensions would bring string theory and particle physics closer together. And string theorists did pay attention, though most of them viewed large extra dimensions as an interesting idea that would never be relevant to string theory. For string theorists the major problem was theoretical: it is very difficult to understand how dimensions could be as large as assumed in the ADD proposal.

Personally, I don't believe that extra dimensions, even if they exist, will turn out to be this large.* Both for theoretical reasons (it's hard to get dimensions that are this large) and for observational ones (it's very tough to get cosmology to work out), the idea seems like a long shot. Even Nima, one of the protagonists, is skeptical at this point. But this was a very important theoretical idea. This new, previously unexplored suggestion highlighted the extent of our ignorance about gravity and the shape of the universe. The ADD paper stimulated a good deal of new thought, and whether or not the idea proves correct, it has had an important impact on physicists' thinking. The large-dimension scenarios have led to many new proposals for extra dimensions and many ideas for experimental tests. After the LHC turns on, theoretical prejudices will be irrelevant in any case, since the implications of hard data will be irrefutable. Who knows? They might turn out to be right.

What's New

- If Standard Model particles are confined to a brane, extra dimensions can be much larger than physicists previously thought: they can be as large as about a tenth of a millimeter.

- Extra dimensions can be so large that they can explain why

---

*If they are flat (see Chapter 22).

gravity is so much weaker than the electromagnetic, weak, and strong forces.

- If large extra dimensions solve the hierarchy problem, higher-dimensional gravity would become strong at about a TeV.

- If higher-dimensional gravity becomes strong at about a TeV, the LHC will produce KK particles at a measurable rate. The KK particles would carry away energy from the collision, so their signature would be events with missing energy.

20

Warped Passage: A Solution to the Hierarchy Problem

What's so small to you,
Is so large to me.
If it's the last thing I do,
I'll make you see.   Suzanne Vega

Athena awoke with a start. She had just revisited her recurring dream, which had again begun with her entering the dreamworld's rabbit hole. In this episode, when the Rabbit announced, "Next stop, TwoD-Land," Athena ignored him and waited to hear the choices that remained.

At the three-spatial-dimensional stop, the Rabbit announced, "If you lived here, you'd be home by now." But he refused to open the doors, despite Athena's pleas that she did indeed live there and very much wanted to return home.

At the next stop, uniformed six-dimensioners tried to enter. But the Rabbit took one look at their inordinately large girth and abruptly closed the doors, saying that they couldn't possibly fit. They quickly departed once the Rabbit threatened to cut them down to size. *

The elevator continued on its extraordinary journey. When it stopped again, the Rabbit announced "Warped Geometry—a five-dimensional world."† He gently pushed Athena towards the door, advising her, "Enter the funhouse mirror—it will take you home."

---

*As we saw in Chapter 18, extra dimensions can be uniform, large, and flat. The Rabbit is skeptical about this idea.
†This counting includes a dimension of time.

Since the Rabbit had mentioned a fifth dimension, Athena found this highly unlikely. But she didn't have any choice but to enter and hope the tricky Rabbit was right.

When you learn a language, the words you remember depend on your particular needs or interests. On a bicycle trip in Italy, for example, I learned to ask for water in many different ways—*acqua di rubinetto, acqua minerale, acqua (minerale) gassata, acqua (minerale) naturale,* etc.* Similarly, when learning about new physical scenarios, each physicist has her own perspective and her own questions, and might therefore notice certain aspects of a system or discover different implications of what is already known. Each of us can hear something different, even when faced with the same words or situation. It makes sense to listen carefully.

Raman and I had each been thinking about the hierarchy problem for years. But we were not searching for a new, better solution to the hierarchy problem when we began our collaboration. We were working on the model of sequestered supersymmetry breaking that I presented in Chapter 17. In the course of that work, we inadvertently discovered a remarkable *warped geometry* of spacetime (a particular type of curved geometry that we'll soon explore) that was bounded by two branes. And because Raman and I were concerned with particle physics and the weakness of gravity, we immediately recognized the warped geometry's potential significance: if the Standard Model of particle physics lies in this spacetime, the hierarchy problem could be solved. I'm not sure that we were the first to study this particular set of Einstein's equations. But we were definitely the first to recognize this startling implication.

The next few chapters explain this and other remarkable possibilities of curved spacetime and how its consequences sometimes violate our expectations. This chapter focuses on a warped five-dimensional world that could help to explain the vast range of masses that are relevant to particle physics. Whereas in four-dimensional quantum field theory, particles are expected to have roughly the same masses,

*Tap water, bottled water, water with gas, water without, etc.

in a warped higher-dimensional geometry this is no longer the case. Warped geometries provide a framework in which very disparate masses emerge naturally, and in which quantum effects are under control.

In the particular geometry described in this chapter, we'll see that space is so strongly warped in the presence of two flat boundary branes that the hierarchy problem of particle physics is automatically solved—without the need for a large dimension, or for any arbitrary large number at all. In this scenario, one brane experiences a large gravitational force but the other does not. Spacetime changes so rapidly along the fifth dimension that it parlays a modest number associated with the separation between the two branes into a huge number (about ten million billion) associated with the relative strength of the gravitational force.

We'll first explain the weakness of gravity on the second brane in terms of the graviton probability function, which determines the graviton's interactions at any particular location in the fifth dimension. But we'll also explain gravity's weakness in different terms, based on the warped geometry itself rather than the graviton interaction strength. We'll see that one of the amazing consequences of warped geometry is that size, mass, and even time depend on position along the fifth dimension. The warping of space and time in this two-brane setup is like the warping of time near the horizon of a black hole. But in this case, time dilates, geometry expands, and on one of the branes particles have a small mass—so the hierarchy problem gets automatically solved.

After discussing the warped geometry and its implications for the hierarchy problem, we'll conclude this chapter with a discussion of the distinctive implications of the theory for future experiments. One of the most exciting aspects of this theory, as with the large extra-dimension models of the previous chapter, is that if it is correct it will very soon have observable consequences at particle accelerators. In fact, we'll see they will be even more dramatic than the missing energy signature we discussed. The KK partners of the gravitons, though visitors from higher-dimensional space, will be distinguishable, visible particles that will decay into familiar particles on our four-dimensional brane.

Warped Geometry and Its Surprising Implications

The geometry that we'll consider in this chapter contains two branes that bound a fifth dimension of space, as illustrated in Figure 78. This setup is similar to the one considered in Chapter 17 in that there are two branes with a fifth dimension that extends between them. However, it really is quite a different theory. The particles and the distribution of energy are different, and the theory is not super-symmetric. Nevertheless, as with that theory, we assume that all of the Standard Model particles, along with a Higgs particle responsible for breaking electroweak symmetry, are confined to one of the two branes.

Also as before, in this setup we'll assume that gravity is the only force that exists throughout the fifth dimension. This means that were

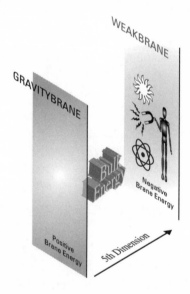

Figure 78. *The warped five-dimensional geometry with a single brane. The universe has five spacetime dimensions, but the Standard Model resides on a brane (the Weakbrane) that has four. Again, the total number of spacetime dimensions in this setup is five, whereas the number of spatial dimensions is four, three of which extend along the branes and one that extends between them.*

it not for gravity, each of the branes would look like a conventional four-dimensional universe. Gauge bosons and particles confined to the branes would communicate forces and interact as if the fifth dimension didn't exist. Standard Model particles would travel only in the three flat spatial brane dimensions, and forces would spread out only along the flat three-dimensional surface of the brane.[35]

Gravity, however, is different since it is not restricted to a brane, but instead exists in the full five-dimensional bulk. The force of gravity would be felt everywhere in the fifth dimension. But this does not necessarily mean that it is felt equally everywhere. Energy on the branes and in the five-dimensional bulk curves spacetime, and this makes an enormous difference to the gravitational field.

The large extra-dimensional theories of the previous chapter took advantage of the fact that branes could trap particles and forces, but neglected the energy that the branes themselves could carry. Raman and I weren't sure that this was always a good assumption, since a central component of Einstein's theory of general relativity is that energy induces a gravitational field, which means that when branes carry energy, they should curve space and time. In a universe with only a single extra dimension, which was what we intended to study, it was not at all clear that one could neglect brane and bulk energy: the gravitational effects of the brane don't dissipate very rapidly, so one would expect distortions of spacetime, even far away from the branes.

We wanted to know how spacetime would curve in the presence of two energetic branes that bounded the extra dimension of space. Raman and I solved Einstein's gravity equations for this two-brane setup, assuming that there was energy both in the bulk and on the branes. We discovered that such energy was indeed very important— the resulting spacetime was dramatically curved.

In some cases, curved spaces are easy to picture. The surface of a sphere, for example, is two-dimensional—you need only latitude and longitude to know your location—but it is nonetheless clearly curved. However, many curved spaces are more difficult to draw because they can't readily be represented in three-dimensional space. The particular warped spacetime that we will now consider is such an example. It is part of a spacetime known as *anti de Sitter space*. Anti de Sitter space has negative curvature, more like a Pringles potato chip than a sphere.

The name comes from the Dutch mathematician and cosmologist Willem de Sitter, who studied a space with positive curvature that is now called *de Sitter space*. Although we don't need the name here, we'll refer to it later on when we connect this theory to a theory of anti de Sitter space that string theorists had been studying.

Although we'll soon explore the interesting way in which the five-dimensional spacetime is curved, let's first focus for a moment on the two branes at the edges of the fifth dimension. These two boundary branes are completely flat. If you were on the brane at either boundary, you would be stuck on a three-plus-one-dimensional world (three dimensions of space and one of time),* which would extend infinitely far in the three spatial dimensions and look like flat spacetime, with no peculiar gravitational effects.

Furthermore, the curved spacetime has the special property that were you to restrict yourself to *any* single slice along the fifth dimension—not just the branes at the ends—you would find that this slice is completely flat. That is, although there aren't branes anywhere in the fifth dimension except at the ends, the geometry of the three-plus-one-dimensional surfaces that you get by restricting yourself to any single five-dimensional point looks flat: it has the same shape as the large flat branes at the boundaries. If you think of the boundary branes as the heels of a loaf of bread, the flat, parallel four-dimensional regions at any location along the fifth dimension of spacetime are like the flat slices of bread from the interior of the loaf.

But the five-dimensional spacetime we are considering is nonetheless curved. That is reflected in the way the four-dimensional flat spacetime slices are glued together along the fifth dimension. I first spoke about this geometry at the Kavli Institute for Theoretical Physics in Santa Barbara, where the string theorist Tom Banks informed me that, technically speaking, the five-dimensional geometry Raman and I found is *warped*. Although many curved spacetimes are colloquially called warped, the technical term refers to geometries in which each slice is flat,† but they are put together with an overall *warp factor*.

---

*I'll sometimes use "three-plus-one" instead of "four" when I want to emphasize the distinction between space and time.

†Really, all the slices have the same geometry; in this case, the slices are all flat.

The warp factor is a function that changes the overall scale for position, time, mass, and energy at each point in the fifth dimension. This fascinating feature of warped geometry is subtle, and I'll explain it further in the following section. The warp factor is also reflected in the graviton's probability function and interactions that we'll soon explore.

A curved space with flat slices is pictured in Figure 79. It is a filled-in funnel. We could slice the funnel into flat sheets with a cleaver, but the surface of the funnel is clearly curved. This is similar in some

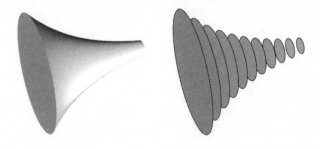

Figure 79. *A filled-in funnel consists of flat slices glued together.*

respects to the curved five-dimensional spacetime we're considering. But the analogy isn't perfect, because the boundary of the funnel, the funnel's surface, is the only place where it's curved, whereas in the warped spacetime the curvature is everywhere. This curvature would be reflected in an overall rescaling of the measuring rod of space and the clock speed for time, which would be different at each point in the fifth dimension.[36]

A simpler way of illustrating the curvature of warped spacetime is through the shape of the graviton's probability function. The graviton is the particle that communicates the gravitational force, and its probability function tells us the likelihood of finding the graviton at any fixed position in space. The strength of gravity is reflected in this function: the larger its value, the stronger the graviton's interactions at that particular point, and the stronger the force of gravity.

For flat spacetime, the graviton would be equally likely to be found anywhere. The probability function for a graviton in flat spacetime would therefore be constant. But for curved spacetime, as in the warped geometry we are considering, this would no longer be the case. Curvature tells us about the shape of gravity. When spacetime is curved, the value of the graviton's probability function is different at different locations in spacetime.

Because each slice of spacetime is completely flat in our warped geometry, the graviton's probability function doesn't vary along the three standard spatial dimensions—it changes only along the fifth dimension.* In other words, even though the graviton's probability function has different values at different places along the fifth dimension, so long as two points are equally far along the fifth dimension, the value will be the same. This tells us that the graviton's probability function depends only on position in the fifth dimension. Nonetheless, it completely characterizes the warped spacetime's curvature. And because that function varies only with a single coordinate, that of the fifth dimension, it is simple to plot.

This graviton's probability function along the fifth dimension is shown in Figure 80. It decreases exponentially quickly, which is to say extraordinarily rapidly, as one leaves the first brane, which we

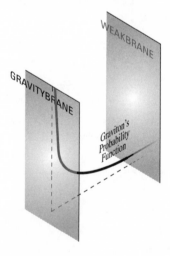

Figure 80. *The graviton's probability function falls off exponentially as it moves away from the Gravitybrane and towards the Weakbrane.*

*Remember that the fifth dimension is the fifth dimension of spacetime and the hypothetical fourth dimension of space.

will call the Gravitybrane, and heads towards the second brane, which we'll call the Weakbrane. The Gravitybrane and the Weakbrane are different because the first carries positive energy, while the second carries negative energy. And this energy assignment makes the graviton's probability function much bigger in the vicinity of the Gravitybrane.

The effect of the plummeting probability function is that the graviton, the physical particle whose exchange generates gravitational attraction, has very little chance to be found near the Weakbrane. The graviton's interactions on the Weakbrane are therefore highly suppressed.

The strength of gravity depends so strongly on position along the fifth dimension that the strengths of gravity experienced on the two branes, which border the opposite ends of this warped five-dimensional world, are extraordinarily different. Gravity is strong on the first brane, where gravity is localized, but feeble on the second, where the Standard Model resides. Because the graviton's probability function is negligibly small on the second brane, the graviton's interactions with Standard Model particles, which are confined there, are extremely weak.

This tells us that in this warped spacetime we would actually expect to find a hierarchy between observed masses and the Planck scale mass. Although the graviton is everywhere, it interacts with far greater strength with particles on the Gravitybrane than with particles on the Weakbrane. The graviton just isn't hanging around there all that much. The graviton's probability function on the Weakbrane is extremely tiny, and if this scenario is the correct description of the world, this tinyness is responsible for the feebleness of gravity in our world.

In this model, feeble gravity on the Weakbrane doesn't require a large separation between the two branes. Once you leave the Gravitybrane, where the graviton's probability function is highly concentrated, gravity becomes exponentially weaker, which makes gravity on the Weakbrane extremely feeble. Because the graviton's probability function is falling precipitously, gravity is highly suppressed on the Weakbrane (where we live). It can be ten million billion times weaker than you would expect without the warping, even if the two branes are fairly close together. This aspect of the theory, the fact that the branes don't need to be separated by very much, makes this model a far more realistic possibility than large extra dimensions. Although large extra dimensions were a tantalizing rephrasing of the hierarchy problem, at the end of the day they still leave an unexplained large number—the extra dimensions' size. In the theory we are now considering, the gravitational force on the Weakbrane is many orders of magnitude weaker than other forces, even when the Weakbrane is only a modest distance away from the first brane (the Gravity-brane).

The distance between branes in this warped geometry need only be a little larger than the Planck scale length. Whereas the large dimensions scenario required the introduction of an extremely large number—namely the size of the dimensions—in the warped geometry, no contrived large number is required to explain the hierarchy. That is because an exponential automatically turns a modest number into an extremely huge number (the exponential) or an extremely tiny one (the inverse of the large exponential). The strength of gravity is smaller on the Weakbrane; it is reduced by a factor of the exponential of the distance between the two branes.* The enormous ratio between the Planck scale mass—the large mass that tells us that gravity is weak—and the mass of the Higgs particle, and therefore the masses of the weak gauge bosons, is expected if the Weakbrane is located at distance of 16 units away,† since the ratio of the different masses is about 10^{16} (ten million billion). That means that a distance between the branes that is only about a factor of sixteen bigger than your most naive guess would suffice to explain the hierarchy. A factor of sixteen might sound big, but it is a lot smaller than ten million billion, the number we are trying to explain.

For years, particle physicists had hoped to find an exponential explanation of the hierarchy. That is, we had hoped to interpret the previously unexplained large number as the consequence of a

---

*The units in which distance is measured are determined by the energy on the brane, which would be determined by the Planck scale mass.
†This number is in units of the curvature, which is in turn determined by the energy on the brane and in the bulk.

naturally occurring exponential function. Now, with extra dimensions, Raman and I had found a way for particle physics to automatically incorporate an exponential hierarchy of masses. The interaction of gravity could be much smaller at the location of our brane, the Weakbrane, than it would be where the graviton's probability function peaks. Because gravity on our brane would be weakened by the warped geometry, if the Standard Model is housed on the Weakbrane, the hierarchy problem would be solved. This was a solution to the hierarchy problem, and it had fallen right into our laps.

Another way to understand this remarkable new feature of the warped geometry is to consider how gravity gets diluted. In Chapter 19, we explained the weakness of gravity in the ADD scenario in terms of the gravitational force lines emitted from a massive object, which were diluted because they were spread throughout large dimensions. If we so chose, we could have described this dilution as a consequence of the graviton's probability function. Remember that the graviton's probability function tells us how gravity is spread out over space. Because gravity in the large extra-dimensions scenario is equally strong everywhere in the extra dimensions, the graviton's probability function in this case is flat. Such a flat graviton probability function would tell us that the graviton, the particle communicating gravity, is spread out over the large region enclosed by the extra dimensions. This flat probability function, that is equally distributed over all of extra-dimensional space, tells us that gravity's influence in four dimensions is vastly diluted.

In the warped five-dimensional spacetime we are now considering, there is an interesting twist. The graviton no longer has equal probability of being in all places in five-dimensional space that lie between its two boundaries, the Gravitybrane and the Weakbrane. The distribution of the graviton is in fact far from democratic as an automatic consequence of the energy carried by the branes and in the bulk. The graviton's probability function varies: it is big in one region and small in all others, and it is this variation that provides the dilution factor responsible for making gravity so weak in our world. Gravity is feeble on the Weakbrane because the graviton's probability function there is so minuscule.

Let us momentarily return to the sprinkler analogy that we used earlier to explain how the strength of gravity decreases with distance. The larger the region over which the sprinkler distributes water (as illustrated in the upper part of Figure 81), the more the water gets diluted. When there are large extra dimensions, gravity is spread over a very big region, and it too gets diluted. Gravity therefore appears to be feeble in the low-energy, effective four-dimensional theory.

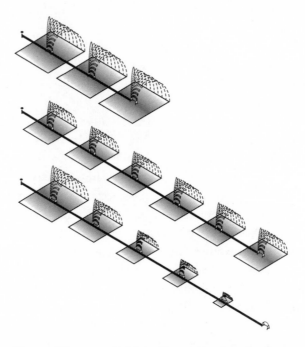

Figure 81. *Three different sprinklers. By comparing the first and second sprinklers, we see that a longer sprinkler delivers less water over any particular region than a shorter one does. The third sprinkler demonstrates that water can be inequitably distributed so that the first garden always gets half the water, the second garden one-quarter, and so on. In that case, the amount of water delivered to the first garden is independent of the sprinkler's length; it always gets half the water.*

The warped geometry, on the other hand, resembles a sprinkler that does not distribute water equally in all directions, but instead delivers it preferentially to one particular region, the region around the

Gravitybrane (see the lower part of Figure 81). With an undemocratic sprinkler, it is obvious that less water will be delivered everywhere aside from the favored region. And, if the amount of water delivered to other regions falls off exponentially from the highly favored location, the water fraction delivered to those other areas will be very tiny indeed, even if they are only a modest distance away. Clearly, the water delivered by the "warped" sprinkler is "diluted" far more than water that is equally distributed to all regions.

The upshot is that if all the Standard Model particles are housed on the Weakbrane, then gravity is so weak compared with the other three forces that the hierarchy problem of particle physics, the question of why gravity is so weak relative to the other forces, is solved. Feeble gravity is a natural consequence of the small amplitude of the graviton's probability function on the Weakbrane, even when it is only a relatively modest distance (about ten times larger than the string-theory-favored Planck scale length) away from the Gravitybrane.

Growing and Shrinking in a Warped Dimension

The previous explanation of the hierarchy in terms of the exponentially falling probability function is entirely adequate for understanding warped spacetime. The intuitive explanation for the weakness of gravity is that the graviton is less likely to be found on the Weakbrane. You are free to accept this explanation and skip to the next section, but you might want to read the following, slightly more rigorous explanation, which delves further into the fascinating properties of warped spacetime.

In this section, we'll see that gravity's weakness on the Weakbrane can also be explained as a consequence of objects getting bigger and lighter as you move away from the Gravitybrane and approach the Weakbrane. Were Athena to move from the Gravitybrane to the Weakbrane (as she will in the story in the next chapter), she would see her shadow on the Gravitybrane increase in size as she moved away. And the amount her shadow would increase is enormous—it would grow by sixteen orders of magnitude!

We will also see that in this geometry, heavy and light particles can peacefully coexist. Even when there are Planck-scale-mass particles on one of the two branes, there are only weak-scale-mass particles on the other. Therefore, there is no longer a hierarchy problem.

To understand how this works, suppose that, like most people (at least those who haven't read this book), you were completely ignorant of the fifth dimension—which is, after all, invisible. Untroubled in the belief that you live in four dimensions, you would know only about four-dimensional gravity, which you would believe was communicated by a conventional four-dimensional graviton. In the four-dimensional effective theory that describes what you see, there would be only one gravitational force, and there could therefore be only a single type of four-dimensional graviton. All particles would interact with that single type of graviton. But that graviton wouldn't contain any information about the location of a particle in the original, higher-dimensional theory.

This reasoning makes it look as though all graviton interactions should be the same—that is, independent of where in the fifth dimension an object originated. After all, you wouldn't know that the object originated in the fifth dimension, or even that there *was* a fifth dimension. Newton's constant of gravitation, which determines the graviton's interaction strength, would be the single quantity that determines the strength of all four-dimensional gravitational interactions. But in the previous section, we saw that gravity's interactions are weaker as you moved from the Gravitybrane towards the Weakbrane. This leaves the question, how can gravity's strength encompass information about an object's fifth-dimensional location?

The resolution to the apparent paradox hinges on the fact that gravitational attraction is also proportional to mass, and mass at different points along the fifth dimension can and must be different. The only way to reproduce the weakened graviton interaction on each successive slice along the fifth dimension is to measure mass differently on each four-dimensional slice.

One of the many remarkable properties of warped spacetime is that as you move from the Gravitybrane to the Weakbrane, energies and momenta shrink. The shrinking energies and momenta (and consistency with quantum mechanics and special relativity) also tell us that distance and time must expand (as shown in Figure 82). In the geometry I am describing, size, time, mass, and energy all depend on

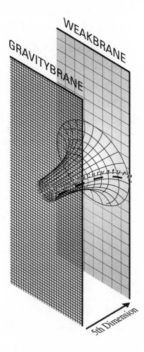

Figure 82. *Sizes increase (and masses and energies decrease) as one moves from the Gravitybrane to the Weakbrane.*

location. Four-dimensional sizes and masses inherit values that depend on their original five-dimensional positions. Physics looks four-dimensional. But the ruler with which length is measured, or the scale with which mass is measured, depends on the original five-dimensional location. Residents of the Gravitybrane or the Weakbrane both see four-dimensional physics, but they would measure different sizes and expect different masses.

The gravitational attraction of masses of particles that originate further away from the Gravitybrane in the original five-dimensional theory is smaller in the four-dimensional effective theory because the masses themselves are smaller. This is because at each position in the fifth dimension, mass and energy get *rescaled* by an amount proportional to the amplitude of the graviton's probability function at that particular point. And the *warp factor*, which is the amount by which you rescale the energies, is smaller further away from the Gravitybrane. In fact, its plot has exactly the same shape as the graviton's probability function. Masses and energy therefore shrink by a different factor at every point along the fifth dimension—and the warp factor determines by how much.

This rescaling might seem arbitrary, but it's not. It is subtle, however, so let's first consider an analogous situation. Suppose that we were to measure time in terms of how long it takes to travel 100 km by train. I will call these units of time TT (train time) units. This is a fine measure of time, except that your determination of time would depend on where you are traveling: are the trains fast there or are they slow? For example, suppose that a movie lasted two hours. If an American train took an hour to travel 100 km, an American viewer would cover 200 km over the course of that movie and say that the movie lasted 2 TTs. A French viewer riding the TGV, on the other hand, would think that the movie lasted 6 TTs, because express trains in France travel about three times faster and the French viewer would need to watch their DVD during a 600-km-long train ride to see how it ends. Because the French viewer's train covers 100 km in 20 minutes, whereas an American's train covers the same distance in an hour, you need to rescale train time if Americans and French are to share common units and agree on the TT length of the movie. To convert from French to American time, you would have to rescale the French train time by a factor of three.

Similarly, on the Weakbrane, where the graviton interaction is far smaller than on the Gravitybrane, the units for the scale used to measure energy must be rescaled to take account of gravity's weakness. At the Weakbrane, the rescaling is by an enormous amount, 10^{16}, ten million billion. What this means is that whereas on the Gravitybrane all fundamental masses are expected to be M_{Pl} (the Planck scale mass), on the Weakbrane they are expected to be only about 1,000 GeV, a factor of 10^{16} smaller. Masses of new particles that live on the Weakbrane might be somewhat larger, perhaps 3,000 or 5,000 GeV, but they shouldn't be much larger than that since all masses have been rescaled enormously.

The hierarchy problem arises when all masses get raised to the largest mass around. If that mass is the Planck scale mass, all masses are expected to be about as big as the Planck scale mass. But owing to the rescaling, if you originally thought that the Planck scale mass was the expected mass for everything on the Gravitybrane, then on the Weakbrane you would conclude that a TeV, sixteen orders of magnitude smaller, is the expected mass.* This means that the mass of the Higgs particle is not at all disturbing: a mass of about a TeV—ten million billion times smaller than the Planck scale mass— is expected, even though gravity is weak. The rescaling, which is essential in this interpretation, solves the hierarchy problem.

By the same reasoning, all new objects on the Weakbrane, including strings, should have mass of about a TeV. This tells us that this model could have dramatic experimental consequences. On the Weakbrane, the extra particles associated with strings would be very much lighter than those on the Gravitybrane—or in a four-dimensional world, for that matter. The Weakbrane presents a fabulous scenario from the perspective of discovering extra dimensions. If this idea is correct, then low-mass particles from extra dimensions should be near at hand. TeV-mass particles would abound on the Weakbrane.

Everything on the Weakbrane is expected to be lighter than the Planck scale mass by a factor of 10^{16}. And according to quantum mechanics, smaller mass means larger size. Athena's shadow would grow as she went from the Gravitybrane towards the Weakbrane. This tells us that strings on the Weakbrane should not be 10^{-33} cm in size. Instead, they should also be sixteen orders of magnitude larger— that is, about 10^{-17} cm.

Although I have focused on a two-brane scenario with a specific warp factor, the features we have considered are likely to be more general than this particular example. With extra dimensions there is

---

*The names Planckbrane and TeV-brane or Weakbrane are the terms commonly used in the physics literature. Gravitybrane will be Branesville in the story in the next chapter. The name Weakbrane refers to the fact that most particles confined to this brane are expected to have a mass about the size of the weak scale mass.

good reason to expect disparate masses. The particle physics intuition that masses should be more or less the same is violated, and a wide range of masses is *expected*. Particles located in different locations would naturally have different masses. Their shadows change as you move around. In our four-dimensional world, the result would be a range of sizes and masses, and that is what we observe.

Further Developments

When our paper explaining the hierarchy in terms of warped geometry appeared in 1999, most of our colleagues didn't recognize that it was a genuinely new theory, very different from the large dimensions idea. Joe Lykken said to me, "Reaction built slowly. Eventually everyone understood that this paper [and another that Chapter 22 will describe] were big and new and generic and opened a whole new arena of ideas, but not at first."

For months after our paper came out, I was asked to give talks about my work on "large extra dimensions." I kept having to object that the beauty of our theory is precisely that the dimensions are not large! In fact, Mark B. Wise, a (very aptly named) Caltech particle theorist, laughed at the title I was assigned for a plenary talk in the closing session of the Lepton-Photon Conference of 2001, the major particle conference where experimenters present important results. The organizers had given my talk a title that referred to all research on extra dimensions except my own!

Mark and his then student, Walter Goldberger, were two of the first to understand the merits of the warped scenario. But they also recognized that Raman and I had left a potential gap in our result that needed to be filled. We had assumed that brane dynamics would naturally lead to branes that are a modest distance apart. However, we had not explicitly said how the distance between the two branes is established. This wasn't just a detail; our theory's role as a solution to the hierarchy problem depended on being able to readily stabilize the two branes a small but finite distance apart. It was possible that the inverse exponential function of the distance (which we wanted to be extremely tiny), rather than the distance itself, would turn out naturally to be a modest number. If so, the predicted hierarchy between the weak scale mass and the Planck scale mass would be a modest number, and not the (much smaller) inverse exponential of that number—and our solution wouldn't work.

Goldberger and Wise did the important research that closed this potentially treacherous loophole in the theory Raman and I had presented. They demonstrated that the distance between the two branes is a modest number, and the inverse of the exponential of that distance is extremely tiny, exactly as was required for our solution to work.

Their idea was elegant, and turned out to be of more general validity than anyone realized at the time. As it happens, any stabilization model is very similar to theirs. Goldberger and Wise suggested that in addition to the graviton, there was a massive particle that lived in the five-dimensional bulk. They assigned properties to this particle that made it act like a spring. In general, a spring has a favored length; any larger or smaller length would carry energy that would make the spring move. Goldberger and Wise had introduced a particle (and associated field) for which the equilibrium configuration for the field and the branes would involve a modest brane separation—again, just what our solution to the hierarchy required.

Their solution relied on two competing effects, one that favors widely separated branes and another that favors nearby branes. The result is a stable compromise position. The combination of the two counteracting effects leads naturally to a two-brane model in which the two branes are a moderate distance apart.

The Goldberger-Wise paper made it clear that the warped two-brane scenario really was a solution to the hierarchy problem. And the fact that the separation between the branes could be fixed was important for another reason. If the distance between the branes was undetermined, the branes could move closer together or further apart as the temperature and energy of the universe evolves. If the brane separation could change, or if different sides of the five-dimensional universe could expand at different rates, the universe would not evolve in the way it's supposed to in four dimensions. Since astrophysicists have tested the expansion of the universe late in its evolution, we know that recently the universe has expanded as if it were four-dimensional.

With the Goldberger-Wise stabilization mechanism, the warped five-dimensional universe agrees with cosmological observations. Once the branes are stabilized with respect to each other, the universe would evolve as if it were four-dimensional, even if it actually has five dimensions. Even though there would be a fifth dimension, the stabilization would rigidly constrain different places along the fifth dimension so that they would evolve in the same way, and the universe would behave as it would in four dimensions. Since the Goldberger-Wise stabilization should happen relatively early on, the warped universe would look four-dimensional for most of its evolution.

Once stabilization and cosmology were understood, the warped geometry solution to the hierarchy problem was in business. Many other interesting developments about this warped geometry soon followed. One of these was unification of forces. All forces, including gravity, might be unified at high energies in the warped geometry we're considering!

Warped Geometry and Unification of Forces

Chapter 13 explained how a major feather in the cap of supersymmetry is that it can successfully accommodate the unification of forces. Extra-dimensional theories that addressed the hierarchy problem seemed to forfeit this potentially important development. Since we have not seen any conclusive experimental evidence for unification, such as proton decay, this is not necessarily a major loss, as we don't yet know for certain that unification is correct. Nonetheless, three lines meeting at a point is intriguing and might presage something meaningful. Even if unification is not yet firmly established, we shouldn't abandon it too hastily.

Alex Pomarol, a Spanish physicist now at the University of Barcelona, observed that unification of forces can also occur in warped geometry. However, the setup he considered is slightly different; the electromagnetic, weak, and strong forces are not confined to a brane, but are instead present in the full five-dimensional bulk. The gauge bosons of the Standard Model—the gluons, the Ws, the Z, and the photon—are not stuck on a three-plus-one-dimensional brane.

According to string theory, gauge bosons could be stuck on a higher-dimensional brane or, along with gravity, they too could be in the bulk. Unlike the graviton, which must arise from a closed string, gauge bosons and charged fermions can correspond to either open or closed strings—it depends on the model. And according to whether they arise from open or closed strings, gauge bosons and fermions will be either stuck on a brane or free to move in the bulk.

In the large extra-dimensional scenario, had nongravitational forces been in the bulk, they would have been far too weak to agree with observations. Bulk forces would have spread throughout an enormous bulk space. Therefore, as with gravity, they too would have been extremely diluted. This would be unacceptable because we have measured the forces' strengths to be much larger than this theory would have predicted.

But if additional dimensions are not large, as is the case in the warped geometry, there is no problem with the nongravitational forces in the five-dimensional bulk. The only thing that can dilute them is the extra dimensions' size, not the warping—and in the warped scenario that size is rather small. This means that the true theory of the world might have all four forces experienced throughout the bulk. In that case, not only particles on the brane, but also particles throughout the bulk, could then feel the electric force, the weak force, and the strong force, as well as gravity.

If gauge bosons in the warped scenario are present in the bulk, they could have energy much bigger than a TeV. The gauge bosons, which would hang out in the bulk, would experience the entire energy range. No longer tethered to the Weakbrane, they could travel anywhere in the bulk, and have energies as high as the Planck scale energy. Only on the Weakbrane does energy have to be less than a TeV. Because the forces would be in the bulk and could therefore operate at high energies, unification of forces would be a possibility. This is exciting because it means that the forces can unify at high energy, even in a theory with an extra dimension. And Pomarol found the very interesting result that unification did indeed occur, almost as if the theory were truly four-dimensional.

But it gets even better. Unification and the warped hierarchy mechanism can be combined. Pomarol showed that forces unify, but he assumed that supersymmetry addressed the hierarchy problem. But the hierarchy problem's solution in warped geometry requires only that the Higgs particle be on the Weakbrane, so that its mass will be about the same as the weak scale energy, between 100 GeV and a TeV. The gauge bosons need not be stuck there.

In the warped geometry, all you need in order to solve the hierarchy problem is that the Higgs particle's mass be low. That is because the Higgs field is responsible for the spontaneous symmetry breaking that is the source of all elementary particle masses. Gauge bosons and fermions won't have a mass unless the weak force symmetry is broken. So long as the Higgs particle has a weak scale mass, the weak gauge boson masses will turn out correct. The warped gravity solution to the hierarchy really only requires the Higgs particle to be on the Weakbrane.

What this all means is that if the Higgs particle is on the Weakbrane, but quarks, leptons, and gauge bosons are in the bulk (see Figure 83), you can have your cake and eat it too. The weak scale would be protected and would be about a TeV, but unification could still occur at very high energies—on the GUT scale. My former student Matthew Schwartz and I showed that supersymmetry isn't the only theory that

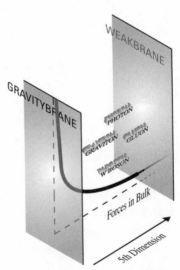

Figure 83. *Nongravitational forces can also be in the bulk. In that case, forces can unify at high energies.*

can be consistent with unification—a warped extra-dimensions theory can be, too!

Experimental Implications

The natural scale on the Weakbrane is about a TeV. Should this warped geometry scenario prove to be a true description of our world, the experimental consequences at the Large Hadron Collider at CERN in Switzerland will be tremendous. Signatures of the warped five-dimensional spacetime could include Kaluza-Klein particles, five-dimensional black holes of anti de Sitter space, and TeV-mass strings.

The KK particles of the warped spacetime are likely to be the most accessible experimental herald of this geometry. As always, KK particles are particles with momentum in the extra dimension. But the new wrinkle in this model is that because the space is curved—not flat—the masses of the KK particles would reflect the idiosyncrasies of the warped geometry.

Since the only particle that we know for certain traverses the bulk is the four-dimensional graviton, let's concentrate on its KK partners. As was true in flat space, the lightest of the KK partners of the graviton will be the one with no momentum at all in the fourth dimension. This particle would be indistinguishable from a particle of genuine four-dimensional origin: it's the graviton that would communicate gravity in what looks like a four-dimensional world and it is the graviton whose probability function we have studied in detail in this chapter. If there were no additional KK particles, the gravitational force would behave in exactly the same way as in a true four-dimensional universe. In this scenario, the universe is secretly five-dimensional, but the particle that acts like a four-dimensional graviton does not reveal this fact. In the absence of heavier KK particles, Athena's world would indeed appear to her to be four-dimensional.

Only the more massive KK particles could communicate the secrets of the five-dimensional theory. But they have to be light enough to be produced. Calculating the KK particles' masses in this theory is a little tricky, however. Because of the distinctive geometry, the KK particles would not have masses proportional to the inverse size of the dimension, as was the case for rolled-up dimensions of flat space. A mass proportional to the inverse size would have been extremely surprising, since, for the small extra dimension we are considering, that would be the Planck scale mass. On the Weakbrane nothing much heavier than a TeV can exist; one certainly wouldn't ever find anything there with the Planck scale mass.

Since a TeV is the mass associated with the Weakbrane, it shouldn't come as too big a surprise that when you do the calculations correctly taking into account warped spacetime, the KK particles turn out to have masses of about a TeV. Both the lightest KK particle, and the difference between the masses of the successively heavier KK particles, turn out to be about a TeV when the fifth dimension ends at the Weakbrane, as we have been assuming. KK particles pile up on the Weakbrane (because their probability function peaks there) and they have all the properties of Weakbrane particles.

This means that there are heavy KK partners of the graviton that are about 1 TeV, 2 TeV, 3 TeV, . . . in mass. And, depending on the ultimate energy reach of the LHC, there is a good chance of finding one or more of them. Unlike the KK partners in the large extra dimensions scenario, these KK partners interact much more strongly than gravity.

These KK particles are not nearly as feebly interacting as the graviton of four dimensions—they have an interaction strength sixteen orders of magnitude bigger. The graviton KK partners interact so strongly in our theory that any KK partner produced at the collider will not simply disappear out of sight, carrying away energy but leaving no visible signal. Instead, they will decay inside the detector into detectable particles, perhaps muons or electrons, which can be used to reconstruct the KK particle from which they originated (see Figure 84).

This is the conventional recipe for discovering new particles: study all the decay products and deduce the properties of what they came from. If what you find isn't something you already know about, it must be something new. If the KK particles decay in the detector, the signal of extra dimensions should be very clean. In our model, rather than simply a missing energy signature, which has no significant labels

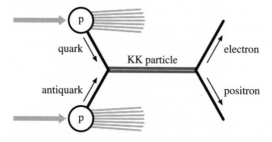

Figure 84. *Two protons collide, and a quark and an antiquark annihilate and produce a KK partner of the graviton. The KK particle can then decay into visible particles, such as an electron and a positron. The gray lines are sprays of particles from the protons.*

that would definitely identify the missing energy's origin and let us distinguish the model from other possibilities, the reconstructed masses and spins of the KK particles should be enormously helpful clues that will tell us quite a lot about the new particles' identities. The spin value of the KK particles—spin-2—will be a virtual ID tag that will tell us that the new particles have something to do with gravity. A spin-2 particle with a mass of about a TeV would be extremely strong evidence for an extra warped dimension. Few other models give rise to such heavy spin-2 particles, and the ones that do would have other distinguishing features.

If we're lucky, in addition to the KK partners of the graviton, experiments might also produce an even richer set of KK particles. In a theory in which most Standard Model particles reside in the bulk, we might also see charged KK partners of quarks and leptons and gauge bosons. Those particles would be both charged and heavy. And they could ultimately give us even more information about the higher-dimensional world.* In fact, the model builders Csaba Csaki, Christophe Grojean, Luigi Pilo, and John Terning have shown that in extra-dimensional warped spacetime with Standard Model particles in the bulk, electroweak symmetry might be broken even without a

---

*Kaustubh Agashe, Roberto Contino, Michael J. May, Alex Pomarol, and Raman Sundrum are among the physicists who have studied detailed models of what might be present.

Higgs particle, and the charged particles that experimenters might then detect could tell us whether this alternative model is true for the world in which we live.

An Even More Bizarre Possibility

I've described quite a few weird properties of extra dimensions. But the most extraordinary possibility is yet to come. We will shortly see that a warped extra dimension can actually stretch infinitely far, yet still be invisible, unlike a flat dimension, which always has to have a finite size to agree with observations.

This result was truly shocking. In Chapter 22, when we discuss this infinite extra dimension, we will focus on the geometry of space, not the hierarchy problem. But I'll briefly mention here how you can solve the hierarchy problem in the infinite-extra-dimensional case as well.

So far, we have considered a model with two branes: the Gravity-brane and the Weakbrane, both of which bound a fifth dimension. However, the Weakbrane doesn't have to be the end of the world (that is, the boundary of the fifth dimension). If the Higgs particle is confined to a second brane placed in the middle of an infinite extra dimension, such a model could also solve the hierarchy problem. The graviton's probability function would be very small on the Weakbrane, gravity would be weak, and the hierarchy problem would be solved just as before when the Weakbrane bounded the extra dimension. The graviton's probability function in the model with an infinite warped dimension would continue beyond the Weakbrane, but that wouldn't affect the solution to the hierarchy problem, which relied only on the small graviton probability function on the Weakbrane.

However, because the dimension is infinite, the KK particles would have different masses and interactions, so the experimental implications of this model would be different than the ones I just described. When Joe Lykken and I first discussed this possibility at the Aspen Center for Physics (an inspirational venue if there ever was one, and also one of the reasons why many theoretical physicists like to hike), we weren't sure whether this idea would actually work. If the fifth dimension didn't end on the Weakbrane, not all the KK particles would be heavy (and have mass of about a TeV). Some KK particles would have very tiny masses. If these particles were detectable but experiments hadn't yet discovered them, the model would be ruled out.

But it turns out that our model is safe. While sitting on a bench surrounded by gorgeous mountain scenery, I worked out the interactions of the KK particles (Joe did the same calculation, but I think he was inside his office at the Center). We calculated a result that told us that, although the KK particles' interactions would be big enough to be interesting for future experiments, they would not be so big as to have already been seen.

In the future, the LHC will have a good chance of producing the KK particles of this model, should they exist. These particles won't look like the ones from the finite-sized extra-dimensional warped model. Instead of nice KK particles that decay inside the detector, the KK particles in this model with an infinite extra dimension will escape into the extra dimension (similar to the KK particles' behavior when there are large dimensions). So if there's an infinite extra warped dimension and a Weakbrane that solves the hierarchy problem, experiments could only hope to discover events with missing energy. Even so, at sufficiently high energies, missing energy should be a sufficiently telling signal that something new is out there.

Black Holes, Strings, and Other Surprises

In addition to KK particles, other remarkable signals of extra dimensions could turn up when the LHC turns on. Although the effects of five-dimensional gravity are minuscule at ordinary energies, five-dimensional gravity will be a major player when colliders create high-energy particles. In fact, when energies reach about a TeV, the effects of five-dimensional gravity would be enormous—they would overwhelm the interactions of the feebly interacting four-dimensional graviton, which has a small probability function on the Weakbrane where we live (and experiments take place).

The enormous strength of five-dimensional gravity means that five-dimensional black holes might be produced, as well as five-dimensional strings. Furthermore, once energies reach about a TeV, everything located on or near the Weakbrane would interact strongly with everything else. That is because the effects of gravity and the extra KK particles would be enormous at TeV energies, and they would conspire to make everything interact with everything else. Such strong interactions among all known particles and gravity wouldn't occur in a four-dimensional scenario; they would be a definite signal of something new. As was the case with large extra dimensions, we don't yet know if the energy will be high enough to see these new objects. But if interactions are strong at energies not too much greater than a TeV, experiments won't miss it.

Coda

The connection between a solution to the hierarchy problem and experimental consequences at TeV energies is robust, but the details of what we will see depend on the model. Different models have distinctive experimental consequences, and that is very reassuring. These distinctive signatures mean that once the LHC is up and running, we have a good chance of identifying which—if any—of these models applies to our world.

What's New

- Spacetime can be dramatically curved in the presence of bulk and brane energy, even when the brane itself is completely flat.

- The model we considered in this chapter has two branes, the Gravitybrane and the Weakbrane, at each of the ends of a finite-sized fifth dimension. Energies in the bulk and on the branes warp spacetime.

- A single extra dimension introduces an entirely new way to solve the hierarchy problem. The fifth dimension in this model is not big, but it is very warped. The strength of gravity depends strongly on where you are in the fifth dimension. Gravity is strong on the Gravitybrane and is feeble on the Weakbrane, where we are located.

- From the perspective of an observer who thinks he is in four dimensions, objects should have different sizes and masses if they originate from different places in the fifth dimension. Objects that are confined to the Gravitybrane should be very heavy (with mass of about the Planck scale mass), whereas objects confined to the Weakbrane should have much lower masses, about a TeV.

- All forces can unify and the hierarchy problem can be solved if the Higgs particle (but not the gauge bosons) is confined to the Weakbrane.

- The Kaluza-Klein partners of the graviton should give rise to very distinctive particle collider events in which they decay into Standard Model particles within the detector.

- In models in which Standard Model particles are in the bulk, other KK particles can be produced and observed.

21

The Warped Annotated "Alice"*

Go ask Alice.
When she's ten feet tall.

Jefferson Airplane

Athena stepped out of the dreamworld's elevator into the warped five-dimensional world and was astonished to see only three spatial dimensions. Was the Rabbit playing games, pretending to take her to a world with four spatial dimensions when in fact there were only three? What a funny way to travel to what looked like an ordinary world!†

With great gallantry, a local received the puzzled new arrival. "Welcome to Branesville,‡ our glorious capital. Permit me to show you around." Athena, who was tired and confused, blurted out, "Branesville doesn't look all that special. Even the mayor looks completely normal," although, she had to confess, she wasn't entirely sure as she had never seen a mayor before.

The mayor to whom Athena referred had arrived accompanied by

---

*This title borrows from Martin Gardner's delightful *Annotated Alice*, in which he explains the wordplay, math riddles, and references in Lewis Carroll's *Alice in Wonderland* and *Through the Looking Glass*.

†The brane itself is large and flat and has only three spatial dimensions. Only gravity makes contact with the additional dimension. Remember that the five-dimensional space has four spatial dimensions (and one of time), whereas the brane has three spatial dimensions. I'll still call time the fourth dimension, and I'll call the additional dimension the fifth.

‡Branesville is the Gravitybrane.

the Cheshire Fat Cat, his Chief Advisor. The Cat's job was keeping tabs on everything in the city, which was greatly facilitated by his skill at catching people unawares—especially surprising in light of the Cat's enormous bulk. The Cat loved to explain that he owed this skill to his ability to disappear into the bulk, but no one ever understood what he meant.*

The Cat materialized next to Athena and asked if she would like to accompany him as he made his rounds. He warned her that she had better be comfortable with bulk, to which Athena eagerly responded that her favorite uncle was in fact very, very fat. The Cat looked skeptical, but agreed to take her along. He offered Athena cream cake with butter frosting, in which she happily indulged. And off they went.

Athena wondered what it was she'd eaten. She now appeared to be on a four-dimensional slice of a five-dimensional world, and as far as she could tell, she was no thicker than this thin four-dimensional slice. She exclaimed, "I am like my paper doll! But whereas Dolly has two spatial dimensions in a three-dimensional world, I have three spatial dimensions in a four-spatial-dimensional world."

The Cat grinned sagely and explained, "You are now conscious of what I like to call The Bulk. You are still in Branesville, but will be leaving (and growing) momentarily. Branesville is in reality part of a five-dimensional universe, but the fifth dimension is warped so discreetly that Branesville residents are completely unaware of its existence. They have no idea that Branesville is the border of a five-dimensional state. You too mistakenly concluded on your arrival that there are only three spatial dimensions. The new Athena, untethered from the brane, is free to travel out into the fifth dimension. May I suggest for our destination another village called Weakbrane, at the other edge of the five-dimensional universe?"

What a strange five-dimensional journey it turned out to be. After leaving Branesville, Athena found herself moving in another dimension, and growing as she did so (as shown in Figure 85).† When

---

*The Fat Cat, unlike Branesville residents, is not confined to the brane.
†Everything is bigger and lighter near the Weakbrane. Athena's shadow over Branesville grew as she got closer to the Weakbrane and further away from the Gravitybrane.

Figure 85. *Alice got bigger as she moved through the bulk from the Weak-brane to the Gravitybrane.*

the observant Cat noticed the confused look on Athena's face, he reassuringly explained, "Weakbrane is close by and we will be there very soon.* It's lovely, but don't be alarmed when you see that, like the Branesville residents you encountered, Weakbrane residents scoff at the notion of four spatial dimensions. You, who can see out into the bulk, will see a huge shadow on Branesville, ten million billion times bigger than the one with which you started. Almost everything else will seem to you and to them to be entirely normal."

But upon her arrival in Weakbrane, Athena noticed one other thing. The four-dimensional graviton had quietly accompanied the travelers on their journey and was softly tapping on her shoulder. He touched her so extremely gently that she had barely noticed.†

But she couldn't ignore the graviton when he launched into a litany of complaints. "Weakbrane would be so exciting, were it not for the

---

*The fifth dimension does not have to be very big in order to solve the hierarchy problem.

†Gravity is feeble on the Weakbrane, where the graviton's probability function is so small.

superior influence of the entrenched hierarchy. The strong, weak, and electromagnetic armed forces on the Weakbrane permit me only the most feeble strength." The graviton whined how everywhere else he was a force to be reckoned with, especially in Branesville, which is ruled by an oligarchy with comparably strong forces.* *Weakbrane, where gravity was the most suppressed, was the graviton's least favorite place.†* The graviton turned to Athena in the hope of enlisting her in his plan to wrest power from the reigning authorities.

Athena thought she had better leave immediately and looked around for the rabbit hole, but couldn't find it. She did find a white rabbit, whom she expected to be an efficient guide. But the Weakbrane rabbit had an alarmingly sluggish gait, and kept repeating how happy he was that his date would wait.‡ Athena realized that this rabbit wasn't going anywhere, so she found a more anxious rabbit she could follow, and worked her way back home. Once she understood the physics implications, Athena enjoyed her dream enormously—though it should be noted that she never again ate cream cake.

*On the Gravitybrane, gravity is no weaker than the other forces.
†The petulant graviton is complaining that on the Weakbrane, gravity is much weaker than the electromagnetic, weak, and strong forces. Gravity would be much stronger (and would have a strength closer to that of the other forces) closer to the Gravitybrane.
‡Things are bigger and time is slower on the Weakbrane. The rabbit's laxness is accounted for by rescaling time.

22

Profound Passage:
An Infinite Extra Dimension

Athena woke up with a start. Her recurring dream had once again taken her down the rabbit hole. This time, however, she asked the Rabbit to take her straight back to the warped five-dimensional world.

Athena arrived back in Branesville (or so she thought). The Cat soon appeared, and she eagerly turned to him, anticipating her dream cake and a delightful excursion to Weakbrane. She was sorely disappointed when the Cat told her there was no such thing as Weakbrane in this particular universe.*

Athena didn't believe the Cat and thought there must be another brane further away. Proud of herself for understanding how, in the warped geometry, further-away branes had weaker gravity, she decided it was probably called "Meekbrane" and asked the Cat whether she could go there.

But once again she was in for a disappointment. The Cat explained, "There is no such place. You are on the Brane; there are no others."

"Curiouser and curiouser," thought Athena. This clearly wasn't

---

*The geometry of this chapter is warped, as in the previous ones, but now there is only a single brane—the Gravitybrane. Although this means that there is an infinite fifth dimension, this chapter will show why this is perfectly fine with the warped spacetime.

exactly the same space as before, since it had only a single brane. But Athena wasn't ready to give up. "May I see for myself that there is no other brane?" she asked in her sweetest tone.

The Cat strongly advised her against it, warning, "Four-dimensional gravity on the Brane is no guarantee of four-dimensional gravity in the bulk. Once I nearly lost everything but my smile there."

Athena was a cautious girl, despite her many adventures, and she took the Cat's warning to heart. But she often wondered what the Cat meant. What did lie beyond the Brane, and how would she ever know?

Curved spacetime has remarkable properties. We explored some of these in Chapter 20, including how mass and size and the strength of gravity can all depend on location. This chapter presents an even more extraordinary feature of curved spacetime: it can appear to have four dimensions, even when there are truly five. By examining the warped spacetime geometry more carefully, Raman and I realized, to our astonishment, that even an infinite extra dimension can sometimes be invisible.

The spacetime geometry that we will consider in this chapter is almost the same as the one described in Chapter 20. But as the above story should suggest, this geometry has a single distinguishing feature: it has only a single brane. But this is a tremendously important distinction: because there is no second boundary brane, a single brane means that the fifth dimension is infinite (see Figure 86).

That is a stupendous difference. For three-quarters of a century after Theodor Kaluza introduced the idea of an extra dimension of space in 1919, physicists believed that extra dimensions were acceptable, but only if they were finite in size, either curled up or bounded between branes. Infinite extra dimensions were supposed to be very easy to rule out because the gravitational force, which would spread infinitely far in these dimensions, would look wrong at all distance scales, even those that we know about already. An infinite fifth dimension was supposed to destabilize everything around us, even the solar system, which is held together by Newtonian physics.

This chapter explains why this reasoning is not always correct.

Figure 86. *Infinite warped spacetime with a single brane. There is a single four-dimensional brane in a five-dimensional universe. The Standard Model resides on this single brane.*

We'll investigate an entirely new reason why extra dimensions might be hidden, which Raman and I discovered in 1999. Spacetime can be so warped that the gravitational field becomes highly concentrated in a small region near a brane—so concentrated that the huge expanse of an infinite dimension is inconsequential. The gravitational force is not lost into the extra dimensions, but remains focused in a small region near a brane.

In this scenario, the graviton, the particle that communicates gravity, is *localized* near a brane, which is the Brane of Athena's story, but which I'll call the Gravitybrane from now on. Athena's dream took her to this warped five-dimensional space, in which the Gravitybrane so radically alters the nature of spacetime that space appears to be four-dimensional, even though it is in reality five-dimensional. Remarkably, a warped higher dimension can have infinite extent, yet nonetheless be hidden, while the three flat infinite dimensions reproduce the physics of our world.

The Localized Graviton

You might recall that when I first introduced branes, I distinguished reluctance to explore distant regions from genuine confinement, which explicitly forbids foreign travel beyond the place where someone or something is confined. Although you've probably never visited Greenland, no law forbids you from going there. But some places are just too much trouble to get to. Even if travel to such places is permitted, and even if they are no further away than some other places you've been, you still might just never go there.

Or imagine someone who has broken his leg. In principle, he could leave the house whenever he likes, but he's much more likely to be found inside the house than outside, even when no bars or locks are keeping him there.

Similarly, the localized graviton has unrestricted access to an infinite fifth dimension. But it is nonetheless highly concentrated in the vicinity of a brane, and has very low probability of being found far away. According to general relativity, everything—including the graviton— is subject to the gravitational force. The graviton is not at all impaired, but it behaves as if it is gravitationally attracted to the brane and therefore remains close by. And because the graviton only rarely travels outside a limited region, the extra dimension can be infinite without giving rise to any dangerous effects that would rule the theory out.

In our work, Raman and I concentrated on gravity in a five-dimensional spacetime with only a single extra dimension of space. We could thereby concentrate on the localization mechanism we are about to discuss, which keeps gravity in a small region of the five-dimensional spacetime. I'll assume that if the universe has ten or more dimensions, some mixture of localization and curling them up hides the rest. Such additional hidden dimensions wouldn't affect the localization phenomenon I'm about to describe, so we'll ignore them and focus on the five dimensions that are critical to our discussion.

In our model a single brane sits at one end of the fifth spacetime dimension. It is reflective, as were the two branes I described in

Chapter 20. Things that hit the brane simply bounce back, so nothing loses energy when it hits this brane. Because the model we are now considering contains only this single brane, we'll assume that the Standard Model particles are confined there; note the distinction from the model I discussed in the previous chapter, in which the Standard Model particles were on the Weakbrane, which no longer exists. The location of the Standard Model particles is not relevant to the spacetime geometry, but it does of course have implications for particle physics.

Although in this chapter we are interested in the single-brane theory, the first clue Raman and I had that an infinite fifth dimension might be legitimate was a curious feature of the warped geometry with two branes. We initially assumed that the second brane served two functions. One was to confine the Standard Model particles; the second was to make the fifth dimension finite. As with flat extra dimensions, a finite fifth dimension guaranteed that at sufficiently large distance, gravity would be that of four-dimensional space-time.

However, a peculiar fact suggested that this latter role for the second brane was a red herring, and that the second brane was not essential for gravity to mimic that of a genuinely four-dimensional universe: the four-dimensional graviton's interactions were virtually independent of the fifth dimension's size. A calculation showed that gravity would have the same strength if the second brane stayed where it was, or if it were twice as far from the Gravitybrane, or if it were ten times as far out into the bulk, further away from the first brane. In fact, four-dimensional gravity persisted even if our model put the second brane at infinity—which is to say, eliminated it altogether. This should not be true if the second brane and a finite dimension were essential for reproducing four-dimensional gravity.

This was our first clue that the intuition that we need a second brane was based on flat dimensions and wasn't necessarily true for warped spacetime. With a flat extra dimension, the second brane is compulsory for four-dimensional gravity. We can see this with the aid of the sprinkler analogy from Chapter 20. A flat extra dimension would correspond to water being distributed equally everywhere along a long straight sprinkler (see Figure 81, p. 396).* The longer the sprinkler, the less water would be sprinkled on any given garden. If we were to extend this reasoning to an infinitely long sprinkler, we would see that the water would be spread so thinly that essentially no water would be sprinkled on any finite-sized garden.

Similarly, if gravity were spread throughout an infinite uniform dimension, the gravitational force would be so attenuated along the extra dimension that it would be reduced to nothing. A geometry with an infinite extra dimension would have to contain some subtlety that goes beyond this simple intuitive picture if gravity is to behave four-dimensionally. And indeed, warped spacetime provides the requisite added ingredient.

To see how this works, let's once again use our sprinkler analogy to identify the loophole in the argument above. Suppose that you have an infinitely long sprinkler, but you don't distribute water equally everywhere. Instead, you have control over how the water is allocated, giving you the option of ensuring that your own garden is well watered. One way of accomplishing this would be to deliver half the water to your plot of land and the remaining half of the water everywhere else. In that case, although the gardens far away would be badly treated, your garden would be guaranteed to receive all the water it needs. Your garden would always receive half the water, even though the sprinkler continues delivering water indefinitely far away. With an inequitable distribution of water, you would get all the water you need. The sprinkler could be infinite but you wouldn't know the distance.

Similarly, the graviton's probability function in our warped geometry is always very big near the Gravitybrane, despite the infinite fifth dimension. As in the previous chapter, the probability function for the graviton peaks on this brane (see Figure 87), and falls off exponentially as the graviton moves away from the Gravitybrane into the fifth dimension. In this theory, however, the graviton's probability function continues indefinitely far but it is inconsequential to the size of the graviton probability function near the brane.

---

*We consider a straight sprinkler, instead of the circular one that we considered before, because it is easier to generalize to the warped scenario.

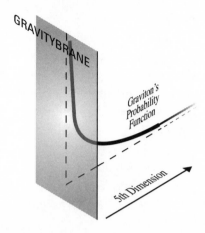

Figure 87. *The graviton's probability function in infinite warped spacetime with a single brane.*

A plummeting probability function of this sort tells us that the likelihood of finding the graviton far from the Gravitybrane is extremely tiny—so tiny that we can generally ignore the distant regions of the fifth dimension. Although in principle the graviton can be anywhere along the fifth dimension, the exponential decrease makes the graviton's probability function very concentrated in the vicinity of the Gravitybrane. The situation is almost, but not quite, as if a second brane confined the graviton to a limited region.

The high probability for the graviton to be found near the Gravity-brane, and the corresponding concentration of the gravitational field there, might also be compared to the high likelihood of greedy ducks in a pond being near the shore. Usually the ducks aren't equally spaced across the pond, but instead congregate near breadcrumbs that bird lovers have tossed in (see Figure 88). So the size of the pond would be essentially immaterial to the distribution of ducks. Similarly, in warped spacetime, gravity attracts the graviton to the Gravitybrane, so the extent of the fifth dimension is irrelevant.

You can also see why the fifth dimension doesn't affect gravity very much by considering the gravitational field surrounding an object on the Gravitybrane. We have seen that in flat spatial dimensions, the force lines emanating from an object spread equally in all directions. And when there were finite extra dimensions, the field lines extended

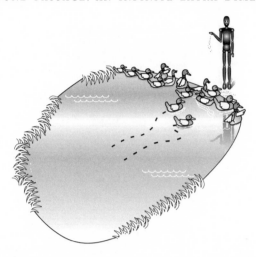

Figure 88. *If ducks are concentrated near the shore, you can count almost all of them by counting the ones nearby.*

in all directions until some reached a boundary and bent around. For this reason, gravitational field lines that are further from an object than the size of the extra dimensions spread only along the three infinite dimensions of the lower-dimensional world.

In the warped scenario, on the other hand, field lines do not distribute themselves equally in all directions; it's only along the brane that they extend equally in all directions. Perpendicular to the brane, they extend very little (see Figure 89). Because the gravitational field lines spread out primarily along the brane, the gravitational field looks almost identical to the field associated with an object in four dimensions. The spread into the fifth dimension is so small (not much more than the Planck scale length, 10^{-33} cm) that we can ignore it. Although the extra dimension is infinite, it is irrelevant to the gravitational field of a brane-bound object.

You can also see how Raman and I resolved the initial puzzle we faced: why the size of the fifth dimension is irrelevant to the strength of gravity. Returning to the sprinkler analogy above, suppose we now specify the distribution of water over the entire sprinkler, so that it resembles the distribution of gravity from the plummeting graviton's probability function: after you take half of the water for your garden, you then send half of the remaining water to the adjacent garden, half

Figure 89. *In the warped scenario, the field lines are equally distributed in all the directions on the brane. However, off the brane the field lines bend back around so that they become essentially parallel to the brane, almost as if the fifth dimension were finite. Even with an infinite dimension, the gravitational field is localized near the brane, and field lines spread essentially as if there were only four (spacetime) dimensions.*

of that amount to the next one, and so on, with everyone in successive gardens receiving half as much water as their neighbor. To mimic a second brane in the fifth dimension, we'll assume that we stop delivering water beyond a certain point, just as a second brane in the fifth dimension would cut off the graviton's probability function at some point along the fifth dimension. And to mimic an infinite fifth dimension, we'll assume that the water delivers water indefinitely along its length.

To show that the size of the fifth dimension is irrelevant to the strength of gravity near the brane, we would want to show that the first few gardens get very nearly the same amount of water, regardless of whether we stop delivering water when we get to the fifth garden or the tenth garden or we don't stop delivering water at all. So let's consider what happens if the sprinkler ended after the first five gardens. Because the sixth garden and beyond would have received so little water, the total amount of water that the sprinkler would send to the first several gardens would differ from that of an infinite sprinkler by only a few percent. And if you stopped the sprinkler after the seventh garden, it would differ by even less. With our distribution of water where nearly all the water is used on the first several gardens, the faraway gardens, which receive only a very tiny fraction of the water, are irrelevant to the amounts of water the first few gardens receive.*

Because I will use the duck analogy again in the next chapter, I'll explain the same thing in terms of counting ducks attracted to the shore where someone has sprinkled crumbs. If you were to first count nearby ducks, and then count ones a little further out, your duck counting would quickly become almost futile. By the time you get a little way away from shore, there would be very few ducks left to count. You don't need to keep counting ducks far from the shore because you've already counted essentially all of them by focusing on the region near the shore (see Figure 88).

The graviton's probability function is simply so small beyond the second brane that a second brane's location would make only a negligible difference to the interaction strength of the four-dimensional graviton. In other words, the extent of the fifth dimension is immaterial to the apparent strength of four-dimensional gravity in this theory, in which the gravitational field is localized near the Gravity-brane.[37] Even if there's no second brane and the fifth dimension is infinite, gravity still looks four-dimensional.

Raman and I called our scenario *localized gravity*. That is because the graviton's probability function is localized near a brane. Although, strictly speaking, gravity can leak out into the fifth dimension because the fifth dimension is indeed infinite, in reality it does not because of the low probability of the graviton being found far away. Space is not truncated, yet everything remains in a concentrated region in the vicinity of the brane. A faraway brane makes no difference to physical processes on the Gravitybrane since very little from the Gravitybrane ventures far away. Anything produced on or near the Gravitybrane remains nearby, in a localized region.

---

*A real-life analogy of this sort would be the Colorado River, where dams and irrigation ensure that water is delivered to the Southwestern United States, but only a small amount of water remains in the river when it reaches Mexico. Putting a dam near the Gulf of California (which would be like putting another brane far from the Gravitybrane) wouldn't affect the amount of water that Las Vegas receives.

Sometimes physicists refer to this model of localized gravity as RS2. The RS stands for Randall and Sundrum, but the 2 is misleading—it refers to the fact that this was the second paper we wrote on warped geometry, not the fact that there are two branes. The scenario with two branes, which addresses the hierarchy problem, is known as RS1. (The names would be less confusing if we had written the papers in the opposite order.) Unlike RS1, the scenario in this chapter is not necessarily relevant to the hierarchy problem, though you can introduce a second brane and solve the hierarchy as well, as we briefly considered towards the end of Chapter 20. But whether or not there is a second brane inside the space to address the hierarchy problem, localized gravity is a radical possibility with important theoretical implications that contradicts the long-held assumption that extra dimensions must be compact.

Kaluza-Klein Partners of the Graviton

The previous section discussed the graviton's probability function, which is heavily concentrated on the Gravitybrane. The particle I was talking about plays the role of the four-dimensional graviton because it travels almost exclusively along the brane and has only a tiny probability of leaking out into the fifth dimension. From the graviton's perspective, space looks as if the fifth dimension is only 10^{-33} cm in size (a size set by the curvature, which is in turn set by the energy in the bulk and on the brane) rather than of infinite extent.

But although Raman and I were rather excited by our discovery, we weren't sure that we had completely solved the problem. Was the localized graviton by itself sufficient to generate a four-dimensional effective theory in which gravity behaved as it would in four dimensions? The potential problem was that Kaluza-Klein partners of the graviton could also contribute to the gravitational force, and could thereby significantly modify gravity.

The reason this seemed so dangerous was that, generally, the larger the size of the extra dimension, the smaller the mass of the lightest KK particle. For our theory with an infinite dimension, this would mean that the lightest KK particle could be arbitrarily light. And because the difference in masses of the KK particles also decreases with the size of the extra dimension, infinitely many types of very light graviton KK partners could be produced at any finite energy. All these KK particles could potentially contribute to the gravitational force law and change it. The problem looked especially bad because even if each KK particle interacted very weakly, if there were too many of them then the gravitational force would nonetheless look quite different from that in four dimensions.

On top of that, since the KK particles are extremely light, they might be easy to produce. Colliders already operate at sufficiently high energy to make them. Even ordinary physical processes, such as chemical reactions, would generate enough energy to create graviton KK partners. If the KK particles carried too much energy to the five-dimensional bulk, the theory would be ruled out.

Fortunately, neither of these concerns turns out to be a problem. When we calculated the probability functions for the KK particles, we found that the graviton KK partners interact extremely weakly on or near the Gravitybrane. Despite the large number of graviton KK partners, they all interact so feebly that there is no danger of producing too many of them or of changing the form of the gravitational force law anywhere. If there is any problem at all, it is that this theory so closely mimics four-dimensional gravity that we don't yet know any way to distinguish it experimentally from a truly four-dimensional world! The graviton KK partners would have such a negligible impact on anything observable that we do not yet know how to tell the difference between four flat dimensions and four flat dimensions supplemented by a fifth, warped one.

You can understand the weakness of the graviton KK partners' interactions from the shape of their probability functions. As with the graviton, these tell us the likelihood of any particle being found at any position along the fifth dimension. Raman and I followed the more or less standard procedure for finding the masses and probability functions of each graviton KK partner in our warped geometry. This involved solving a quantum mechanics problem.

For a flat fifth dimension, the quantum mechanics problem,

described in Chapter 6, was to find the waves that fit around the rolled-up dimension and thereby quantize the allowed energies.* For our warped, infinite fifth-dimensional geometry, the quantum mechanics problem looked rather different, since we needed to take into account the energy on the brane and in the bulk that warped spacetime. But we were able to modify the standard procedure to suit our setup. The results were fascinating.

The first KK particle we found was the one with no momentum in the fifth dimension. The probability function of this particle is heavily concentrated on the Gravitybrane and decreases exponentially away from it. This shape should sound familiar: it is the probability function for the same four-dimensional graviton we have already discussed. This massless KK mode is the four-dimensional graviton that communicates Newton's four-dimensional force law.

The remaining KK particles are very different, however. None of these other KK particles are likely to be found near the Gravitybrane. Instead, what you find is that for any value of mass between zero and the Planck scale mass, there exists a KK particle with that particular mass, and the probability function for each of those particles peaks at a different place along the fifth dimension.

In fact, there is an interesting interpretation for the locations of the different peaks. We saw in Chapter 20 that in warped spacetime, in order to put all particles on the same footing in the four-dimensional effective theory so that they all interact with gravity in the same way, we rescaled all distances, times, energies, and momenta differently along the fifth dimension. As one travels out away from the brane, each point gets associated with an exponentially smaller energy. That was why particles on the Weakbrane were expected to have a mass of about a TeV. The shadow of Athena traveling out into the fifth dimension became bigger, and Athena became lighter, as she moved from the Gravitybrane toward the Weakbrane.

Each point along the fifth dimension can be associated with a particular mass in the same way; the mass is related to the Planck

---

*The curled-up space is still mathematically "flat." That is because you can unroll the dimension to something you would recognize as flat; that is not true of a sphere, for example.

scale mass by the rescaling at that point. And the KK particle whose gravity function peaks at a particular point has approximately that rescaled Planck scale mass. As you travel out into the fifth dimension, you encounter successively lighter KK particles whose probability functions peak there.

In fact, you might say that the Kaluza-Klein spectrum exhibits a highly segregated society. Heavy KK particles are banished from the regions of space where the rescaled energy is too small to produce them. And light KK particles are rarely found in those regions that contain very energetic particles. KK particles concentrate as far from the Weakbrane as they can, given their mass. Their locations are like the size of teenage boys' pants, which are as baggy as they can be without falling down. Fortunately, the physical laws that determine the KK particles' locations are easier to understand than the far more perplexing rules of teenage fashion.

For us, the most important feature of the probability functions for the light KK particles is that they are extremely small on the Gravitybrane. That means there is only a small probability of finding light KK particles on or near there. Because light KK particles shy away as much as possible from the Gravitybrane, light particles (aside from the exceptional graviton whose probability function peaks on the Gravitybrane) would only rarely be produced there. Furthermore, light KK particles don't significantly modify the gravitational force law because they tend to stay away from the Gravitybrane and therefore don't interact much with brane-bound particles.

Putting everything together, Raman and I decided that we had found a theory that worked. The Gravitybrane-localized graviton is responsible for the appearance of four-dimensional gravity. Despite the abundance of KK graviton partners, they interact so weakly on the Gravitybrane that their effect is not at all noticeable. And despite the existence of an infinite fifth dimension, all physical laws and processes, including that of gravity, appear to agree with what is expected of a four-dimensional world. In this highly warped space, an infinite extra dimension is permissible.

As mentioned earlier, if anything, this model is frustrating from an observational vantage point. Amazing as it may seem, this five-dimensional model mimics four dimensions so extraordinarily well that it will be extremely difficult to tell them apart. Particle physics experimenters will certainly have a hard time.

Physicists have, however, begun to explore astrophysical and cosmological features that might distinguish the two worlds. Many physicists[*] have considered black holes in the warped spacetime and they continue to investigate whether there exist distinguishing features that we can use to determine which type of universe we actually live in.

As of now, we know that localization is a new and fascinating theoretical possibility for extra dimensions in our universe. I eagerly look forward to further developments that could ultimately determine whether it's a true feature of our world.

What's New

- A dimension can be infinitely long, yet be invisible, if spacetime is suitably warped.

- Gravity can be localized, even if it is not strictly confined to a finite region.

- In localized gravity, the massless KK particle is the localized graviton. It is concentrated close to the Gravitybrane.

- All other KK particles are concentrated far from the Gravitybrane; the shape of their probability function and the locations where they peak depend on their mass.

[*]They include Juan Garcia-Bellido, Andrew Chamblin, Roberto Emparan, Ruth Gregory, Stephen Hawking, Gary T. Horowitz, Nemanja Kaloper, Robert C. Myers, Harvey S. Reall, Hisa-aki Shinkai, Tetsuya Shiromizu, and Toby Wiseman.

23

A Reflective and Expansive Passage

Someday girl I don't know when
We're gonna get to that place
Where we really want to go.

Bruce Springsteen

Ike XLII was ready to live large. He wanted to test the Alicxvr's ultra-high settings of many megaparsecs, with which he could explore places beyond the Galaxy and the known universe and experience distant regions no one had ever seen before.

Ike was thrilled when the Alicxvr took him to distances 9, 12, and even 13 billion light-years away. But his excitement diminished when he tried to go farther and his signal strength fell precipitously. When he aimed for 15 billion light-years, his exploration aborted completely: he no longer received any information at all. Instead, he heard, "Message 5B73: The Horizon customer you are trying to reach is beyond your calling area. If you need assistance, please contact your local long-distance operator."

Ike couldn't believe his ears. It was the thirty-first century, yet his Horizon service still provided only limited coverage. When Ike tried to contact the operator, a recording said, "Please stay on the brane. Your call will be answered in the order in which it was received." Ike suspected that the operator would never respond, and was wise enough not to wait.

The previous chapter explained why warping can emancipate an extra dimension and allow it to be infinite, yet unseen. But the infinite extra dimension is not the end of the physics story: things get even more bizarre. This chapter will explain how four-dimensional gravity (that is, with three dimension of space and one of time) can be truly a local phenomenon—gravity might look very different far away. We'll see that not only could space appear to be four-dimensional when there are truly five dimensions, but we might be living in an isolated pocket with four-dimensional gravity inside a five-dimensional universe.

The model we'll now consider demonstrates that, remarkable as it might seem, different regions of space can appear to have different numbers of dimensions. The physicist Andreas Karch and I found a model for spacetime in which this was the case in the course of investigating some perplexing features of localized gravity. The new and radical scenario we ended up with suggests that the reason we don't see additional dimensions could be much more peculiar to our environment than anyone had ever believed. We could be living in a four-dimensional sinkhole in which three spatial dimensions is merely an accident of location!

Reflections

When I look back at the e-mail record from the time during which Raman and I had been collaborating, I find it a little mind-boggling how we completed our work in the midst of so many other distractions. When we began our research, I was in the process of moving from MIT to Princeton, where I was about to take a position as a professor, and I was also planning a six-month workshop in Santa Barbara for the following year. Raman, who had had several post-doctoral positions, was concerned about getting a faculty offer, so he was busy preparing talks and job applications. It was difficult to believe. He had done great work, yet I and others were trying to convince him that it would work out in the end and he shouldn't abandon physics and search for another career. Raman was clearly meant to continue with physics and strongly deserved an excellent faculty position, yet he was having trouble finding a job.

The e-mails from the time illustrate the chaos: interesting physics issues alternate with requests for letters of recommendation, scheduling talks, arranging my Princeton housing, and some Santa Barbara conference organization. There were also a few e-mail exchanges with other physicists about our work. But not many. Although the RS2 paper was ultimately cited thousands of times and became well accepted, the work's initial reception was mixed. It took a while before most physicists understood and believed us. A colleague tells me that at first people were waiting for someone else to find the loophole so they wouldn't have to pay attention. Certainly at Princeton, the reaction to a talk Raman gave could at best be described as tepid.

Even those who did listen didn't necessarily believe us right away. A conversation we had with the string theorist Andy Strominger was very enlightening, even though he now laughs at how he initially didn't believe a word we said. Fortunately, he hadn't been too skeptical to listen and talk.

In the physics community, there were a few who understood and believed what we were doing right from the start. We were lucky that Stephen Hawking was among them, and that he did not hesitate to share his enthusiasm with physics audiences. I remember Raman excitedly telling me how Hawking's prestigious Loeb Lectures at Harvard concentrated heavily on our work.

Several others also worked on some related ideas. But the following fall, several months after our paper was published (and many months after we had begun talking about it), was when the theoretical physics community at large started paying attention. It turned out to be good fortune that David Kutasov, a University of Chicago physicist from Israel, and Misha Shifman, a Russian-born particle theorist from the University of Minnesota, and I had organized a six-month workshop in the fall of 1999 at the Kavli Institute for Theoretical Physics in Santa Barbara. The original goal of this workshop had been to bring together string theorists and model builders, and profit from an incipient convergence of research interests on topics such as supersymmetry and strongly interacting gauge theories. We had started planning the workshop well in advance, before the concept of branes and extra dimensions had created such a stir. Although we had hoped for some positive synergy between string theorists and model builders, we didn't know at the time we started the organization that we'd be thinking about extra dimensions when the conference actually happened.

But the timing proved to be fortuitous. The workshop provided an excellent opportunity to flesh out ideas about extra dimensions, and for model builders, string theorists, and general relativists to share expertise. Many exciting discussions took place, and warped geometry was one of the chief topics. In the end, both model builders and string theorists took warped five-dimensional geometry seriously. In fact, the distinction between the two fields blurred as people teamed up to work on similar problems on warped geometry and other ideas.

Many physicists later worked on other aspects of warped geometries, establishing connections and exploring subtleties that made localized gravity even more interesting. Although string theorists had originally dismissed RS1 (the warped geometry with two branes) as just a model, once they began to search they found ways to realize the RS1 scenario in string theory. Questions about black holes, time evolution, related geometries, and the connections to ideas from string theory and particle physics have also been fertile areas of research. Localized gravity has now been investigated in various contexts, and new ideas continue to emerge.

After our theory was accepted and no longer thought incorrect, some physicists actually went overboard in a different direction, claiming our theory was nothing new. One string theorist even went so far as to conclude that a string theory calculation of the impact of Kaluza-Klein modes was the "smoking gun" that proved our theory was the same as a version of string theory that string theorists had already been studying. This conformed to the joking adage in science that a new theory goes through three stages before being accepted: first it's wrong, then it's obvious, and finally somebody claims that someone else did it first. In this case, however, the smoking gun went up in smoke when physicists realized that the string theory calculation was subtler than they had thought, and the purported string theory answer actually hadn't been right.

The truth was that the intersection with work in string theory was very exciting to all of us, and led to important new insights. Localized gravity turned out to have strong overlaps with the most important string developments of the time: both our work and the research of string theorists involved a similarly warped geometry. In fact, perhaps because our research didn't directly challenge string theory models, the string theory community actually accepted and recognized the significance of our work sooner than the model building community. Although it had initially seemed coincidental, maybe this was some indication that we were all on the right track. And happily, Raman had no trouble getting job offers afterward. (He's now a professor at Johns Hopkins.)

However, some skeptics remained. The precise model Raman and I considered led to interesting questions that no one could answer right away. Did localization depend on the form of spacetime very far away? When people tried to find examples of the type of geometry that Raman and I had suggested in supergravity theories, the form of gravity far from the localizing brane seemed to be the stumbling block. But were those conditions essential? Another question we wanted to answer was, did spacetime necessarily look four-dimensional everywhere? Localized gravity made the entire five-dimensional universe behave as if there were four-dimensional gravity. Did this always happen, or could some regions look four-dimensional and some regions behave differently? And what happens when the Gravitybrane isn't completely flat? Does localization work the same way for a brane with a different geometry? These are some of the questions that *locally localized gravity*, the theory that Andreas and I developed, could address.

Locally Localized Gravity

How many dimensions of space are there? Do we really know? By now, I hope you will agree that it would be overreaching to claim that we know for certain that extra dimensions do not exist. We see three dimensions of space, but there could be more that we haven't yet detected.

You now know that extra dimensions can be hidden either because they are curled up and small, or because spacetime is warped and gravity so concentrated in a small region that even an infinite dimension is invisible. Either way, whether dimensions are compact or localized, spacetime would appear to be four-dimensional everywhere, no matter where you are.

This might be a little less obvious in the localized gravity scenario, in which the graviton's probability function becomes smaller and smaller as you go out into the fifth dimension. Gravity acts as it does in four dimensions if you're near the brane. But what about everywhere else?

The answer is that in RS2, the influence of four-dimensional gravity is inescapable, no matter where you are in the fifth dimension. Although the graviton's probability function is largest on the Gravitybrane, objects everywhere can interact with one another by exchanging a graviton, and therefore all objects would experience four-dimensional gravity, independently of location. Gravity everywhere looks four-dimensional because the graviton's probability function is never actually zero—it continues on for ever. In the localized scenario, objects far from a brane would have extremely weak gravitational interactions, but weak gravity would nonetheless behave in a four-dimensional manner. So, for example, Newton's inverse square law would hold, no matter where you were along the fifth dimension.

The small but nonzero graviton probability function away from the Gravitybrane was essential to the solution to the hierarchy problem I presented in Chapter 20. The Weakbrane, located away from the Gravitybrane in the bulk, experiences gravity that appears to be four-dimensional, even if it experiences that gravity only extremely feebly. Like water far from your own garden in the sprinkler analogy, there is always some water, just not a lot.

But suppose we reflect even further and ask what we really know with certainty about the dimensions of space. We do not know that space everywhere looks three-dimensional, only that space *near us* looks three-dimensional. Space appears to have three dimensions (and spacetime to have four) at the distances *that we can see*. But space might extend beyond that, into inaccessible territory.

After all, the speed of light is finite, and our universe has existed for only a finite amount of time. That means that we can only possibly know about the surrounding region of space within the distance that light could have traveled since the universe's inception. That is not infinitely far away. It defines a region known as the *horizon*, the dividing line between information that is and is not accessible to us.

Beyond the horizon, we don't know anything. Space needn't look like ours. The Copernican Revolution is repeatedly updated and revised as we see further into the universe and realize not everywhere is necessarily the same as what we see. Even if the laws of physics are the same everywhere, that doesn't mean that the stage on which they are played out is the same. It could be that nearby branes induce a different gravitational force law in our vicinity than would be seen elsewhere.

How can we claim to know the dimension of the universe outside our purview? There would be no contradiction if the universe beyond exhibited more dimensions—maybe five, maybe ten, maybe more. By thinking about the bare essentials, rather than assuming that everywhere, even inaccessible regions, is made up of spacetime that looks like ours, we can deduce what is really fundamental and what is ultimately conceivable and legitimate.

All we know is that the space we experience appears to be four-dimensional. It might be overstepping the mark to assume that all other regions of the universe must be four-dimensional as well. Why should a world extremely far from ours, which might not interact with us at all—or perhaps only via extremely weak gravitational signals—have to see gravity and space the way we do? Why can't it have a different type of gravity?

The marvelous thing is that it can. Our braneworld could experience three-plus-one dimensions, while outside regions do not. To our amazement, in 2000, Andreas Karch and I developed a theory in which space looks four-dimensional on or near the brane, but most of the space far from the brane appears higher-dimensional. This idea is schematically illustrated in Figure 90.

We named our scenario *locally localized gravity* because localization produces a graviton that communicates four-dimensional gravitational interactions only in a local region—the rest of space doesn't look four-dimensional. A four-dimensional* world exists only

*This model is also known as "KR," after the initials of our last names.

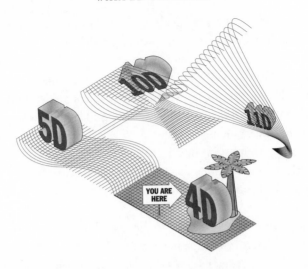

Figure 90. *We could be living in a four-dimensional sinkhole in a higher-dimensional space.*

on a gravitational "island." The dimensionality you see depends on your location in the five-dimensional bulk.

To understand local localization, let's return to our ducks in a pond. You might have disagreed when I said that the size of the pond doesn't matter. If the pond were truly enormous, ducks on the opposite side of the pond wouldn't congregate with the ducks on your side. In fact, it would be very strange if you could influence ducks that were very far away. The distant ducks wouldn't notice your bread, and would obliviously paddle about in a remote part of the lake.

The basic idea underlying locally localized gravity is very similar. Localization of gravity on a brane shouldn't necessarily depend on what is happening in distant regions of space. Although the model I studied with Raman had a graviton whose probability function decayed exponentially but was never quite zero—and that four-dimensional gravity would be experienced everywhere—gravity's behavior far away should not be essential to determining whether four-dimensional gravity exists in the vicinity of the brane.

That is the essence of locally localized gravity. A graviton can be localized and generate a four-dimensional gravitational force in the vicinity of a brane without affecting the gravitational force far away. Four-dimensional gravity can be a completely local phenomenon, relevant only to some portion of space.

Ironically, Andreas, who is an excellent physicist and a very nice guy, had first started thinking about the model that showed that this was possible while he was working on a research project with one of my former MIT colleagues, who had intended to challenge Raman's and my work. (Happily for us, their collaboration did a beautiful job of showing that our work was right.) In the course of his project, Andreas identified a model that was closely related to the one Raman and I had developed, but which had some very peculiar properties. When Andreas visited Princeton, he came to talk to me about it. Eventually we figured out that this model has some startling implications. At first, Andreas and I collaborated via e-mail and on visits to each other's institutions, and afterward, more easily, when I was back in Boston. And what we found was quite remarkable.

This model was very similar to the one I had studied with Raman; it had a single brane in five-dimensional warped space. But the difference in this case was that the brane was not exactly flat. This was because it carried a tiny amount of negative vacuum energy. In general relativity, as we have seen, not only relative energy but also the total amount of energy is meaningful. The total energy tells spacetime how to curve. For example, constant negative energy in five-dimensional spacetime gives rise to the warped spacetime that we have been discussing in the last few chapters. However, in that case the branes themselves were flat. Here, negative energy on the brane makes the brane itself slightly curved.

The negative energy on the brane leads to an even more interesting theory. However, we weren't actually interested in the negative energy itself—if we live on a brane, our brane should actually have a tiny positive energy to agree with observations. Andreas and I decided to study this model solely because of its fascinating implications for dimensionality.

To understand what we found, let's briefly return to a setup with two branes, with the understanding that afterward we will remove the second one. When the second brane was sufficiently far away, we found that there were two *different* gravitons, one localized near each of the two branes. Each of the graviton probability functions peaked near one of the two branes, and decreased exponentially quickly as you left it.

Neither of the gravitons was responsible for four-dimensional gravity over the entire space. They produced four-dimensional gravity only in the region adjacent to the brane on which they were localized. The gravities experienced on the different branes were different. They could even have very different strengths. And objects on one brane didn't interact gravitationally with objects on the other.

The setup with two widely separated branes can be compared to a situation in which someone on the opposite, very distant shore is also feeding ducks. Those ducks could even be of a different type; perhaps you are attracting mallards but, on the opposite shore, someone is attracting wood ducks. In that case, there would be a second concentration of ducks along the opposite shoreline, analogous to the second graviton probability function that is localized near a second brane.

The appearance of two different particles that both look like the four-dimensional graviton was a big surprise to us. General physical principles were supposed to ensure that there is only a single theory of gravity. And indeed, there is a single five-dimensional theory of gravity. However, five-dimensional spacetime turns out to contain two distinct particles that each communicate a gravitational force that acts as if it is four-dimensional, each in a distinct region of five-dimensional space. Different regions of space look like they both contain four-dimensional gravity, but the graviton communicating the four-dimensional gravitational force in those theories is different.

But there was a second surprise as well. According to general relativity, the graviton is massless. Like the photon, it should travel at the speed of light. But Andreas and I discovered that one of the two gravitons has a nonzero mass and didn't travel at this speed. This was truly surprising—but also disturbing. The physics literature said that no graviton with mass would ever produce a gravitational force that matched all observations. In fact, just as we discussed in the case of a heavy gauge boson in Chapter 10, a graviton with mass would have more polarizations than a massless one. And physicists demonstrated, by comparing different measured gravitational processes, that no effects of any extra graviton polarizations have ever been seen. This puzzled us for some time.

But the model outsmarted conventional wisdom. Once we had discovered this model, Massimo Porrati, a physicist at New York University, and Ian Kogan, Stavros Mousopoulos, and Antonios Papazouglou at Oxford University, found that in certain cases the graviton could in fact have mass and still yield correct gravitational predictions. They analyzed technicalities in the theory and demonstrated the loophole in the logic of why a graviton with mass should not agree with observed gravitational processes.

And the model has even weirder implications. Let's think now about what happens when we eliminate the second brane. Physical laws will then still appear to be four-dimensional on the remaining brane, the Gravitybrane, despite the infinite extra dimension. Gravity near the Gravitybrane is virtually identical to that in the RS2 model. For things on the Gravitybrane, the single graviton communicates the force of gravity, and gravity appears to be four-dimensional.

However, there is an important distinction between this model and RS2. In this model, which is different only because of the negative energy on the brane, the graviton that is localized near the brane does not dominate the gravitational force over the entire space. The graviton does not interact with objects anywhere in the space; it yields four-dimensional gravity only on or near the brane. Far from the brane, gravity no longer looks four-dimensional![38]

This might seem to contradict what I said earlier, that gravity must exist everywhere in the higher-dimensional bulk. This is not a false statement; five-dimensional gravity is everywhere. However, unlike the other extra-dimensional theories we have so far considered, in which physics always has a four-dimensional interpretation, this theory looks four-dimensional only for things that are on or near the brane. Newton's gravitational force law applies only on or near the brane. Everywhere else, the gravitational force is five-dimensional.

In this setup, four-dimensional gravity is a completely local phenomenon, experienced only in the vicinity of the brane. The dimensionality you would deduce from the behavior of gravity would depend on where you are in the fifth dimension. If this model is correct, we would have to live on the brane to experience four-dimensional gravity. If we were anywhere else, gravity would look five-dimensional. The brane is a four-dimensional gravity sinkhole— a four-dimensional gravitational island.

Of course, we don't yet know whether locally localized gravity applies in the real world. We don't even know whether extra dimensions exist or—if they do—what has become of them. However, if string theory is right, there are extra dimensions. And if so, they could be hidden by either compactification or localization (or local localization) or by some combination of the two. Many string theorists continue to believe that compactification is the answer, but because there are so many puzzles about the gravity that emerges from string theory, no one can be sure. I view localization as a new option. When gravity is localized, physical laws behave as if the dimensions weren't there, just as with rolled-up dimensions. Localized gravity therefore supplements our model building toolkit and increases the chances of discovering a realization of string theory that agrees with observations.

I like the way locally localized gravity concentrates on what we can explicitly verify. It says only that the universe has to look four-dimensional where we can test it—not that it has to *be* four-dimensional. Our three spatial dimensions could be a mere accident of our location. This idea has yet to be fully explored. But it is not out of the question that different regions of space could appear to have different numbers of dimensions. After all, new physics is revealed each time we probe shorter distances beyond what had previously been seen. Maybe the same thing is true about large distances: if we live on a brane, who knows what lies beyond?

What's New

- Localized gravity is a local phenomenon. It doesn't depend on distant regions of spacetime.

- Gravity can behave as if the world has different dimensions in different regions, since a localized graviton does not necessarily extend over all of space.

- We could be living in an isolated pocket of space that appears to be four-dimensional.

24

Extra Dimensions:
Are You In or Out?

> But I still haven't found what I'm looking for.   U2

Athena's dreams about OneDLand, branes, and five dimensions were passed down for generations. When Ike XLII heard them, he wanted to check whether there was any truth to her stories. So he took out his Alicxvr and went down to a very small scale—not so small that strings would appear, but sufficiently small to check whether there was a fifth dimension. The Alicxvr answered Ike's question by sending him off to a five-dimensional world.

But Ike was not completely satisfied. He remembered the bizarre things that had happened earlier on when he had fooled around with the hyperdrive option. So he once again cranked up the hyperdrive lever—and once again, everything changed drastically. Ike couldn't identify a single familiar object. He could tell only one thing: the fifth dimension had disappeared.

Ike was mystified, so he searched the spacernet to see what it could tell him about "dimensions." He waded through numerous sites that he recognized from his more embarrassing spam, but soon realized that he'd have to refine his search. When he still couldn't find anything definitive, he conceded that he wouldn't know the fundamental origin of dimensions any time soon. So he decided to turn his attention to time travel instead.

Physics has entered a remarkable era. Ideas that were once the realm of science fiction are now entering our theoretical—and maybe even experimental—grasp. Brand-new theoretical discoveries about extra dimensions have irreversibly changed how particle physicists, astrophysicists, and cosmologists now think about the world. The sheer number and pace of discoveries tells us that we've most likely only scratched the surface of the wondrous possibilities that lie in store. Ideas have taken on a life of their own.

Nonetheless, many questions have yet to be fully answered, and our journey is far from over. Particle physicists still want to know why we see the particular forces we see, and are there any more. What is the origin of the masses and properties of familiar particles? We also want to know whether string theory is right. And if it is, how does it connect to our world?

Recent observations of the cosmos point to even more mysteries we want to address. What composes most of the energy and matter in the universe? Was there a brief phase of explosive expansion early on in the universe's evolution, and if so, what caused it? And everyone wants to know what the universe looked like when it started.

We now know that gravity can behave very differently on different length scales. At very short distances, only a quantum theory of gravity such as string theory will describe gravity. On larger scales, general relativity applies admirably well, but recent observations across the universe at very big distances pose cosmological puzzles, such as what accelerates its expansion. And at longer distances still, we reach the cosmological horizon beyond which we know nothing.

One of the intriguing aspects of extra-dimensional theories is that they naturally have different consequences on different scales. Gravity in such theories exhibits behavior at distances smaller than curled-up dimensions, or where the curvature is too small to have an effect, that is different from the behavior at larger distances where dimensions might be invisible, or warping can be important. This gives us reason to believe that extra dimensions might eventually shed light on some of the mysterious features of the cosmos. If we do live in a multidimensional world, we certainly won't be able to neglect its cosmological implications. Some research has already been done on this subject, but I'm sure many more interesting results await us.

Where do I expect physics to go from here? There are too many possibilities to enumerate. But let me describe a few intriguing observations that suggest that more important theoretical surprises lie in store—ones that might come closer to resolution some time soon. These mysteries center on a question that at this point might sound shocking, namely:

What Are Dimensions, Anyway?

How can I ask such a question this late in the game? I've already spent much of this book discussing the meaning of dimensions and some of the potential implications of proposed extra-dimensional worlds. But now that I've told you what we understand about dimensions, allow me to return briefly to this question.

What does the number of dimensions really mean? We know that the number of dimensions is defined as the number of quantities that you need to locate a point in space. But I also presented examples in Chapters 15 and 16 showing that ten-dimensional theories sometimes have the same physical consequences as eleven-dimensional theories.

Such duality suggests that our notion of dimension isn't quite as firm as it looks—there's a plasticity in the definition that eludes the conventional terminology. Dual descriptions of a single theory tell us that no single formulation is necessarily the best one. The formulation and even the number of dimensions in the best description might depend on the strength of the string coupling, for example. Because no single theory is always the best description, the question of the number of dimensions doesn't always have a simple answer. This ambiguity in the meaning of dimensions and the apparent emergence of an additional dimension in strongly interacting theories are among the most important theoretical physics observations of the last decade. Let me now list a few more intriguing recent theoretical discoveries that indicate that the notion of dimension is somewhat fuzzier than we'd maybe like to believe.

I. Warped Geometry and Duality

In Chapters 20 and 22, I explained some of the consequences of the warped spacetime geometry that Raman Sundrum and I developed. In that geometry, the masses and sizes of objects depend on location along a fifth dimension and, furthermore, gravity is localized in the vicinity of a brane. But there is still one more amazing feature of this warped spacetime, known technically as anti de Sitter space, that I have yet to tell you about—one that leads to further questions about dimensionality.

The remaining remarkable feature of anti de Sitter space is the existence of a dual four-dimensional theory. Theoretical clues tell us that everything that happens in five-dimensional anti de Sitter space can be described using a dual four-dimensional framework in which there are extremely strong forces that have special properties. According to this mysterious duality, everything in the five-dimensional theory has an analog in the four-dimensional theory. And vice versa.

Although mathematical reasoning tells us that a five-dimensional theory in anti de Sitter space is equivalent to a four-dimensional one, we don't always know the precise particle content of that four-dimensional dual theory. However, Juan Maldacena, an Argentinian-born string theorist now at the Institute for Advanced Study in Princeton, triggered a string theory frenzy in 1997 when he derived an explicit example of a similar duality in string theory. He realized that a version of string theory with a large number of overlapping D-branes on which strings interact strongly can be described either with a four-dimensional quantum field theory or with a ten-dimensional gravitational theory in which five of the ten dimensions are rolled up and the remaining five are in anti de Sitter space.

How can a four-dimensional and a five- (or ten-)dimensional theory have the same physical implications? What is the analog of an object traveling through the fifth dimension, for example? The answer is that an object moving through the fifth dimension would appear in the dual four-dimensional theory as an object that grows or shrinks. This is just like Athena's shadow on the Gravitybrane, which grew as she moved away from the Gravitybrane across the fifth dimension. Furthermore, objects moving past each other along the fifth dimension correspond to objects that grow and shrink and overlap in four dimensions.

Once you introduce branes, the consequences of the duality are even stranger. For example, five-dimensional anti de Sitter space with gravity but without branes is equivalent to a four-dimensional theory without gravity. But once you include a brane in the five-dimensional theory, as Raman and I did, the equivalent four-dimensional theory suddenly contains gravity.

Does this duality mean that I was cheating when I said that the warped geometries were higher-dimensional theories? Absolutely not. The duality is intriguing, but it doesn't really change anything I've told you. Even if someone finds the precise dual four-dimensional theory, such a theory will be extremely difficult to study. It has to contain an enormous number of particles and such extremely strong interactions that perturbation theory (see Chapter 15) wouldn't apply.

Theories in which objects strongly interact are almost always impossible to interpret without an alternate, weakly interacting description. And in this case, that tractable description is the five-dimensional theory. Only the five-dimensional theory has a simple enough formulation to use for computation, so it makes sense to think of the theory in five-dimensional terms. Nonetheless, even if the five-dimensional theory is more tractable, duality still makes me wonder what the word "dimensions" really means. We know that the number of dimensions should be the number of quantities you need to specify the location of an object. But are we always sure we know which quantities to count?

II. T-duality

Another reason to question the meaning of dimensions is an equivalence between two superficially different geometries that is known as *T-duality*. Even before string theorists discovered any of the dualities I've discussed, they discovered T-duality, which exchanges a space with a tiny rolled-up dimension for another space with a huge rolled-up dimension.[39] Odd as it may seem, in string theory, extremely small and extremely large rolled-up dimensions yield the same physical consequences. A minuscule tiny volume of rolled-up space has the same physical consequences as an extremely large one.

T-duality applies in string theory with curled-up dimensions because there are two different types of closed string in spacetime compactified on a circle, and these two strings get interchanged when a space with a tiny rolled-up dimension is exchanged for a space with a large one. The first type of closed string oscillates up and down as it circles the closed dimension, similar to the behavior of the Kaluza-Klein particles we looked at in Chapter 18. The other type wraps around the curled-up dimension. It can do so once, twice, or any number of times. And T-duality operations, which interchange a small rolled-up space for a large one, exchange these two types of string.

In fact, T-duality was the first clue that branes had to exist: without them, open strings wouldn't have had analogs in the dual theory. But if T-duality does apply and a minuscule rolled-up dimension yields the same physical consequences as an enormous rolled-up dimension, it would mean that, once again, our notion of "dimension" is inadequate.

That is because if you imagine making the radius of one rolled-up dimension infinitely large, the T-dual rolled-up dimension would be a circle of zero size—there would be no circle at all. That is, an infinite dimension in one theory is T-dual to a theory with one dimension fewer (since a zero-size circle doesn't count as a dimension). So T-duality also shows that two apparently different spaces could appear to have a different number of large extended dimensions, yet make identical physical predictions. Once again, the meaning of dimension is ambiguous.

III. Mirror Symmetry

T-duality applies when a dimension is rolled up into a circle. But an even weirder symmetry than T-duality is *mirror symmetry*, which sometimes applies in string theory when six dimensions are rolled up into a Calabi-Yau manifold. Mirror symmetry says that six dimensions can be curled up into two very different Calabi-Yau manifolds, yet the resulting four-dimensional long-distance theory can be the same. The mirror manifold of a given Calabi-Yau manifold could look entirely different: it might have different shape, size, twisting, or even number of holes.* Yet when there exists a mirror to a given Calabi-Yau manifold, the physical theory where six of the dimensions are curled up into either one of the two will be the same. So with mirror manifolds as well, two apparently different geometries give rise to the same predictions. Once again, spacetime has mysterious properties.

IV. Matrix Theory

Matrix theory, a tool for studying string theory, provides still more mysterious clues about dimensions. Superficially, matrix theory looks like a quantum mechanical theory that describes the behavior and interactions of Do-branes (pointlike branes) moving through ten dimensions. But even though the theory doesn't explicitly contain gravity, the Do-branes act like gravitons. So the theory ends up having gravitational interactions, even though the graviton is superficially absent.

Furthermore, the theory of Do-branes mimics supergravity in eleven dimensions, not ten. That is, the matrix model looks as if it contains supergravity with one more dimension than the original theory seems to describe. This suggestive behavior (along with other mathematical evidence) has led string theorists to believe that matrix theory is equivalent to M-theory, which also contains eleven-dimensional supergravity.

One especially bizarre feature of matrix theory is Edward Witten's observation that when Do-branes come too close to each other, you can no longer know exactly where they are. As Tom Banks, Willy Fischler, Steve Shenker, and Lenny Susskind—the originators of matrix theory—said in their paper, "Thus for small distances there is

---

*Manifolds can have different numbers of holes; for example, a sphere has no holes, whereas a torus—a donut-like shape—has one.

no representation of the configuration space in terms of ordinary positions."* That is, the location of a Do-brane is no longer a meaningful mathematical quantity when you try to define it too precisely.

Although such strange properties make matrix theory very tantalizing to study, it is presently very difficult to use it for computations. The problem is that—like nearly all other theories containing strongly interacting objects—no one has yet found a way to solve many of the most important questions that will help us better to understand what is really going on. Even so, because of the emergence of an extra dimension and the disappearance of dimensions when Do-branes come too close together, matrix theory is one more reason to wonder what dimensions really mean.

What to Think?

Although physicists have mathematically demonstrated these mysterious equivalences between theories with different numbers of dimensions, we are clearly still missing the big picture. Do we know with certainty that these dualities apply, and if so, what they tell us about the nature of space and time? Moreover, no one knows what the best description would be when a dimension is neither very big nor very small (relative to the extraordinarily tiny Planck scale length). Perhaps our notion of spacetime breaks down altogether once we try to describe something so small.

One of the strongest reasons for believing that our spacetime description is inadequate at the Planck scale length is that we don't know any way, even in theory, to examine such a short distance. We know from quantum mechanics that it takes a lot of energy to investigate small length scales. But once you put too much energy into a region as small as the Planck scale length, 10^{-33} cm, you get a black hole. You then have no way to know what's happening inside. All that information is trapped within the black hole's event horizon.

On top of that, even if you were to try to cram more energy into

*T. Banks, W. Fischler, S.H. Shenker, and L. Susskind, "M theory as a matrix model: a conjecture," *Physical Review* D, vol. 55, pp. 5112–28 (1997).

that tiny region, you wouldn't succeed. Once you've put that much energy inside the Planck scale length, you can't add any more without the region expanding. That is, the black hole would grow if you added energy. So rather than making a nice tiny probe to study that distance, you would blow the region up into something bigger and never get to study it while it's small. It would be like trying to study delicate artifacts in a museum with a fine laser beam that instead burns them up. Even in physics thought experiments, you simply never see a region that is very much tinier than the Planck scale length. The rules of physics that we know break down before you get there. Somewhere in the vicinity of the Planck scale, conventional notions of spacetime almost certainly do not apply.

Facts so bizarre cry out for a deeper explanation. One of the most important lessons of the perplexing discoveries of the last decade is likely to be that space and time have more fundamental descriptions. Ed Witten succinctly summarized the problem when he said that "space and time may be doomed." Many leading string theorists agree: Nathan Seiberg asserted, "I am almost certain that space and time are illusions"; whereas David Gross imagines that, "Very likely, space and perhaps even time have constituents; space and time could turn out to be emergent properties of a very different-looking theory."* Unfortunately, no one yet has any idea what the nature of this more fundamental description of spacetime will turn out to be. But a deeper understanding of the fundamental nature of space and time clearly remains one of the biggest and most intriguing challenges for physicists in the coming years.

*The quotes are from K.C. Cole's article, "Time, space obsolete in new view of universe," *Los Angeles Times*, November 16, 1999.

25

(In)Conclusion

It's the end of the world as we know it (and I feel fine).

REM

Icarus Rushmore XLII used his time machine to visit the past and warn Icarus III of the disaster that awaited him should he continue driving his Porsche. Ike III was so astounded by his visitor from the future that he heeded Ike XLII's warning. He traded in his Porsche for a Fiat and subsequently led a full, contented, and slower-paced life.

Athena was ecstatic to be reunited with her brother, and Dieter was happy to see his friend, though both of them were confused since it seemed as if Ike had never left. Athena and Dieter realized that the time travel that Ike reported to them was pure fiction. Even in dreams, the Cat never looped through time, the Rabbit never reached a stop with extra time dimensions, and the quantum detective refused to contemplate such odd behavior of time. But Athena and Dieter preferred happy endings. So they suspended disbelief and accepted Ike's fantastic story all the same.

Despite the impressive physics developments of the last few years, we don't yet know how to harness the force of gravity or teleport objects across space, and it's probably too soon to invest in property in extra dimensions.[40] And because we don't know how to connect universes in which you could loop through time to the one in which we live, no one can create a time machine, and most likely no one will do so any time soon (or in the past).

But even if ideas like these remain in the realm of science fiction, we live in a wonderful and mysterious universe. Our goal is to learn how its pieces fit together and how they've evolved into their current state. What are the connections that we haven't yet figured out? What are the answers to questions like those I asked in the previous chapter?

Even if we have yet to understand the ultimate origin of matter at the deepest level, I hope I've convinced you that we do understand many aspects of its fundamental nature on the distance scales we have experimentally studied. And even if we don't know the most basic elements of spacetime, we do understand its properties for distances far away from the Planck scale length. In those regimes, we can apply physical principles we understand and deduce the sorts of consequences I've described. We've encountered many unexpected features of extra dimensions and branes, and those features might play a critical role in solving some of the puzzles of our universe. Extra dimensions have opened our eyes and our imaginations to amazing new possibilities. We now know that extra-dimensional setups can come in any number of shapes and sizes. They could have warped extra dimensions, or they could have large extra dimensions; they might contain one brane or two branes; they might contain particles in the bulk and other particles confined to branes. The cosmos could be larger, richer, and more varied than anything we imagined.

Which, if any, of these ideas describes the real world? We'll have to wait for the real world to tell us. The fantastic thing is that it probably will. One of the most exciting properties of some of the extra-dimensional models I've described is that they have experimental consequences. I can't overemphasize the significance of this remarkable fact. Extra-dimensional models—with new features that we might have thought were either impossible or invisible—could have consequences that we might see. And from these consequences, we might be able to deduce the existence of extra dimensions. If we do, our vision of the universe will be irrevocably altered.

There might be tests of extra-dimensional spacetime in astrophysics or cosmology. Physicists are now developing detailed theories of black holes in extra-dimensional worlds, and have found that although they are similar to their properties in four dimensions, there are subtle differences. The properties of extra-dimensional black holes could turn out to be sufficiently distinctive that we will be able to discern recognizable differences.

Cosmological observations might also ultimately tell us more about the structure of spacetime. Observations today probe what the universe looked like billions of years ago. Many agree with predictions, but several important questions remain. If we live in a higher-dimensional universe, it must have been very different earlier on. And some of those differences might help to explain perplexing features of observations. Physicists are now studying the implications of extra dimensions for cosmology. We might learn about dark matter hidden on other branes, or cosmological energy stored by hidden higher-dimensional objects.

But one thing is certain. Within the next five years, the Large Hadron Collider particle accelerator at CERN will turn on and probe physical regions no one has ever observed before. My colleagues and I are eagerly awaiting that time. The LHC is a great bet—for scientists it doesn't get much better. Experiments at the LHC will almost certainly discover particles whose properties will give us new insights into physics beyond the Standard Model. The exciting thing is that no one yet knows what those new particles will be.

During the time I've been doing physics, the only new particle discoveries have been particles that theoretical considerations told us we were pretty sure to find. Not to undermine those discoveries— they were impressive accomplishments—but finding something genuinely new and unknown will be far more thrilling. Until the LHC starts running, no one can be really certain where to best concentrate their efforts. Results from the LHC are likely to change the way we view the world.

The LHC will have enough energy to produce the new types of particle that promise to be so revealing. These particles could turn out to be superpartners or other particles that four-dimensional models predict. But they might also be Kaluza-Klein particles—particles that traverse extra dimensions. If and when we see those KK particles will depend entirely on the size and shape of the cosmos in which we live. Do we live in a multidimensional universe? And will the size or shape of that universe make KK particles visible?

All of the models that address the hierarchy problem have visible weak-scale consequences. The signatures of the warped geometry that addresses the hierarchy problem are particularly amazing. If this theory is right, we will detect KK particles and measure their properties from the clues they leave behind. If, instead, other extra-dimensional models describe the universe, energy will disappear into extra dimensions and we'll ultimately detect those dimensions through the resulting unbalanced energy accounting.

We certainly don't yet know all the answers. But the universe is about to be pried open. Astrophysical observations will explore the cosmos earlier, further away, and in more detail than ever before. Discoveries at the LHC will tell us about the nature of matter at distances smaller than any physical process ever observed. At high energies, truths about the universe should start to explode.

Secrets of the cosmos will begin to unravel. I, for one, can't wait.

Glossary

action at a distance The hypothetical instantaneous effect of objects on other, distant objects.

aether A hypothesized invisible substance (now debunked) whose vibrations were supposed to be electromagnetic waves.

alpha particle A helium nucleus (consisting of two protons and two neutrons).

anarchic principle The statement that all interactions that are not forbidden by symmetries will occur.

anomaly Symmetry violation that arises from quantum contributions to a physical interaction, but which is not present in the corresponding classical theory (in which quantum contributions are not taken into account).

anomaly-free theory A theory for which the symmetries of the classical theory are also symmetries of the theory with quantum contributions included.

anomaly mediation Communication of supersymmetry breaking by quantum effects.

anthropic principle The reasoning that says, out of many possible universes, we could live only in a place where **structure** could have formed.

anti de Sitter space Spacetime with constant negative curvature.

antiparticle A particle with the same mass as another particle, but opposite charge.

atom A building block of matter consisting of electrons orbiting a positively charged nucleus.

beta decay Radioactive decay in which a neutron splits into a proton, an electron, and a neutrino.

black hole A compact object that is so dense that nothing can escape from its surrounding gravitational field.

blackbody An idealized object that absorbs all heat and energy and radiates it back in a manner determined solely by its temperature.

blackbody radiation Radiation emitted by a blackbody.

boson A particle with integer spin—1, 2, etc. (one of two categories of particle established by quantum mechanics, the other being the **fermion**); photons and the Higgs particle are examples of bosons.

bottom quark A short-lived, heavier version of the down and strange quarks.

brane A membrane-like object in higher-dimensional space that can carry energy and confine particles and forces.

braneworld A physical setup in which matter and forces are confined to branes.

bulk Full higher-dimensional space.

Calabi-Yau manifold A six-dimensional compact space, defined by its particular mathematical properties, that plays an important role in string theory.

CERN The Conseil Européen pour la Recherche Nucléaire (now called the Organisation Européenne pour la Recherche Nucléaire or, in English, the European Organization for Nuclear Research), a high-energy accelerator facility in Switzerland; future home of the Large Hadron Collider (LHC).

charm quark A short-lived, heavier version of the up quark.

chirality The **handedness** of a particle with spin.

classical physics Physical laws that takes neither quantum mechanics nor relativity into account.

closed string A **string** that loops around and has no ends.

collapse of the wavefunction Reduction of the quantum state after a precise measurement fixes the value of the measured quantity.

compact space A finite space.

compactified A compactified space is one that is rolled-up into a finite size.

Compton scattering The scattering of a photon off an electron.

cosmological constant The value of a constant background energy density that isn't carried by matter.

cosmology The science of the evolution of the universe.

coupling constant The number that determines interaction strength.

curvature A quantity that describes bending or curving of an object, space, or spacetime.

D-brane A brane in string theory on which open strings end.

dark energy The measured vacuum energy in the universe that constitutes about 70% of the universe's energy but is not carried by any form of matter.

dark matter The nonluminous matter that carries about 25% of the energy in the universe.

de Sitter space Spacetime with constant positive curvature.

deep inelastic scattering The experiment that discovered quarks by scattering electrons off protons and neutrons.

desert hypothesis The assumption that there are no particles, aside from those included in the Standard Model, that can be produced at energies below the unification energy.

dimension An independent direction in space or time.

dimensionality The number of quantities required to uniquely pin down a point.

dimensionality of a brane The number of dimensions in which brane-bound particles are permitted to travel.

down quark One of the elementary quarks that compose the proton and the neutron.

dual theories Two equivalent descriptions of a single theory that might be superficially quite different.

effective field theory A quantum field theory defined at a particular energy that describes those particles and forces relevant to the energies to which it applies.

effective theory A theory describing those elements and forces that are in principle observable at the distance or energy scales over which it applies.

Einstein's equations The equations of general relativity with which you determine the **metric** (and hence the gravitational field) from the distribution of matter and energy.

electromagnetism One of the four known forces; electromagnetism describes both electricity and magnetism.

electron A very light elementary particle with a negative charge.

electroweak theory The theory incorporating both electromagnetism and the weak force; an essential component of the Standard Model of particle physics.

equivalence principle The principle that uniform acceleration and gravity are indistinguishable.

eV (electronvolt) The energy required to move an electron against a potential difference of 1 volt.

external particles Real physical particles that can enter and leave an interaction region.

family See **generation**.

Fermi interaction An interaction that is generated by the exchange of one of the massive weak gauge bosons.

Fermilab A collider facility in Illinois; home of the **Tevatron**.

fermion A particle with half-integer spin—$\frac{1}{2}$, $\frac{3}{2}$, etc. (one of two categories of particle established by quantum mechanics, the other being the **boson**); quarks and electrons are examples of fermions.

Feynman diagram A diagram that schematically illustrates allowed particle-physics interactions.

field A physical quantity that exists and has a particular value for each point in space. Examples include the classical electric field and quantum fields.

fine-tuning Fudging by adjusting a parameter to a very specific (and unlikely) value.

flavor A label that distinguishes different types of quark or lepton (often used to distinguish quarks and leptons from different generations).

flavor problem (of supersymmetry) The overly high prediction for flavor-changing processes (due to virtual squarks and sleptons) that plagues most models of supersymmetry breaking.

flavor symmetry Symmetry that interchanges different flavors of a particular particle category.

frame of reference An observational vantage point or a set of coordinates for describing events in space or spacetime.

gauge boson A particle that communicates an elementary force.

gaugino The superpartner of a force-carrying gauge boson.

gaugino mediation Communication of supersymmetry breaking by gauginos.

general relativity The theory of gravity that describes the gravitational field due to any source of matter and energy, including that stored by the gravitational field itself, in any frame of reference; general relativity encapsulates the gravitational field in the curvature of spacetime.

generation Each of the three sets of the full complement of particle types (left- and right-handed charged lepton, up-type quark, down-type quark and left-handed neutrino).

geodesic In space, the shortest path between two points; in spacetime, the path a free-falling observer (one with no forces acting on him) would follow.

GeV (gigaelectronvolt) A unit of energy equal to one billion eV.

gluon The elementary particle that communicates the strong force.

Grand Unified Theory (GUT) A proposed theory in which the three known nongravitational forces fuse into a single force at high energy.

gravitational lensing The splitting of light into multiple images as it bends around a massive object.

gravitino The superpartner of the graviton.

graviton The particle that communicates the force of gravity.

hadron A strongly bound object with constituent quarks and/or gluons.

handedness The direction of spin, to the left or to the right.

heterotic string theory A version of string theory in which the oscillation modes that travel clockwise are different from the modes that travel counterclockwise.

hierarchy problem The question of the weakness of gravity or, equivalently, of why the Planck scale mass that characterizes gravity's strength is sixteen orders of magnitude greater than the weak scale mass associated with the weak force.

Higgs field The field that participates in the Higgs mechanism, that is responsible for breaking the symmetry associated with the electroweak force.

Higgs mechanism The spontaneous breaking of the electroweak symmetry that allows gauge bosons and other elementary particles to acquire mass.

Hořava-Witten theory The strongly coupled heterotic string version of string theory, or equivalently (by duality) a version of string theory with two branes that are separated by an eleventh dimension in which the two branes house the forces of the heterotic string.

horizon A region beyond which nothing can escape.

hypercube A generalization of a cube to more than three dimensions.

inertial frame of reference A reference frame that moves at fixed velocity with respect to a fixed reference frame, such as the one at rest.

intermediate (internal) particles Virtual particles whose exchange mediates interactions among other particles.

internal symmetry A symmetry in which physical laws do not change for a set of transformations that does not change the geometric position of particles, but only some internal properties or labels.

intrinsic spin (spin) A number that characterizes how a particle behaves—as if it were spinning. Spin can take integer or half-integer values.

inverse square law The rule that describes those forces whose strength decreases with separation as the square of the distance between them; classical gravitational and electric forces obey inverse square laws.

ion A charged bound state of a nucleus and electrons; an atom with too few or too many electrons.

jet An energetic cluster of strongly interacting particles surrounding an energetic quark or gluon that moves in a particular direction.

Kaluza-Klein (KK) mode A four-dimensional particles with a higher-dimensional origin; KK modes are distinguished by their extra-dimensional momenta.

kinetic energy Energy due to motion.

lepton An elementary fermionic particle that does not experience the strong force.

LHC (Large Hadron Collider) A high-energy particle collider that will bang together 7 TeV proton beams and produce particles with mass up to a few TeV.

local interaction An interaction between adjacent or coincident objects.

localized gravity A high concentration of the gravitational field in a particular region of space; gravity appears to be lower-dimensional since it isn't diluted into an extra dimension.

locally localized gravity A theory in which four-dimensional gravity is not experienced everywhere, but only in the region of space where the probability function of the particle acting like a four-dimensional graviton is concentrated.

longitudinal polarization Wave oscillation along the direction of motion.

M-theory A hypothesized all-embracing theory that unifies all known versions of ten-dimensional string theory and eleven-dimensional supergravity.

matrix theory A ten-dimensional quantum-mechanical theory that might be equivalent to string theory.

mediate To communicate a particle's influence (by an intermediate particle).

metric A quantity or quantities that establish the measurement scale that determines physical distances and angles.

model A candidate theory.

molecule A bound state of two or more atoms in which electrons are shared between them.

multiverse A hypothetical generalization of a universe containing regions that don't interact or interact only extremely weakly.

muon A short-lived, heavier version of the electron.

neutral object An object that is immune to a force; neutral objects have net charge equal to zero.

neutrino A fundamental elementary particle that interacts only via the weak force.

neutron An ingredient of the atomic nucleus in which two down quarks and an up quark are tightly bound to each other.

Newton's gravitational constant The overall coefficient that determines the strength of gravitational attraction in Newton's law of gravity; it is inversely proportional to the square of the Planck scale mass.

Newton's gravitational force law The classical law of gravity that says that the strength of gravity between two massive objects is proportional to their masses and inversely proportional to the square of their separation.

nucleon A proton or neutron.

nucleus The hard, dense central component of an atom.

old quantum theory The predecessor to quantum mechanics that postulated quantization rules but didn't systematically determine them or describe the evolution of a quantum state through time.

open string A string with two ends.

p-brane A solution to Einstein's equations that expands infinitely far in some spatial directions, but in the remaining dimensions acts as a black hole, trapping objects that come too close.

particle accelerator A high-energy physics facility that accelerates particles to high energy.

particle collider A high-energy accelerator that smashes together particles to create enormous amounts of energy.

particle physics The study of the most elementary building blocks of matter.

Pauli exclusion principle The statement that two identical fermions cannot occupy the same position.

perturbation A small modification to a known theory.

perturbation theory When the theory you are interested in is distinguished from a solvable (usually non-interacting) theory by a small parameter (which could be a small interaction strength, for example), perturbation theory allows you to extrapolate from the solvable theory to the theory of interest through a systematic expansion in that small parameter. The results are expressed as power expansions in the corresponding parameter, usually the coupling constant.

photino The superpartner of the photon.

photon The elementary particle that communicates the electromagnetic force; the quantum of light.

Planck scale energy The energy at which gravity becomes a strong force and quantum-mechanical contributions need to be taken into account.

Planck scale length The length scale at which gravity is strong and quantum effects must be included in gravitational predictions.

Planck's constant A quantum-mechanical quantity that relates energy to frequency and momentum to wavelength.

polarization The direction of oscillation of a wave.

positron The positively charged antiparticle of the electron.

potential energy Stored energy that can be released as kinetic energy.

probability function The square of the absolute value of the wavefunction that determines the probability of finding a particle at a given location.

projection A definite prescription for creating a lower-dimensional representation of an higher-dimensional object.

proton An ingredient of the atomic nucleus in which two up quarks and a down quark are tightly bound to each other.

QCD (quantum chromodynamics) The quantum field theory of the strong force.

QED (quantum electrodynamics) The quantum field theory of electromagnetism.

quantum A discrete unbreakable unit of a measurable quantity; the smallest unit of that quantity.

quantum contribution A contribution to a physical process due to virtual particles.

quantum field theory The theory used to study particle physics with which one can calculate the rates for processes in which particles can interact, be created, or be destroyed. According to quantum field theory, fluctuations of fields manifest themselves as particles.

quantum gravity A theory of gravity that incorporates both quantum mechanics and general relativity.

quantum mechanics The theory that is based on the assumption that all matter consists of discrete elementary particles that have associated wavefunctions.

quark An elementary fermionic particle that experiences the strong force.

quasicrystal A solid material whose crystalline structure is derived from higher dimensions.

redshift The lowering of frequency of a wave when the object emitting the wave is either moving away (Doppler redshift) or is slowed down by a strong gravitational field (gravitational redshift).

relativity One of Einstein's two theories of spacetime: **special relativity**, which unifies space and time, and **general relativity**, which explains gravity as the curvature of spacetime.

renormalization group A calculation technique for relating quantities that apply in different energy or distance regimes.

rotational invariance The independence of the results of experiments on orientation (or direction).

selectron The superpartner of the electron.

sequestering The physical separation of different elementary particle types in extra dimensions.

singularity A region where a mathematical description of an object breaks down because some quantity becomes infinite.

slepton The superpartner of a lepton.

spacetime The concept that unifies space and time into a single framework; the mathematical formulation of the region where physical processes can occur.

special relativity The theory of gravity that describes motion in inertial frames of reference.

spectral lines Discrete frequencies at which non-ionized atoms emit or absorb light.

spectrum A function that gives the spread of energy emitted across all frequencies.

spin See **intrinsic spin**.

spontaneously broken symmetry Symmetry that is preserved by physical laws but broken by the actual physical state of a system.

squark The superpartner of a quark.

Standard Model (of particle physics) The effective theory that describes all known particles and nongravitational forces and the interactions among them.

strange quark A short-lived, heavier version of the down quark.

string A one-(spatial)dimensional extended object whose oscillations constitute elementary particles.

string coupling A quantity that determines the strength of the interaction between strings.

string theory The theory that posits that the ingredients of the universe are fundamental strings and which should consistently incorporate quantum mechanics and general relativity.

strong force One of the four known forces; the strong force is responsible for binding quarks in a proton or neutron, for example.

structure Constituents of matter.

substructure More elementary ingredients of constituents of matter.

supergravity A supersymmetric theory that includes gravity.

superpartner (of a particle) The particle that supersymmetry pairs with another particle; if the original particle is a boson, the superpartner is a fermion, and vice versa.

superspace An abstract space that incorporates the familiar four dimensions as well as theoretical fermionic dimensions.

superstring theory The supersymmetric version of string theory without **tachyons** that includes fermions in addition to gravity and gauge bosons.

supersymmetry A symmetry that interchanges partnered bosons and fermions.

symmetry A property of an object or a physical law such that certain physical operations are undetectable.

symmetry transformation A manipulation of a physical system that does not change its properties or behavior; the action that transforms different configurations that are related by symmetry into each other.

T-duality An equivalence between physical phenomena in a universe with a small rolled-up dimension and another universe with a large one (the size of the radius of a curled-up dimension is exchanged with its inverse).

tachyon A particle that signals an instability and superficially appears to have negative squared mass.

tau A short-lived particle with identical charge to the electron and muon but heavier than either.

tension Resistance to being stretched that determines how readily a string will oscillate and produce heavy particles.

TeV (teraelectronvolt) A unit of energy equal to one trillion eV.

Tevatron The high-energy collider currently in operation at Fermilab that collides beams of TeV-energy protons with TeV-energy antiprotons.

theory A definite set of elements and principles with rules and equations for predicting how those elements interact.

thought experiment An imagined physics experiment through which you can evaluate the consequences of a given set of physical assumptions.

top quark A short-lived, heavier version of the up quark; the heaviest known quark.

translational invariance The independence of physical laws of location in space.

transverse polarization Wave oscillation perpendicular to the direction of motion.

ultraviolet catastrophe An infinite energy emitted at high frequencies that is predicted by the classical theory of a blackbody.

uncertainty principle The basic principle underlying quantum mechanics that restricts the accuracy with which pairs of quantities (such as position and speed) can be simultaneously measured.

up quark One of the elementary quarks that compose the proton and the neutron.

vacuum The state of the universe with the lowest possible energy and no particles.

vacuum energy The energy carried by the vacuum, the state in which particles are absent; also known as the **cosmological constant**.

velocity The quantity that specifies both speed and direction of motion.

virtual particle An ephemeral particle allowed only by quantum mechanics; virtual particles carry the same charge as the corresponding true physical particles but have the wrong energy.

warp factor The overall scaling of a metric that varies with respect to one coordinate.

warped spacetime geometry Spacetime that would be flat (more generally, each slice would have the same shape) except for an overall scaling that varies with the position in a particular direction.

wavefunction A quantum-mechanical function that determines the relative likelihood of the corresponding object being at any point in space.

weak force One of the four known forces; the weak force is responsible for beta decay of neutrons into protons, for example.

weak gauge boson An elementary particle (with three varieties, W^+, W^-, and Z) that communicates the weak force.

weak scale energy The energy at which the symmetry associated with the weak force is spontaneously broken. The weak scale energy determines the masses of elementary particles.

weak scale length The length, 10^{-16} cm, or one ten thousand trillionth of a centimeter, that corresponds (via quantum mechanics and special relativity) to the weak scale energy. It is the range of the weak force—the maximum distance between particles that can influence each other through this force.

weak scale mass The mass that is related to the weak scale energy (of 250 GeV) through the speed of light. In conventional mass units, the weak scale mass is 10^{-21} grams.

Math Notes

1. This isn't really a math note, but the Saturday Night Baby is three-dimensional. [Figure M1]

Figure M1. *Saturday Night Baby.*

2. A metric on space can take the form $ds^2 = a_x dx^2 + a_y dy^2 + a_z dz^2$, where x, y, z are the three coordinates of space, and a_x, a_y, and a_z can be numbers or they can be functions of x, y, and z. The metric determines lengths, distances, and angles between lines. For example, the length of a vector pointing from the origin to the point with coordinates (x, y, z) is $\sqrt{(a_x x^2 + a_y y^2 + a_z z^2)}$. If $a_x = a_y = a_z = 1$, we have flat space, and distances and lengths would be measured in the familiar manner. For example, the length of a vector pointing from the origin to (x, y, z) would be $\sqrt{(x^2 + y^2 + z^2)}$. More complicated metrics can have cross terms, such as $dxdy$. In that case, the

metric must be described by a tensor with two indices that tells the coefficients a_{ij} of each term in the metric of the form $dx_i dx_j$. Later, when we discuss relativity, the metric will also have a term dt^2 and could also have terms of the form $dt dx_i$.

3. A hypersphere is defined by $x_1^2 + x_2^2 + \ldots + x_n^2 = r^2$. Here x_i refers to the ith coordinate (the location in the ith dimension) and r is the radius of the hypersphere. The cross-section of the hypersphere when it crosses a fixed location in the nth dimension, $x_n = d$, is described by the equation $x_1^2 + x_2^2 + \ldots + x_{n-1}^2 = r^2 - d^2$. This is the equation of a hypersphere of one lower dimension and radius $\sqrt{(r^2 - d^2)}$. So, for example, when $n = 3$ and a sphere crosses Flatland, Flatlanders would see circles. (They would see disks if they saw the circles and their interiors, which would be mathematically described with an inequality.)

4. Calabi-Yau manifolds are not the only possible stringy hidden manifolds. We now know that others, such as ones called G2 holonomy manifolds, might also give acceptable models.

5. In string theory, we also sometimes use the word "brane" to mean space-filling branes that have the same number of dimensions as the higher-dimensional space. Here, however, we will concern ourselves only with branes that have fewer dimensions than the full higher-dimensional space, so I will restrict myself to the use of the term as described in the text.

6. A brane that extends in the dimensions $x_1, \ldots, x_j$ is described by the $n - j$ equations $x_{j+1} = c_{j+1}, x_{j+2} = c_{j+2} = \ldots, = x_n = c_n$, where x_i are coordinates, n is the number of dimensions of space, and c_i are fixed constants describing the location of the brane. More complicated branes that curve in the given coordinate system are described by more complicated equations that describe the surface.

7. In equation form, Newton's law says that the gravitational force is $G\, m_1 m_2/r^2$, where G is Newton's constant of gravitation, m_1 and m_2 are the two masses that are attracted to each other, and r is the distance between them.

8. Newtonian gravity respects Euclidean geometry. In Euclidean geometry, $x^2 + y^2 + z^2$, the length of a vector pointing to the point with coordinates (x, y, z), is independent of the coordinate frame. That is, you can rotate your coordinates but the distance to any point won't change, even though the individual coordinates will. Special relativity puts time into the picture. It says that $x^2 + y^2 + z^2 - c^2t^2$ is independent of your choice of inertial reference frame. Notice that this invariant quantity involves both space and time, but time is treated differently because of the minus sign in front of the term c^2t^2. Also notice that for this quantity to be independent of inertial frame, changes

in reference frame must mix the values of the space and time coordinates. If one reference frame moves at speed v with respect to the other in the x direction, the transformation of coordinates from (t, x, y, z) to (t', x', y', z') would be $x' = \gamma x - c\beta\gamma t$, $t' = \gamma t - \beta\gamma x/c$, $y' = y$, $z' = z$, where $\beta = v/c$, c is the velocity of light, and $\gamma = 1/\sqrt{(1 - \beta^2)}$.

9. Einstein's equations tell us how to derive the metric $g_{\mu\nu}$ from a known distribution of matter and energy: $R_{\mu\nu} - \frac{1}{2} g_{\mu\nu} R = 8\pi G T_{\mu\nu}/c^4$. $R_{\mu\nu}$ is the Ricci curvature tensor and is related to the metric $g_{\mu\nu}$, $T_{\mu\nu}$ is the stress-energy tensor describing the matter-energy distribution, G is Newton's constant of gravitation, and c is the speed of light. For example, for matter of mass density ϱ at rest, $T_{00} = \varrho$ while all the other components of the tensor are 0.

10. The energy per unit frequency emitted by a blackbody of temperature T depends on the frequency, f, as $f^3/(e^{hf/kT} - 1)$, where $k = 1.3807 \times 10^{-16}$ erg K^{-1} is Boltzmann's constant, which converts temperature to energy. Notice that at low frequencies the energy increases with frequency. But at frequencies where the energy of a quantum, hf, is large compared with kT, the spectrum drastically cuts off; the emitted energy is exponentially smaller at higher frequencies.

11. A wavefunction is actually a complex-valued function. This is the source of many of quantum mechanics' strange properties. When you add two complex functions together and then square the sum, you generally get a different result from when you first square and then add. That results in interference phenomena. For example, in the double-slit experiment the probability that is recorded on a screen results from the interference of the waves that describe the electron's two possible paths.

12. More precisely, it's the product of Planck's constant and the absolute value of the commutator of the two quantities divided by 2.

13. Special relativity tells us that a stationary object with rest mass m_0 carries energy $E = m_0 c^2$. More generally, an object that is moving with velocity v (with $\beta = v/c$ and $\gamma = 1/\sqrt{(1 - \beta^2)}$) will carry energy $E = \gamma m_0 c^2$. The rest mass is also known as the *invariant mass* (independent of reference frame). That is because, according to the transformation laws of special relativity, the quantity $E^2 - p^2 c^2 = m_0 c^4$ is the same in any reference frame. Notice that you always need an energy at least equal to $m_0 c^2$ to produce an object of mass m_0. Also notice that when an object has low mass compared with its energy (really, energy/c^2), the energy and momentum are related approximately by $E = pc$. That is why, at high energy, energy and momentum are roughly interchangeable.

14. Maxwell's Equations (in c.g.s. units) are

$$\nabla \cdot \mathbf{E} = 4\partial\varrho$$

$$\nabla \times \mathbf{E} = -\frac{1}{c}\frac{\partial \mathbf{B}}{\partial t}$$

$$\nabla \cdot \mathbf{B} = 0$$

$$\nabla \times \mathbf{B} = \frac{4\partial}{c}\mathbf{J} + \frac{1}{c}\frac{\partial \mathbf{E}}{\partial t},$$

where $\mathbf{E}$ is the electric field, $\mathbf{B}$ is the magnetic field, ϱ is the charge, and $\mathbf{J}$ is the current. These are first-order differential equations; by combining two of them you can derive a second-order differential equation involving only the electric or magnetic field. This equation takes the form of a wave equation—that is, its solutions are sinusoidal waves.

15. Actually, according to special relativity's underlying principles, there could have been a fourth polarization as well, one that would oscillate in the time direction. But that one doesn't exist either, and the same internal symmetry that eliminates the third (longitudinal) polarization eliminates "time polarization" as well. Since it plays no role in the discussion in this or the following chapter, we won't consider it any further.

16. The true symmetries associated with all the forces are actually more subtle and rotate fields, which are complex quantities, into each other. The symmetries don't merely interchange fields, they turn one field into a linear superposition of the others. The force associated with electromagnetism rotates a single complex field, whereas the weak force rotates two complex fields into each other, and the strong force rotates three.

17. To make a Higgs model work, at least one of the Higgs fields must be forced to take a nonzero value. This would be true if the minimum energy configuration occurs when the value of at least one of the Higgs fields is nonzero. One way this can happen is illustrated in Figure M2, which shows the so-called Mexican hat potential, a plot of the energy the system would take for any combination of values of the two Higgs fields, where the two lower axes are the absolute values of the two Higgs fields and the height of the three-dimensional surface represents the energy of that particular configuration. This particular potential takes the form $\lambda(|H_1|^2 + |H_2|^2 - v^2)^2$, where λ determines how bowed up the potential is and v determines the value that $|H_1|^2 + |H_2|^2$ will take when the potential is at its minimum. The key feature of this potential is that when both fields have zero value, it is at a local maximum. Therefore, energy considerations tell us that the Higgs fields will not both be zero. Instead, they will take values that put them at the bottom of the circular basin surrounding the origin.

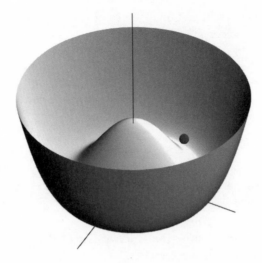

Figure M2. *The "Mexican hat" potential for the Higgs field.*

18. A more accurate way to describe the weak force symmetry would be to say that it rotates fields, rather than interchanging them.

19. This actually simplifies the symmetry breaking. Even if both x and y were nonzero—if both x and y were 5, for example—the rotational symmetry would be broken since a particular direction is picked out, the direction pointing from $x = 0$, $y = 0$ to the point where $x = 5$ and $y = 5$. A similar "rotational" symmetry applies to Higgs$_1$ and Higgs$_2$, but I have simplified and described the symmetry simply as an interchange symmetry. In the true description, even if both Higgs fields take the same value, the weak interaction symmetry would be broken—in much the same way as the point $x = 5$, $y = 5$ spontaneously breaks rotational symmetry.

20. Although this model starts with two complex Higgs fields, there is only a single Higgs particle in the end. That is because the three other (real) fields become the three additional fields that are required to turn three massless particles with two physical polarizations into massive particles with three polarizations. Three of the Higgs fields become the third polarizations of the three heavy weak gauge bosons—the two Ws and the Z. The fourth remaining Higgs field should create true physical Higgs particles. If this model is right, the LHC should produce them.

21. The strength of each of the forces is determined by a numerical coefficient. Renormalization group calculations show that the values of these quantities change logarithmically with energy.

22. Whereas the weak force symmetry mixes pairs of fields and the strong

force symmetry mixes three fields, the Georgi-Glashow Grand Unification symmetry group mixes five fields. Some of the symmetry transformations associated with the forces of the GUT coincide with the weak symmetry and the strong symmetry transformations. The forces are unified because a single symmetry group of transformations includes all the symmetry transformations of the Standard Model.

23. This connection to space and time is actually most manifest when two supersymmetry transformations are performed in succession, first in one order and then in the other, and then subtracted one from the other. In that case, fermions remain fermions and bosons remain bosons, but the system is moved; the net result of the transformation is exactly the same as a conventional spacetime transformation. The commutator of the two supersymmetry transformations, which perform exactly the same operation as a single spacetime symmetry transformation, decisively demonstrates that supersymmetry transformations must be connected to the symmetries that act on space and time and move things around.

24. The trajectory of a particle is a worldline that gives the particle's position as a function of time. The trajectory of a string is a surface that describes the position of the entire string as it moves through time. The worldsheet represents the motion of an open string, whereas the worldtube represents the motion of a closed string. This is shown in Figure M3, which illustrates the motion through time and the "softer" interactions of strings.

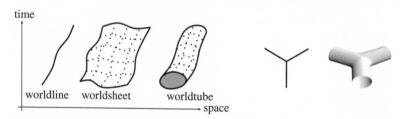

Figure M3. *(Left figure) Wordline of a particle, worldsheet of an open string, worldtube of a closed string. (Right figure) Interactions of three particles and three strings.*

25. The string tension is not always as high as you would guess from the Planck scale energy. It depends on how strongly strings interact. Joe Lykken and others have considered the possibility that it is much smaller, in which case the additional particles from string theory could be much lighter.

26. Actually, according to the duality we learn about in this chapter, even the probes used to study a given version of string theory change character when the coupling becomes strong. So if Ike really was part of the string world, he, too, would change.

27. They can also extend in zero dimensions, in which case they are new kinds of particle called D0-branes, or in one dimension, in which case they are new types of string called D1-branes.

28. Branes don't necessarily interact via ordinary charges. They interact via a higher-dimensional generalization of charges.

29. The symmetry actually rotates branes into each other, but this is beyond the technical reach of this book (and would make Igor's head spin).

30. Ordinarily, the gaugino masses fall in a ratio of about $1:3:30$, where the photino is the lightest, winos are next (though the zino might be a little heavier or lighter than the winos), and the gluinos are heaviest. In sequestered models the ratio is $1:2:8$, where the winos are the lightest, the photino is heavier, and the gluino is again the heaviest.

31. The wavefunctions of the Kaluza-Klein modes are the modes that occur in the generalized Fourier decomposition of the higher-dimensional wavefunction.

32. This also assumes that there are no singularities in the spacetime geometry—that is, no place where the space shrinks to zero size.

33. D. Cremades, S. Franco, L. Ibanez, F. Marchesano, R. Rabadan, and A. Uranga also suggested an interesting alternative. Their idea is that particles are not confined on an individual brane, but are instead confined to the intersections of multiple branes. As with separated parallel branes, strings extending between branes will generally be heavy. But light or massless particles arise from zero-length strings, which in this case would be confined to the region where the branes intersect.

34. We can also show this in a slightly different way with a more mathematical argument. When there are curled-up dimensions, the force lines emanating from a massive object behave according to the gravitational law of the higher-dimensional theory at short distances and according to four-dimensional gravity at long distances. The only way to reconcile the two force laws and switch smoothly from one to the other is by noting that at about the distance corresponding to the extra dimensions' sizes, the force lines spread as if there were only four dimensions, but with a reduced strength because of the extra volume of the curled-up space. Beyond the size of the extra dimensions, gravity behaves four-dimensionally but with its strength suppressed by the spreading out over the extra-dimensional volume.

Newton's law of gravitation says that when there are three spatial dimensions, the force is proportional to $1/M_{Pl}^2 \times 1/r^2$. If there are n additional dimensions, the force law would be $1/M^{n+2} \times 1/r^{n+2}$, where M sets the strength of higher-dimensional gravity, similarly to the way in which M_{Pl} sets the strength of four-dimensional gravity. Notice that the higher-dimensional force law varies more quickly with r since the force lines would spread over a hypersphere whose surface would have $n + 2$ dimensions (as opposed to the two-dimensional surface of a sphere that gives rise to the force law of three-dimensional space). However, when the extra-dimensional volume is finite and the n extra dimensions have size R, the force law will be $1/M^{n+2} \times 1/R^n \times 1/r^2$ when r is greater than R, and the force lines can no longer spread in the extra dimensions. This is the form of a three-spatial-dimensional force law if we make the identification $M_{Pl}^2 = M^{n+2} R^n$. Since R^n is the volume of the higher-dimensional space, we find that the strength of gravity decreases with volume, or equivalently (because gravity's strength is weaker when the Planck scale energy is bigger), the Planck scale energy is big if the volume is big.

35. A flat metric with three spatial dimensions is $ds^2 = dx^2 + dy^2 + dz^2 - c^2 dt^2$. Because there are no spatial or time-dependent coefficients, measurements are independent of where you are or which direction you point in; that is to say, spacetime is completely flat. All three spatial coordinates as well as the time coordinate (up to the minus sign that always singles time out) are treated the same; that is, the coefficients of the terms in the metric are completely independent of time and spatial location.

36. The metric in the warped geometry is $ds^2 = e^{-k|r|}(dx^2 + dy^2 + dz^2 - c^2 dt^2) + dr^2$, where r is the coordinate of the fifth dimension. That tells us that at any fixed location in the fifth dimension, which corresponds to fixed r, spacetime is completely flat. However, the overall r-dependent factor tells us that how we measure size changes according to the position of an object in the fifth dimension. The exponential falloff of the coefficient, which is the warp factor, is the reason that the graviton's probability function falls exponentially, and is also why we need to rescale mass, energy, and size to make a single, four-dimensional effective theory.

37. Because space is not flat, the extra-dimensional volume that enters when we calculate M_{Pl} in four dimensions is not simply $M_{Pl}^3 R$, as it would be when space is flat. Instead, the value of M_{Pl} depends on the curvature. If the metric has the form $ds^2 = e^{-k|r|}(dx^2 + dy^2 + dz^2 - c^2 dt^2) + dr^2$, where r is the coordinate of the fifth dimension, then, roughly, $M_{Pl}^2 = M^3/k$. In other words, the size of the space is largely irrelevant. This makes sense because the curvature of space—not the extra dimension's size—determines how the field lines spread in the extra dimension and hence the strength of four-dimensional gravity. In fact, there is small dependence on R: the real formula is $M_{Pl}^2 =$

$M^3/k(1 - e^{-kR})$, but when kR is big, the exponential term is largely irrelevant and can be neglected.

38. The warp factor in the locally localized gravity model that Andreas Karch and I developed is the sum of a decreasing exponential function (like the warped geometries we have already considered) and an increasing exponential function. It is proportional to $\cosh(kc - k|r|)$, where k is related to the bulk energy and c is related to the brane energy. Like the localized gravity warp factor we have already considered, this warp factor falls exponentially as you leave the brane. But unlike the previous case, the warp factor turns around and then exponentially increases. The four-dimensional graviton is localized in the region between the brane and this "turnaround" point. Beyond that distance, four-dimensional gravity no longer applies.

39. Under T-duality, the compactification radius, r, is interchanged with its inverse, $1/r$ (with distances measured in units of the string length).

40. The physicists Csaba Csaki, Joshua Erlich, and Christophe Grojean have, however, made the interesting observation that the speed of light and the speed of gravity can be different (the speed of gravity can actually be faster) if there is an *asymmetrically warped* spacetime in which the scalings of time and spatial coordinates along a fifth dimension are different from each other.

Permissions

Index

Numbers in italics indicate Figures.